Robust Design in Geotechnical Engineering

Geotechnical design confronts uncertainties from incomplete subsurface knowledge, natural variability, geomaterial properties, construction errors, and differing solution models. Traditional approaches use conservative factors of safety or pre-calibrated partial factors to address uncertainties indirectly. Reliability-based design incorporates uncertainties directly but requires full statistical characterization, which is often unattainable in practice. Robust geotechnical design offers an alternative, by optimizing safety, cost, and robustness when uncertainties cannot be accurately quantified—where reliability-based and load and resistance factor design prove less effective. This first comprehensive book on robust geotechnical design presents detailed methodologies, design examples, and reviews existing methods, benefiting researchers and engineers interested in geotechnical and geological design under uncertainty.

Robust Design in Geotechnical Engineering

Wenping Gong, C. Hsein Juang,
Lei Wang, and Chao Zhao

CRC Press
Taylor & Francis Group
Boca Raton London New York

CRC Press is an imprint of the
Taylor & Francis Group, an **informa** business

Designed cover image: Shutterstock

First edition published 2027
by CRC Press
2385 NW Executive Center Drive, Suite 320, Boca Raton FL 33431

and by CRC Press
4 Park Square, Milton Park, Abingdon, Oxon, OX14 4RN

CRC Press is an imprint of Taylor & Francis Group, LLC

ISBN: 978-1-032-91424-4 (hbk)
ISBN: 978-1-032-91684-2 (pbk)
ISBN: 978-1-003-56445-4 (ebk)

DOI: 10.1201/9781003564454

Typeset in Times
by Apex CoVantage, LLC

Contents

Preface

Geotechnical engineering, as a discipline grounded in geological complexities, inherently involves uncertainty. In the face of spatial variability in subsurface conditions, limited site characterization efforts, incomplete human knowledge, and construction variability, the geotechnical engineering community needs a new design approach that is not only technically sound but also robust to various sources of uncertainty. Uncertainties in geotechnical design can arise from geotechnical parameters, solution models, and construction errors. Yet, modern infrastructure demands designs that are safe, economical, and resilient under various operational conditions. Bridging this divide requires a new design methodology that acknowledges uncertainties while delivering safe, economical, and robust design. The latter is the essence of robust geotechnical design (RGD).

This book is the first on RGD, a relatively new and novel approach for addressing uncertainties in geotechnical design. It presents a comprehensive framework for robust geotechnical design, integrating new robustness measures, probabilistic methods, computational modeling, and robustness-oriented design optimization. The robust geotechnical design methodology presented in this book can easily be integrated with existing geotechnical design methods, such as deterministic factor of safety (FS)-based design, reliability-based design (RBD), and load and resistance factor design (LRFD). In this book, we presented detailed methodologies and provided examples and associated computer codes for the application of robust design methods to geotechnical design problems such as foundations, slopes, retaining walls, tunnels, and deep excavations. This book reflects both theoretical underpinnings and practical applications, offering tools for engineers, researchers, and students seeking to design various geotechnical systems in the face of uncertainties.

The chapters are structured to build progressively from fundamental concepts in geotechnical uncertainties and existing geotechnical design approaches to theoretical frameworks for robust geotechnical design, followed by advanced topics such as multi-objective optimization, robust geotechnical design methodology, and different robustness measures for implementing robust geotechnical design. Case studies and design examples are embedded throughout to demonstrate how a robust geotechnical design methodology can be effectively implemented across diverse geotechnical applications.

The materials in this book are primarily based on the outcomes of RGD research conducted by Prof. Juang and his coworkers over the last 15 years, with financial support mainly from the National Science Foundation. This work is a collaborative effort among leading researchers who believe that geotechnical engineering must evolve to meet the demands of a changing world with transformative design tools. We hope that this book contributes meaningfully to that evolution, equipping readers with the methodologies and practical examples necessary to create safer, more reliable, and robust geotechnical infrastructure.

This book is intended for graduate students, researchers, and practicing engineers who seek to advance their understanding and skills in designing robust geotechnical systems. We express our gratitude to our academic and industry collaborators whose insights and feedback have shaped this work, and we look forward to the continued advancement of robust geotechnical engineering through research, practice, and innovation.

Wenping Gong, C. Hsein Juang
Lei Wang, and Chao Zhao

About the Authors

Wenping Gong is a professor at China University of Geosciences in Wuhan. He received his PhD in civil engineering (geotechnical) from Clemson University, USA. His research focuses primarily on probabilistic site characterization, uncertainty analysis and design methods in geotechnical and geological engineering, and landslide risk recognition and assessment. He has published more than 100 papers in peer-reviewed international journals. Dr. Gong has already obtained several research projects from the Natural Science Foundation of China. He has received multiple awards, including the Richard Wolters Prize Runner-Up (IAEG), the GEOSNet Young Researcher Award (GEOSNet), and the Bright Spark lecture (ISSMGE). He is the editor-in-chief of *Engineering Geology* and also serves on the editorial boards of several journals, such as *Bulletin of Engineering Geology and the Environment* and *Marine Georesources & Geotechnology.*

C. Hsein Juang is an emeritus professor at Clemson University, USA, after 36 years of academic service (1982–2018). Professor Juang earned his BS (1974) and MS (1976) degrees from National Cheng Kung University, Taiwan, and a PhD degree (1981) from Purdue University, USA. Dr. Juang has received many awards, including the GEOSNet Distinguished Award 2022, the Wilson Tang Lecturer 2019, the GEOSNet Award 2015, and the TK Hsieh Award 2001 by the Institution of Civil Engineers. Among his career achievements, Dr. Juang is most pleased to have been selected by his students as the Outstanding Teacher and to have graduated 22 doctoral students at Clemson University. Dr. Juang has served the geotechnical profession in various capacities, including as chair of the ASCE/GI RAM committee (2009–2012), conference chair for ASCE Georisk 2011, and editor-in-chief of *Engineering Geology* (2012–2024), among others. Dr. Juang has made many original contributions to soil liquefaction, supported excavation, reliability-based methods, and robust design.

Lei Wang is an associate professor of civil engineering at the University of Cincinnati, USA. He received his PhD degree from Clemson University, USA, in 2013. Dr. Wang has an extensive research experience in geotechnical engineering, including geotechnical risk and reliability, underground excavation and tunneling, soil/rock-structure interaction, numerical and centrifuge modeling, earth dams and levees, foundation engineering, and geotechnical earthquake engineering. He has published over 70 refereed journal papers. He was a recipient of many honors and awards, including the Best Paper Award by the Taiwan Geotechnical Society. He serves on editorial boards for international journals, including *Engineering Geology* (Elsevier), *Computers and Geotechnics* (Elsevier), *Journal of Computing in Civil Engineering* (ASCE), *Journal of Cold Regions Engineering* (ASCE), *Bulletin of Engineering Geology and the Environment* (Springer), and *Marine Georesources & Geotechnology* (Taylor & Francis Group).

Chao Zhao is an associate professor at Yangtze University, China. He received his PhD in geological engineering from China University of Geosciences (Wuhan) in 2022. His main research interests include probabilistic site characterization and risk assessment of geohazards. He has published more than 20 papers in peer-reviewed international journals. He has already obtained several research projects from the Natural Science Foundation of China and the China Postdoctoral Science Foundation.

Acknowledgments

he publication of this book, *Robust Design in Geotechnical Engineering*, would not ave been possible without the invaluable contributions and unwavering support of any individuals.

We would like to express our deepest appreciation to the graduate students, aoyin Xing, Liang Wang, Wenxing Tang, Yiyang Wang, and Ting Xiong, for their gnificant efforts in preparing the illustrative examples and figures presented in hapters 1–5. Their diligence, technical proficiency, and meticulous work greatly nhanced the clarity and quality of the examples provided.

Our sincere gratitude is also extended to Dr. Liang Zhang, who made substantial ontributions to the case studies, figure preparation, and reference compilation in hapter 6. His rigorous approach and attention to detail were instrumental in finalizg the chapter to a high scholarly standard.

We are grateful to the people at CRC Press (Taylor & Francis Group) for the wonerful support in ensuring the timely processing of the manuscript and the prompt ublication of this book for the readers.

Finally, we want to extend our thanks to all colleagues and students whose disussions, feedback, and encouragement inspired the refinement of the methodologies nd examples in this book.

1 Uncertainty Characterization and Probabilistic Analysis for Geotechnical Design

1.1 INTRODUCTION

Uncertainty is a distinct feature of geotechnical analysis and design. Uncertainties in geotechnical engineering can be induced by various factors, such as the inherent variability of geotechnical properties, measurement error due to imperfect testing, statistical uncertainty due to limited test data, abstraction adopted in solution model development, construction errors, and incomplete knowledge of the problem. In the face of uncertainties, the performance or safety of a geotechnical system can not be ascertained. Because of these inevitable uncertainties, the selection of an optimal design (among the candidate designs) in geotechnical practice can be a challenge. To cope with this challenge, the geotechnical engineer may select one or more design approaches, such as the deterministic factor of safety (FS)-based design, reliability-based design (RBD), and load and resistance factor design (LRFD), as the reference design.

In the context of the FS-based design, the uncertainties are not explicitly characterized and included in the performance analysis and design; instead, the uncertain parameters involved are captured by their conservative estimates, and the design safety is guaranteed through the selection of a conservative target FS. Within the context of the RBD, the behavior of a candidate design is studied with probabilistic methods that can explicitly consider the uncertainties in the parameters and solution models, and the design safety is guaranteed by a target reliability index (or the equivalent probability of failure). In the context of the LRFD, partial factors such as the resistance and load factors, which have previously been calibrated to achieve the target reliability index or the target FS approximately over a range of design scenarios covered by the design code, are employed to cope with the uncertainties. From a user's perspective, the partial factors for a given geotechnical design using a given solution model are specified in the code, and the main task involved in the design is a proper selection of the characteristic values of uncertain parameters.

It should be noted that the design approaches mentioned above are not perfect, and there is space for improvement. For example, the FS-based deterministic design usually leads to an over-design with an unknown and/or inconsistent safety level, although under-design is also a possibility. RBD is often conducted using inexact or assumed statistics, as the uncertainties of the parameters and solution models in

DOI: 10.1201/9781003564454-1

geotechnical engineering are usually hard to characterize accurately because of the limited availability of site-specific data. Therefore, the results of RBD are influenced by the adopted (or assumed) statistics of the parameters and model errors. The standard LRFD often involves fixed partial factors, which are easy to use (and conveniently codified). However, the fixed partial factors cannot cover all design scenarios involving different levels of variation of the parameters and model errors. Thus, for a given design scenario, the use of the standard LRFD code might lead to a geotechnical design that deviates from the target reliability index or FS by an unknown quantity.

As can be seen, the dilemma of whether to over-design for safety or under-design for cost efficiency has not been entirely overcome, even though the geotechnical design approach has evolved from the FS-based design method to RBD and LRFD methods. In this regard, the concept of *robust design* has recently been adapted for geotechnical designs. In the context of robust design, a candidate design is deemed robust if the design performance is insensitive to (or robust against) the uncertainties in the "hard-to-control" uncertain input parameters, known as *noise factors* (Taguchi 1986). Note that the robust design concept originated from the fields of quality and industrial engineering (e.g., Taguchi 1986; Tsui 1992; Phadke 1989; Beyer and Sendhoff 2007), which is aimed at deriving an optimal design that is insensitive to the variation of noise factors by adjusting the "easy-to-control" input parameters, known as *design parameters*. Robust geotechnical design (RGD), a robust design method that is adapted specifically for geotechnical engineering, has been introduced by Juang and his coworkers (e.g., Juang and Wang 2013; Juang et al. 2013a, 2013b, 2014; Wang et al. 2013, 2014; Gong et al. 2014a, 2014b, 2014c). The parameters and solution models involved in geotechnical design are highly uncertain, which can lead to high variation in the performance of the design; thus, it is desirable to reduce this variation in the performance prediction.

The variation of the performance (or response) of the geotechnical system can be reduced by decreasing the variation of noise factors (e.g., uncertain parameters and solution models). Alternatively, this objective may be achieved by adjusting the design parameters (e.g., the geometry and other parameters that the designer can easily control). The latter is indeed the essence of RGD. According to the characterized levels of uncertainties in geotechnical design, several robustness measures have been proposed for ease of application of the RGD.

It should be noted that although the focus of this book is on the general principle and implementation procedures of RGD in various design scenarios, uncertainty characterization and probabilistic analysis in geotechnical engineering are fundamental elements of the RGD and the existing geotechnical design approaches (i.e., FS-based design method, RBD, and LRFD). Thus, essential background on the uncertainty characterization and probabilistic analysis for geotechnical design will be briefly introduced in this chapter. It starts with a discussion of the sources of uncertainties. Then, basic concepts of probability are introduced. Next, basic methods for characterizing uncertainties and those for probabilistic analysis are briefly introduced. Finally, the summary and concluding remarks are drawn.

1.2 SOURCES OF UNCERTAINTIES IN GEOTECHNICAL DESIGN

Various types of uncertainties can be involved in geotechnical design. The uncertainties involved are mainly from geotechnical design parameters, solution models, and construction errors. These sources of uncertainties are further elaborated.

1.2.1 Uncertainties from Geotechnical Design Parameters

The properties of geomaterials are the most vital input parameters in geotechnical analysis and design. The geomaterials (i.e., soil and rock) at a site are natural materials. The properties of geomaterials are strongly dependent upon the geologic processes that produce and continually modify the geomaterials in situ. The inherent variability involved in the natural and uncertain geologic processes inevitably leads to the inherent spatial variability of the geotechnical properties. However, engineers cannot know the geologic processes and inherent spatial variability of geotechnical properties with certainty. In such a situation, site investigation (e.g., on-site borehole investigation and laboratory tests of collected samples) is often conducted to derive the geotechnical properties at the concerned site. Nevertheless, site investigation effort that can be afforded in a specified project is often limited, and only a limited number of boreholes (or samples) can be conducted (or tested). Thus, geotechnical properties can only be known at sparse borehole positions; in contrast, the geotechnical properties at other positions cannot be known and must be characterized (or interpreted) from those known at sparse borehole positions. Because of the inherent variability of geotechnical properties and the incomplete knowledge that engineers are equipped with, essential assumptions or simplifications must be made in the interpretation of the geotechnical properties. That is to say, uncertainty inevitably exists in the characterized geotechnical properties during the geotechnical design.

Further, testing errors (i.e., the discrepancies between the true properties and the tested properties) and transformation uncertainties (i.e., uncertainties involved in the transformations from the tested properties to those directly applicable to the geotechnical analysis and design) may not be avoided. In general, the testing error can be affected by the testing apparatus and the experience of the engineer, and geotechnical properties obtained from on-site tests could be different from those obtained from laboratory tests of collected samples because of disturbances involved in soil and/or rock sample preparation and in-situ stress release. The transformation uncertainties are mainly related to the transformations from the tested properties to those applicable to the geotechnical analysis and design through some empirical models. For example, the blow counts obtained from on-site standard penetration tests (SPT) may be transformed into the friction angle before they can be applied to solving geotechnical problems such as slope stability or bearing capacity of foundations. Error or uncertainty cannot be avoided in such a transformation.

In summary, there are three primary sources of uncertainties in a geotechnical design parameter: inherent variability, measurement error, and transformation uncertainty, as shown in Figure 1.1 (Phoon and Retief 2016).

Note that apart from the geotechnical properties, uncertainties may also exist in other parameters associated with the geotechnical design. For example, the loading

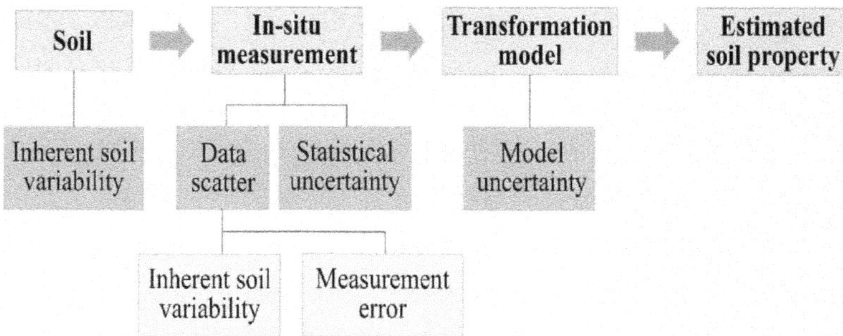

FIGURE 1.1 Three primary sources of uncertainties for a geotechnical design parameter (modified from Phoon and Retief 2016).

applied to the geotechnical system may also involve uncertainties. Because of the uncertainties involved in the dead and live loads, the loadings applied to the infrastructure or building foundation are uncertain and vary with time. For geotechnical systems such as retaining walls, the loading may have to be estimated based on the geotechnical properties with empirical or analytical equations, and the uncertainties involved in the geotechnical properties and solution models can propagate to the loading estimated. Further, the loading applied to the geotechnical system may also be affected by possible earthquakes and nearby engineering activities. The earthquake and engineering activities in the future cannot be accurately forecasted or evaluated; consequently, the resulting loading in the geotechnical design cannot be accurately determined.

1.2.2 Uncertainties from Solution Models

Most solution models adopted for the performance analysis of geotechnical systems are only abstractions or simplifications of the real world. Discrepancies between model predictions and field observations are often the norm rather than the exception (Ang and Tang 1984; Cheung and Tang 2005). Thus, error or uncertainty typically exists in the solution model adopted for a geotechnical design. With advances in mathematics and computational power, scientists are afforded the opportunity to develop complex models that seek to improve model fidelity. However, even with sophisticated and complex models, discrepancies between model predictions and field observations cannot be eliminated. On the other hand, simple and robust solution models are usually preferred in engineering practice, as the more sophisticated and complex models are difficult to apply, yielding less robust predictions, especially in the face of uncertainties. In short, a dilemma often exists between the choice of a complex model for fidelity and that of a simple model for usability and robustness (Gong et al. 2018).

Note that various factors can induce uncertainty in a solution model in geotechnical engineering. For ease of capturing the influences of the geomaterial composition, particle-size distribution, water content, and stress-strain history, any constitutive models have been proposed for characterizing the stress-strain lationships of geomaterials. However, no single constitutive model is perfect, d the effectiveness of the adopted constitutive model can be degraded by the nproper calibration of model parameters (due to the limited data availability). hese issues are further complicated by the fact that the performance prediction of geotechnical design may also be affected by the associated geometry and bound-y conditions. Finally, abstractions or simplifications of the constitutive model d boundary conditions often have to be adopted in the derivation of the solution odel.

In general, the uncertainty of a solution model may be characterized by a bias ctor or residual factor as follows.

$$d = \lambda \cdot f(\boldsymbol{x}) \tag{1.1}$$

$$d = f(\boldsymbol{x}) + \varepsilon \tag{1.2}$$

here $f(\boldsymbol{x})$ is the prediction obtained from the solution model $f(\cdot)$ with input param-ers $\boldsymbol{x}$, in which $\boldsymbol{x}$ is a vector of geotechnical design parameters; d is the observed erformance of the geotechnical system; λ is the bias factor of the solution model; nd, ε is the residual factor or model error. For an unbiased solution model, the mean f the bias factor λ should be 1.0, while the mean of the residual factor ε should be 0. he statistical information (i.e., mean, standard deviation, and type of distribution) of e bias factor λ or the residual factor ε in geotechnical practice is calibrated based n the collected discrepancies between predictions and field observations.

.2.3 Construction Errors

is well known that the construction variability of a geotechnical system is more omplex than that of a building or infrastructure on the ground. For example, the iaphragm wall is constructed in the slurry, the tunnel is built under the influence f surrounding geomaterials and water, and some mining or geotechnical activities ay even be conducted in an environment with toxic gas. Because of the collapse f the geomaterials during the borehole drilling or excavation, the on-site geome-y of a geotechnical structure, such as a drilled shaft, can be much different from at designed in the office. Due to the negative influence of the slurry, the strength f the concrete cast on-site may be lower than the designed strength. Further, the onstruction quality and duration of a geotechnical system such as a tunnel can e affected by some unforeseen geological conditions such as faults. In short, the onstruction quality of a geotechnical structure may be much lower than that of uildings constructed on the ground or precast in factories. Thus, apart from the eotechnical design parameters and solution models mentioned above, another ource of uncertainties in geotechnical design arises from the construction. For ase of geotechnical analysis and design, uncertainties caused by the geotechni-al structure construction are often captured through a term called "construction rror," the definition and calibration of which are similar to those from solution odels.

1.3 BASIC CONCEPTS OF PROBABILITY AND RELIABILITY

Probability theory is most commonly used to quantify or characterize uncertai ties in geotechnical design, and these uncertainties are incorporated explicitly i the analysis that yields a probability of failure. The computed probability of failu is used to capture the safety (or performance) of a given geotechnical design (or designed system). Thus, basic concepts of probability are briefly introduced below.

1.3.1 Three Axioms of Probability

Probability informs how often an event will happen after many repeated trials; i other words, probability can capture the likelihood or possibility of the event. Thre axioms that are widely adopted for the probability calculation are summarized belo

Axiom 1: The probability of an event A, denoted as $P(A)$, will always be non-negative, expressed as $P(A) \geq 0$.

Axiom 2: The probability of the sample space S, denoted as $P(\boldsymbol{S})$, is 1.0, expressed as $P(\boldsymbol{S}) = 1.0$. If event A is only a subset of the sample space $\boldsymbol{S}$, then the probability of event A should be no greater than 1.0, expressed as $0 \leq P(A) \leq 1.0$.

Axiom 3: For two mutually exclusive events, A_1 and A_2, the probability of their union is equal to the summation of their individual probabilities, expressed as $P(A_1 \cup A_2) = P(A_1) + P(A_2)$. Mutual exclusivity means that the intersection of the two events is a null set, the probability of which is 0, expressed as $P(A_1 \cap A_2)$ or, herein, simply, $P(A_1A_2) = 0$.

1.3.2 Conditional Probability and Total Probability

Conditional probability $P(A_1A_2)$ defines the probability of the occurrence of eve A_1, given that the other event A_2 has occurred. In a conditional probability calculatio the sample space can be reduced by conditioning it with respect to the occurrence c the other event. For example, the probability of the intersection of events A_1 and A denoted as $P(A_2A_1)$, can be calculated using the following multiplication rule.

$$P(A_1A_2) = P(A_1 \mid A_2)P(A_2) = P(A_2 \mid A_1)P(A_1) \quad (1.\text{:}$$

where $P(A_2A_1)$ defines the conditional probability of occurrence of event A_2 give that event A_1 has occurred; and, $P(A_1)$ and $P(A_2)$ represent the probability of occu rence of event A_1 and that of event A_2, respectively. In a situation where the occu rence of event A_1 does not depend on the occurrence of event A_2, these two even are referred to as statistically independent events. If events A_1 and A_2 are two statis tically independent, then $P(A_1 \mid A_2) = P(A_1)$. In such situations, the probability c the intersection of events A_1 and A_2 is calculated as follows.

$$P(A_1A_2) = P(A_1)P(A_2) \tag{1.4}$$

Suppose that the occurrence of an event A depends on the occurrence of other events $E_i (i = 1, 2, 3, ..., n)$, and events E_i are mutually exclusive and collectively exhaustive, in which n represents the number of events. According to the total probability theorem, the probability of the occurrence of event A, denoted as $P(A)$, can be calculated as follows.

$$P(A) = \sum_{i=1}^{n} P(A \mid E_i)P(E_i) \tag{1.5}$$

where $P(A \mid E_i)$ denotes the conditional probability of occurrence of event A given that event $E_i (i = 1, 2, 3, ..., n)$ has occurred, and $P(E_i)$ denotes the probability of occurrence of event E_i.

1.3.3 Discrete Random Variables and Continuous Random Variables

A random variable is a variable whose value cannot be predicted with certainty before experiments. Random variables are widely used to statistically represent uncertain events or phenomena such that they can be further analyzed and processed using probability theory.

Discrete random variables are random variables that can only take finite different values in the sample space. Let X denote a discrete random variable, and let $x_i (i = 1, 2, 3, ..., n)$ indicate the possible value of X. For the discrete random variable X, its probability distribution can be described by its probability mass function (PMF), denoted as $P(X = x_i)$ $(i = 1, 2, 3, ..., n)$. The expected value or mean value of the discrete random variable X, denoted as m_X, is computed with the following equation.

$$\mu_X = \sum_{i=1}^{n} x_i P(X = x_i) \tag{1.6}$$

The variance represents the expectation of the squared deviation of a random variable from its mean, which could measure the dispersion of the random variable relative to its mean. For the discrete random variable X, its variance, denoted as σ_X $Var(X)$, is computed as follows.

$$Var(X) = \sum_{i=1}^{n} (x_i - \mu_X)^2 P(X = x_i) \tag{1.7}$$

The standard deviation of a random variable indicates the square root of its variance, the unit of which is the same as that of the mean. As such, the standard deviation is a more convenient indicator or measure of the dispersion of the random variable. For the discrete random variable X, its standard deviation, denoted as σ_X, is computed from its variance $Var(X)$ as follows.

$$\sigma_X = \sqrt{Var(X)} \tag{1.8}$$

The coefficient of variation (COV) represents the normalized standard deviation of a random variable relative to its mean. Thus, the COV can be more convenient than the standard deviation for comparing the degree of dispersion of different random variables. Let COV_X denote the COV of the random variable X, and it can be computed as follows.

$$\text{COV}_X = \frac{\sigma_X}{\mu_X} \tag{1.9}$$

A random variable (Y) is a continuous random variable if its possible values comprise either a single interval on the number line or a union of disjoint intervals. The distribution of a continuous random variable can be represented by its probability density function (PDF). Let $f(y)$ denote the PDF of a continuous random variable Y. Then, its expectation or mean value, denoted as μ_Y, can be computed with the integration below.

$$\mu_Y = \int_{y_L}^{y_U} yf(y)dy \tag{1.10}$$

where y_L and y_U represent the lower and upper bounds of the continuous random variable Y in its sample space, respectively. The variance of the continuous random variable Y, denoted as $Var(Y)$, can be computed with the following equation.

$$Var(Y) = \int_{y_L}^{y_U} (y - \mu_Y)^2 f(y)dy \tag{1.11}$$

For the continuous random variable Y, the probability of being smaller than y can be represented by its cumulative distribution function (CDF), denoted as $F(y)$, and the CDF can be calculated from its PDF as follows.

$$F(y) = \int_{y_L}^{y} f(y)dy \tag{1.12}$$

1.3.4 Common Distribution Types Employed in Geotechnical Practice

The distribution types commonly employed in geotechnical engineering, including the normal distribution, lognormal distribution, and Binomial distribution, are presented below.

The normal distribution is also called the Gaussian distribution. The PDF and CDF of a normal variable X, denoted as $f(x)$ and $F(x)$, respectively, are formulated as follows.

$$f(x) = \frac{1}{\sqrt{2\pi}\sigma_X} \exp\left[-\frac{(x-\mu_X)^2}{2\sigma_X{}^2}\right] \tag{1.13}$$

$$F(x) = \int_{-\infty}^{x} \frac{1}{\sqrt{2\pi}\sigma_X} \exp\left[-\frac{(t-\mu_X)^2}{2{\sigma_X}^2} \right] dt \tag{1.14}$$

where μ_X and μ_X are the mean and standard deviation of the normal variable X, respectively. An example of the PDF of a normal variable is shown in Figure 1.2. The PDF is centered around the mean value μ_x, and the spread of the PDF is controlled by its standard deviation σ_X.

For a case in which $\mu_X = 0$ and $\sigma_X = 1.0$, the normal distribution is known as the standard normal distribution, and its PDF, denoted as $\varphi(u)$, is known as follows.

$$\varphi(u) = \frac{1}{\sqrt{2\pi}} \exp\left(-\frac{u^2}{2} \right) \tag{1.15}$$

And, the CDF of the standard normal distribution is designated by $\Phi(u)$, which is computed as:

$$\Phi(u) = \int_{-\infty}^{u} \frac{1}{\sqrt{2\pi}} \exp\left(-\frac{u^2}{2} \right) du \tag{1.16}$$

A general normal random variable (X) with a mean of μ_X and a standard deviation of σ_X can be related to a standard normal variable (u) with the following equation.

$$X = \mu_X + u \cdot \sigma_X \tag{1.17}$$

The random variable X follows a lognormal distribution if its logarithm ln X follows the normal distribution. According to this definition, the lognormal variable X cannot be negative. In geotechnical practice, many geotechnical parameters physically cannot take negative values. Thus, the lognormal distribution is often adopted in geotechnical engineering to model the non-negative geotechnical

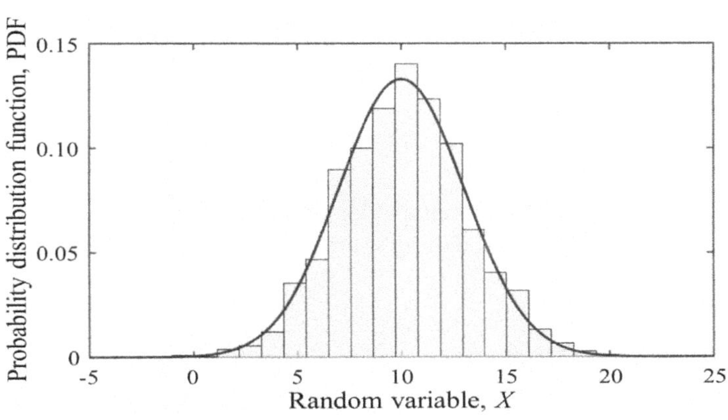

FIGURE 1.2 The PDF for a normal variable $X \sim N(10, 3^2)$.

parameters. The PDF of a lognormal random variable X, denoted as $f(x)$, is formulated as follows.

$$f(x)=\frac{1}{\sqrt{2\pi}\xi_X x}\exp\left[-\frac{(\ln x-\lambda_X)^2}{2\xi_X{}^2}\right] \tag{1.18}$$

where ξ_X and λ_X are the mean and standard deviation of the logarithm $\ln X$, respectively, and which can be estimated from the mean and standard deviation of X with the following equations.

$$\lambda_X=\ln(\mu_X)-\frac{1}{2}\xi_X{}^2 \tag{1.19}$$

$$\xi_X=\sqrt{\ln\left[1+\left(\frac{\sigma_X}{\mu_X}\right)^2\right]} \tag{1.20}$$

The Bernoulli process refers to a sequence of repeated trials or experiments based on the following assumptions: (1) only two possible outcomes exist in each trial, (2) the probability of occurrence of the event in each trial is constant, and (3) the trials are statistically independent. Let p denote the occurrence probability of an event in a Bernoulli sequence. The probability of observing k a occurrences of the event out of n trials during the Bernoulli sequence is governed by the Binomial distribution with the following PMF.

$$P(X=k)=C_n^k p^k(1-p)^{(n-k)} \quad (k=0,1,2,\cdots,n) \tag{1.21}$$

where $C_n{}^k$ represents the Binomial coefficient. The mean and variance of the Binomial random variable X, denoted as μ_X and $Var(X)$, respectively, are calculated as follows.

$$\mu_X=np \tag{1.22}$$

$$Var(X)=np(1-p) \tag{1.23}$$

As can be seen, compared to normal and lognormal distributions, the Binomial distribution is more suitable for modelling some discrete random variables in geotechnical design.

1.3.5 Multivariate Distribution

Covariance measures the joint variability of two random variables, and the covariance between two random variables X_1 and X_2, denoted as $\mathrm{COV}(X_1,X_2)$, is defined as follows.

$$\mathrm{COV}(X_1,X_2)=E\left[\left(X_1-\mu_{X_1}\right)\left(X_2-\mu_{X2}\right)\right]=E(X_1X_2)-E(X_1)E(X_2) \quad (1.24)$$

here $E(\cdot)$ denotes the expectation of a random variable. When the random variables and X_2 are statistically independent, $E(X_1X_2) = E(X_1)E(X_2)$,the covariance etween these two variables X_1 and X_2 is equal to 0, expressed as $COV(X_1,X_2) = 0$. o eliminate the effect of the magnitude of the variables in the interpretation of the ovariance, the normalized covariance, in terms of the correlation coefficient, is pre- rred in engineering practice. Mathematically, the correlation coefficient between riables X_1 and X_2, denoted as $\rho_{X1,X2}$, is defined as follows.

$$\rho_{X_1,X_2}=E\left[\left(\frac{X_1-\mu_{X_1}}{\sigma_{X_1}}\right)\left(\frac{X_2-\mu_{X2}}{\sigma_{X_2}}\right)\right]=\frac{\mathrm{COV}(X_1,X_2)}{\sigma_{X_1}\sigma_{X_2}} \quad (1.25)$$

he correlation coefficient is also a dimensionless measure of the linear dependence etween two random variables, the value of which ranges from −1.0 to 1.0. In situa- ons where multiple random variables are involved, the covariance for these random ariables may be formulated as a matrix, which is known as the covariance matrix. ikewise, the correlation coefficients for these multiple random variables can be xpressed as a correlation matrix.

Often, multiple random variables are involved in a geotechnical design problem. this scenario, the joint distribution of multiple random variables can be modeled ith multivariate distributions. The multivariate normal distribution is a general- ation of the normal distribution for the case of multivariate normal variables. If ultiple random variables $X = [X_1, X_2, X_3, \ldots, X_n]^T$ follow a multivariate normal dis- ibution, their joint PDF, denoted as $f(\mathbf{x})$, can be written as:

$$f(\boldsymbol{x})=\frac{1}{(2\pi)^{\frac{n}{2}}\sqrt{|\boldsymbol{C}|}}\exp\left[-\frac{(\boldsymbol{x}-\boldsymbol{\mu}_X)^T\boldsymbol{C}^{-1}(\boldsymbol{x}-\boldsymbol{\mu}_X)}{2}\right] \quad (1.26)$$

here $\boldsymbol{C}$ is the covariance matrix of multiple random variables $m_X = [m_{X1},$ $_{X2}, m_{X3}, \ldots, m_{Xn}]^T$ is the mean vector of multiple random variables $\boldsymbol{X}$; and, $|\cdot|$ is the eterminant of the matrix.

Similarly, the multivariate lognormal distribution is a generalization of the log- ormal distribution for the case of multiple random variables. If the logarithms of e multiple random variables $\boldsymbol{X}$, expressed as $lnX = [lnX_1, lnX_2, lnX_3, \ldots, lnX_n]^T$, llow a multivariate normal distribution, multiple random variables $\boldsymbol{X}$ follow a mul- variate lognormal distribution and each element of $\boldsymbol{X}$ follows a lognormal distribu- on. The joint PDF of the multivariate lognormal variables $X = [X_1, X_2, X_3, \ldots, X_n]^T$, enoted as $f(\boldsymbol{x})$, is mathematically formulated as follows.

$$f(\boldsymbol{x})=\frac{1}{x_1x_2x_3\cdots x_n(2\pi)^{\frac{n}{2}}\sqrt{|\boldsymbol{C}_{\ln}|}}\exp\left[-\frac{(\ln\boldsymbol{x}-\boldsymbol{\lambda})^T\boldsymbol{C}_{\ln}^{-1}(\ln\boldsymbol{x}-\boldsymbol{\lambda})}{2}\right] \quad (1.27)$$

where $\lambda = [\lambda_1, \lambda_2, \lambda_3, \ldots, \lambda_n]^T$ is the mean vector of $lnX = [lnX_1, lnX_2, lnX_3, \ldots, lnX_n]$ and C_{ln} is the covariance matrix of $\mathrm{lnX} = [\mathrm{lnX}_1, \mathrm{lnX}_2, \mathrm{lnX}_3, \ldots, \mathrm{lnX}_n]^T$. It is noted th the correlation coefficient between logarithms lnX_i and lnX_j, denoted as $r_{i,j}$, can b different from that between variables X_i and X_j, denoted as $\rho_{i,j}$, the relationsh between them is expressed as follows.

$$\rho_{i,j} = \frac{\ln\left(1 + r_{i,j}\dfrac{\sigma_i \sigma_j}{\mu_i \mu_j}\right)}{\sqrt{\ln\left[1 + \left(\dfrac{\sigma_i}{\mu_i}\right)^2\right]}\sqrt{\ln\left[1 + \left(\dfrac{\sigma_j}{\mu_j}\right)^2\right]}} \quad (1.2$$

where μ_i and σ_i represent the mean and standard deviation of variable X_i respectivel and μ_j and σ_j represent the mean and standard deviation of variable X_j, respectivel

Note that the construction of the correlation matrix of multiple random variabl can be a challenge in geotechnical engineering due to the limited data availabilit and the marginal distribution can vary among the multiple random variables. In suc a situation, the construction of the joint PDF of the multivariate distribution cou also be a challenge. To overcome this problem, scholars have adopted the Copula the ory for constructing the multivariate distribution and dependence structure amor dependent variables (e.g., Tang and Phoon 2016). Copula functions can conne the marginal distributions (of multiple random variables) to their joint distributio through which linear or nonlinear relationships among the multiple random variabl can be modeled. The readers are referred to Tang and Phoon (2016) for more info mation on the Copula theory-based multivariate distribution characterization.

1.4 METHODS FOR CHARACTERIZING UNCERTAINTIES IN GEOTECHNICAL PRACTICE

The performance of a geotechnical design can be affected by the uncertaintie involved in the parameters and solution models. As such, characterization of thes uncertainties is a vital component in the analysis and design of geotechnical system Here, some basic methods that are often adopted for uncertainty characterization i geotechnical practice are introduced.

1.4.1 Three-Sigma Bounds or Its Variant

Because of the limited availability of site-specific data, uncertainty characteriz tion has long been a significant challenge in geotechnical engineering. For exar ple, at a site with minimal site-specific data, only nominal values of the uncertai variables may be estimated with confidence. It should be mentioned that althoug the data obtained directly from the site investigation can be limited in a specifi project, more data may be available from the literature, through local exper ence, and by engineering judgment. As such, the potential distribution ranges c the concerned uncertain variables, captured by their highest conceivable values (HCVs) and lowest conceivable values (LCVs), might be estimated. These intervals between HCVs and LCVs provide a significant level of knowledge of the uncertainty of these variables. For example, with the adoption of three-sigma bounds or its variant (e.g., two-sigma bounds), the mean and standard deviation of an uncertain variable X can be approximated from its HCV and LCV as follows (e.g., Duncan 2000; Christian 2004).

$$\mu_X = \frac{X_{HCV} + X_{LCV}}{2} \tag{1.29}$$

$$\sigma_X = \frac{X_{HCV} - X_{LCV}}{6} \text{ or } \frac{X_{HCV} - X_{LCV}}{4} \tag{1.30}$$

where X_{HCV} and X_{LCV} represent the HCV and LCV of the uncertain variable X, respectively.

1.4.2 Moment Method

With the size increase of the site-specific data, the statistics of the uncertain variable X can also be estimated with the moment method. In the context of the moment method, the mean and standard deviation of the uncertain variable X are approximated with its sample mean and sample standard deviation, respectively, adopting the mathematical formulations below.

$$\mu_X \approx \frac{1}{n}\sum_{i=1}^{i=n} X^{(i)} \tag{1.31}$$

$$\sigma_X \approx \sqrt{\frac{1}{n-1}\sum_{i=1}^{i=n}\left(X^{(i)} - \mu_X\right)^2} \tag{1.32}$$

where n is the number of samples collected for the uncertain variable X, and $X^{(i)}$ represents the ith sample. Mathematically, the mean and standard deviation of an uncertain variable could be accurately estimated when the sample size is large enough. In general, the number of samples required for an accurate estimate of the standard deviation is greater than that for the mean (Gong et al. 2017). Nevertheless, when the number of samples ranges from 30 to 100, both mean and standard deviation may be estimated with considerable accuracy. With this moment method, the correlation coefficient or Pearson product-moment correlation coefficient between two random variables X_1 and X_2 can be estimated as follows.

$$\rho_{12} \approx \frac{\frac{1}{n-1}\sum_{i=1}^{i=n}\left(X_1^{(i)} - \mu_{X1}\right)\cdot\left(X_2^{(i)} - \mu_{X2}\right)}{\sqrt{\frac{1}{n-1}\sum_{i=1}^{i=n}\left(X_1^{(i)} - \mu_{X1}\right)^2} \times \sqrt{\frac{1}{n-1}\sum_{i=1}^{i=n}\left(X_2^{(i)} - \mu_{X2}\right)^2}} \tag{1.33}$$

where n represents the number of collected pairs for the uncertain variables X_1 and X_2; $X_1^{(i)}$ and $X_2^{(i)}$ represent the ith samples of uncertain variables X_1 and X_2, respectively; and, μ_{X1} and μ_{X2} represent the means of uncertain variables X_1 and X_2, respectively.

Note that apart from the statistics mentioned above, the predicted performance of the geotechnical system can also be influenced by the distribution type of uncertain variables. To determine the distribution type of a variable, the histogram of this uncertain variable is first plotted, and then the distribution type is determined by comparing the plotted histogram with the PDFs of candidate distribution types. Alternatively, the distribution type of the variable can be determined utilizing the Chi-square or Kolmogorov-Smirnov test.

1.4.3 Maximum Likelihood Estimate (MLE) Method

Given a prespecified distribution type and a set of site-specific data, the statistics of an uncertain variable can also be estimated with the maximum likelihood estimate (MLE) method (Gong et al. 2014a; Juang et al. 2015). Let θ denote the statistical parameters (in terms of the mean and standard deviation) of the uncertain variable to be estimated, and let $\boldsymbol{D}$ denote the observed data of the uncertain variable, which is a vector. Let $f(\boldsymbol{D}/\boldsymbol{\theta})$ denote the joint PDF of $\boldsymbol{D}$ given $\boldsymbol{\theta}$, or equivalently, the likelihood of observing $\boldsymbol{D}$ given $\boldsymbol{\theta}$. When the joint PDF $f(\boldsymbol{D}/\boldsymbol{\theta})$ is taken as a function of observation $\boldsymbol{D}$, a likelihood function, $l(\boldsymbol{\theta}/\boldsymbol{D})$, is formed. The maximum likelihood principle states that the optimal values of statistical parameters $\boldsymbol{\theta}$ may be determined by maximizing the likelihood function $l(\boldsymbol{\theta}/\boldsymbol{D})$ (e.g., Givens and Hoeting 2012). Thus a series of candidate values of $\boldsymbol{\theta}$ may be assumed, and the corresponding values of the likelihood function $l(\boldsymbol{\theta}/\boldsymbol{D})$ are computed; then, the value of θ that yields the highest likelihood value has the best chance to observe the site-specific data $\boldsymbol{D}$. It is noted that maximization of the likelihood function $l(\boldsymbol{\theta}/\boldsymbol{D})$ is equivalent to the maximization of the log-likelihood function $\ln[l(\boldsymbol{\theta}/\boldsymbol{D})]$. As the maximum value of the log-likelihood function is often easier to evaluate, the log-likelihood function is more commonly used for statistical estimation of parameters using the MLE principle (Givens and Hoeting 2012; Juang et al. 2015).

The MLE method may also be applied to estimate the statistical parameters of a set of correlated uncertain variables (in terms of the mean values and covariance matrix). The only difference is the formulation of the likelihood function $l(\boldsymbol{\theta}/\boldsymbol{D})$. The likelihood function $l(\boldsymbol{\theta}/\boldsymbol{D})$ in this situation should be based on the concept of the joint PDF.

Let $\boldsymbol{d}_1$, $\boldsymbol{d}_2$, $\boldsymbol{d}_3$, . . . , and d_n denote n set of the site-specific data. Assuming the site-specific data are statistically independent, and then the likelihood function $l(\boldsymbol{\theta}/\boldsymbol{D})$, which is the chance to observe the site-specific data $\boldsymbol{D} = \{\boldsymbol{d}_1, \boldsymbol{d}_2, \boldsymbol{d}_3, \ldots, \boldsymbol{d}_n\}$, can be formulated as follows.

$$l(\theta \mid \boldsymbol{D}) = \prod_{i=1}^{n} f(\boldsymbol{d}_i \mid \theta) \tag{1.34}$$

where $\prod_{i=1}^{n} f\left(d_i \mid \theta\right)$ indicates the joint probability of observing $\boldsymbol{D}$ given $\boldsymbol{\theta}$, it is the product of n probability terms $f(\boldsymbol{d}_i \mid \boldsymbol{\theta})$ since $\boldsymbol{d}_1, \boldsymbol{d}_2, \boldsymbol{d}_3, \ldots$, and $\boldsymbol{d}_n$ are statistically independent. Based on the maximum likelihood principle, the optimal values of θ may be obtained by maximizing the likelihood function or the logarithm of the likelihood function as follows.

$$\ln\left[L(\theta \mid \boldsymbol{D})\right] = \sum_{i=1}^{n} \ln f\left(\boldsymbol{d}_i \mid \theta\right) \tag{1.35}$$

The same optimal values of $\boldsymbol{\theta}$ can be obtained regardless of whether the likelihood function or the log-likelihood function $\ln\left[(\theta|\boldsymbol{D})\right]$ is maximized. However, maximizing the latter is more computationally efficient. It must be stressed that when the site-specific data $\boldsymbol{d}_1, \boldsymbol{d}_2, \boldsymbol{d}_3, \ldots$, and $\boldsymbol{d}_n$ are correlated, the correlation among these data must be explicitly considered in the formulation of the likelihood function. This situation occurs when the MLE method is applied to estimate the autocorrelation structure of a random field, in which the site-specific data are spatially correlated because of the inherent spatial correlation. Indeed, an accurate formulation of the likelihood function $l(\theta \mid \boldsymbol{D})$ is a premise in the estimate of the statistical parameters of the uncertain variables using the MLE method. The readers are referred to Edwards (1974), Aldrich (1997), Stigler (2007), and Juang et al. (2015) for more information on the MLE method.

1.4.4 Bayesian Inference Method

Unlike the MLE method, which considers only site-specific data, the Bayesian inference method can consider additional information, such as the prior statistical information of uncertain variables. In the Bayesian inference method, the posterior PDF of the statistical parameters θ of the uncertain variables, denoted as $f\left(\theta|\boldsymbol{d}\right)$, can be computed as follows (Gong et al. 2017).

$$f\left(\theta|\boldsymbol{D}\right) = kl\left(\boldsymbol{D}|\theta\right)f\left(\theta\right) \tag{1.36}$$

where $f\left(\theta\right)$ represents the prior PDF of the statistical parameters $\boldsymbol{\theta}$, which could be estimated based on prior knowledge such as local experience, literature, and engineering judgment (Cao et al. 2015); $\boldsymbol{l}\left(\boldsymbol{D}|\boldsymbol{\theta}\right)$ represents the likelihood of observing the site-specific data $\boldsymbol{D}$, given the statistical parameters θ; and k represents a normalization factor that guarantees a unity for the cumulative probability over the entire sample space of the statistical parameters $\boldsymbol{\theta}$.

The posterior PDF of the statistical parameters $\boldsymbol{\theta}$ formulated in Eq. (1.36) could be solved utilizing the direct integration method, numerical integration method (Zhang and Chu 2009a, 2009b), or sampling method (Juang et al. 2013b; Wang and Cao 2013). The Markov chain Monte Carlo (MCMC) simulation-based sampling method is commonly adopted in geotechnical engineering for deriving the posterior PDF

$f(\boldsymbol{\theta}|\boldsymbol{d})$; and, the Metropolis-Hastings algorithm (Metropolis et al. 1953; Hastings 1970; Zhang et al. 2010) is often adopted for sampling in the MCMC simulation. The procedures for implementing the MCMC-based sampling of the statistics parameters $\boldsymbol{\theta}$ using the Metropolis-Hastings algorithm are summarized below.

Step 1: At stage $m = 1$, sample a starting point $\boldsymbol{\theta}_1$ in the Markov chain. This point can be randomly generated from the prior distribution or taken as the mean of the prior distribution.

Step 2: At the next stage (m starts from 2), sample a candidate point $\boldsymbol{\theta}^*$ from a proposal distribution $\boldsymbol{J}(\theta^* | \boldsymbol{\theta}_{m-1})$, which is assumed to be a multivariate normal distribution, in which the mean is set to the current sample point $\boldsymbol{\theta}_{m-1}$ and the covariance matrix is set to $s\mathbf{C}_\theta$, where s is a scaling factor, and $\mathbf{C}_\theta$ is the covariance matrix of the prior distribution of $\boldsymbol{\theta}$.

Step 3: Calculate the density ratio of the un-normalized posterior PDF of θ^* over that of θ_{m-1}, denoted as ζ, as follows.

$$\zeta = \min\left[\frac{l(\boldsymbol{D}|\boldsymbol{\theta}^*)f(\boldsymbol{\theta}^*)}{l(\boldsymbol{D}|\boldsymbol{\theta}_{m-1})f(\boldsymbol{\theta}_{m-1})}, 1\right] \tag{1.37}$$

Step 4: Sample a random variable, denoted as ξ, from a uniform distribution $U(0, 1)$.

Step 5: Determine whether the candidate point $\boldsymbol{\theta}^*$ can be acceptable using the following acceptance rule: if $\zeta \le \zeta$ then the candidate point θ^* is acceptable and set $\theta_m = \theta^*$; otherwise, the candidate point $\boldsymbol{\theta}^*$ is not acceptable and set $\theta_m = \theta_{m-1}$. It is noted that when the number of the observation data $\boldsymbol{D}$ is large, the likelihood function $l(\boldsymbol{D}|\boldsymbol{\theta}^*)$ becomes fairly small, and the mathematical operation of the density ratio ζ can be a challenge. To overcome such an issue, the logarithm of the density ratio ζ, denoted as $\ln\zeta$, can be alternatively computed for assessing the acceptability of the candidate point $\boldsymbol{\theta}^*$, and Eq. (1.37) can be modified as follows.

$$\ln\zeta = \min\left\{\ln\left[l(D|\theta^*)\right] + \ln\left[f(\theta^*)\right] - \ln\left[l(\boldsymbol{D}|\boldsymbol{\theta}_{m-1})\right] - \ln\left[f(\boldsymbol{\theta}_{m-1})\right], 0\right\} \tag{1.38}$$

Accordingly, the acceptance rule should be modified as: if $\ln\zeta \le \ln\zeta$, then accept the candidate point $\boldsymbol{\theta}^*$ and set $\theta_m = \theta^*$; otherwise, reject the candidate point $\boldsymbol{\theta}^*$ and set $\theta_m = \theta_{m-1}$.

Step 6: Repeat **Steps 2–5** until the target number of samples, in terms of the length of the Markov chain, is achieved.

MCMC sampling at the mth step could approach the target distribution of $[l(\boldsymbol{D}|\theta) \cdot l(\theta)]$ as the number of samples m approaches infinity (Gelman et al. 2004; Zhang et al. 2010). Thus, once a sufficient number of MCMC samples of the statistics of the uncertain variables are obtained, the posterior PDF of the statistics could be derived based on statistical analyses of the MCMC samples. In general, the initial samples obtained from MCMC are not valid since the Markov chain has not been stabilized. Thus, the initial MCMC samples are generally taken as burn-in samples, which should be discarded in the derivation of the posterior PDF of the statistics of the concerned geotechnical properties (Zhang et al. 2010; Juang et al. 2013b).

.5 BASIC METHODS FOR GEOTECHNICAL RELIABILITY ANALYSIS

the face of uncertainties, the performance of a geotechnical design may be ana-zed with probabilistic methods, with which the uncertainties in parameters, solu-on models, and construction errors can be explicitly considered. The outcome of the obabilistic analysis is a probability of failure, denoted as P_f, expressed as:

$$P_f = \Pr[g(\boldsymbol{x}) \le 0] = \int_{g(\boldsymbol{x})\le 0} f(\boldsymbol{x})d\boldsymbol{x} \tag{1.39}$$

here $\boldsymbol{x} = [x_1, x_2, x_3, .., x_n]^T$ is a vector of uncertain variables $x_i (i = 1, 2, 3,...,n)$, in hich n is the dimension of uncertain variables; $g(\boldsymbol{x})$ is the performance function or mit state function, which is formulated such that $g(\boldsymbol{X}) \le 0$ denotes the failure of the gineering design; and $f(\boldsymbol{x})$ is the joint PDF of uncertain variables $\boldsymbol{x}$. Difficulty in riving the closed-form solution of the multifold probability integral in Eq. (1.39) as led to various probabilistic analysis methods. The probabilistic analysis methods at are commonly adopted in geotechnical practice, such as the mean-value first-der second-moment (MVFOSM) method, advanced first-order reliability method FORM), and sampling-based method, are introduced below.

.5.1 Mean-Value First-Order Second-Moment Method

or a nonlinear performance function $z = g(\boldsymbol{x})$, the Taylor series expansion may be nployed to approximate the performance function in the vicinity of a given point. he Taylor series expansion of the performance function around the mean value is stablished as follows.

$$\begin{aligned} z &= g(x_1, x_2, x_3, \cdots, x_n) \\ &= g(\boldsymbol{\mu}_x) + \sum_{i=1}^{i=n} \frac{\partial g}{\partial x_i}(x_i - \mu_{xi}) + \frac{1}{2}\sum_{i=1}^{i=n} \frac{\partial^2 g}{\partial x_i \partial x_j}(x_i - \mu_{xi})(x_j - \mu_{xj}) + \cdots \end{aligned} \tag{1.40}$$

here the partial derivative $\frac{\partial g}{\partial x_i} (i = 1, 2, 3,\ldots,n)$ is derived at the mean values of e random variables $\boldsymbol{x} = [x_1, x_2, x_3, .., x_n]^T$, and $\boldsymbol{\mu}_x = [\mu_{x1}, \mu_{x2}, \mu_{x3}, .., \mu_{xn}]^T$ are the ean values of the random variables $\boldsymbol{x}$. By truncating the Taylor series at the linear rms, the first-order approximate mean and variance of the performance function z an be obtained as follows.

$$\mu_g \approx g(\boldsymbol{\mu}_x) = g(\mu_{x1}, \mu_{x2}, \mu_{x3}, \cdots, \mu_{xn}) \tag{1.41}$$

$$\sigma_g^{\ 2} = \sum_{i=1}^{i=n}\sum_{j=1}^{j=n} \frac{\partial g}{\partial x_i}\frac{\partial g}{\partial x_j} \mathrm{COV}(x_i, x_j) \tag{1.42}$$

here μ_g and σ_g represent the mean and standard deviation of the performance func-on $g(\boldsymbol{x})$, respectively, and $\mathrm{COV}(x_i, x_j)$ is the covariance between uncertain variables x_i and $x_j (i, j = 1, 2, 3,...,n)$. Based on the estimated mean and standard deviation of the performance function $g(\boldsymbol{x})$, the reliability index (β), which is defined as tl ratio of the mean over the standard deviation of the performance function $g(\boldsymbol{x})$, ca be obtained.

$$\beta = \frac{\mu_g}{\sigma_g} = \frac{g(\mu_{x1}, \mu_{x2}, \mu_{x3}, \cdots, \mu_{xn})}{\sqrt{\sum_{i=1}^{i=n}\sum_{j=1}^{j=n} \frac{\partial g}{\partial x_i}\frac{\partial g}{\partial x_j} \mathrm{COV}(x_i, x_j)}} \quad (1.43)$$

The reliability index (β) can be related to the probability of failure (P_f) as follows

$$P_f = \Phi(-\beta) \quad (1.44)$$

where $\Phi(\cdot)$ denotes the CDF of the standard normal variable. Because the perfo mance function is linearized at the mean values of the random variables, the methc mentioned above is known as the MVFOSM method. If all the uncertain variabl are normally distributed and the performance function is linear, the reliability inde obtained is accurate; otherwise, the reliability index obtained is only an approxima estimate.

1.5.2 Advanced First-Order Reliability Method

Within the AFORM, the mean and standard deviation of the performance functic are estimated as follows (e.g., Ang and Tang 2004).

$$\mu_g \approx g(\boldsymbol{\mu}_x) + (\boldsymbol{\mu}_x - \boldsymbol{x})\left(\nabla g\big|_{x=x'}\right) \quad (1.45)$$

$$\sigma_g \approx \sqrt{\left(\nabla g\big|_{x=x'}\right)\mathbf{C}_x\left(\nabla g\big|_{x=x'}\right)^T} \quad (1.46)$$

where C_x is the covariance matrix among the uncertain variables $\boldsymbol{x} = [x_1, x_2, x_3, .., x_n]^T$ and $\nabla g\,|_{x=x^*}$ represents the gradient of the performance function to the uncertai variables at the design point of $\boldsymbol{x}'$, which can be formulated as follows.

$$\nabla g\big|_{x=x'} = \left\{ \frac{\partial g(\boldsymbol{x})}{\partial x_1}\bigg|_{x=x'}, \frac{\partial g(\boldsymbol{x})}{\partial x_2}\bigg|_{x=x'}, \frac{\partial g(\boldsymbol{x})}{\partial x_3}\bigg|_{x=x'}, \cdots, \frac{\partial g(\boldsymbol{x})}{\partial x_n}\bigg|_{x=x'} \right\} \quad (1.47)$$

With the assumption that the performance function $z = g(\boldsymbol{x})$ follows a normal distr bution, the reliability index (β) of a geotechnical design can be estimated as follow

$$\beta = \frac{\mu_g}{\sigma_g} = \frac{g(\boldsymbol{\mu}_x) + (\boldsymbol{\mu}_x - \boldsymbol{x})\left(\nabla g\big|_{x=x'}\right)}{\sqrt{\left(\nabla g\big|_{x=x'}\right)\mathbf{C}_x\left(\nabla g\big|_{x=x'}\right)^T}} \quad (1.48)$$

In reference to Eq. (1.48), the reliability index (β) obtained varies with the selection of the design point $\boldsymbol{x}'$. Nonetheless, a critical point $\boldsymbol{x}^*$ yielding an accurate reliability index of the system can be found by solving the following non-linear constrained optimization problem:

$$\underset{\boldsymbol{x}'}{\mathrm{Min}}(\beta) \quad \text{s.t.} \quad g(\boldsymbol{x}') = 0 \tag{1.49}$$

The critical design point $\boldsymbol{X}^*$ obtained may be interpreted as the most probable combination of the uncertain variables $\boldsymbol{X}^*$ that yields unsatisfactory performance of the geotechnical system, and the related reliability index yields the accurate reliability of the geotechnical system.

Alternatively, the reliability index (β) could be solved in the sample space of the correlated standard normal variables with the following constrained optimization.

$$\beta = \underset{\boldsymbol{n}}{\mathrm{Min}}\left(\sqrt{\boldsymbol{n}^T \boldsymbol{R}^{-1} \boldsymbol{n}}\right) \quad \text{s.t.} \quad g(\boldsymbol{n}) = 0 \tag{1.50}$$

where $\boldsymbol{n} = [n_1, n_2, n_3, .., n_n]^T$ represents the normalized variables of the uncertain variables $\boldsymbol{x}$, and the element n_i is related to the uncertain variable x_i $(i = 1, 2, 3, \ldots, n)$ as follows.

$$\Phi(n_i) = F(x_i) \tag{1.51}$$

where $\Phi(\cdot)$ is the CDF of the standard normal variable, and $F(x_i)$ denotes the CDF of the normal variable x_i. The symbol of $\boldsymbol{R}$ in Eq. (1.50) represents the correlation matrix of the normalized uncertain variables $\boldsymbol{n} = [n_1, n_2, n_3, .., n_n]^T$.

1.5.3 Sampling-Based Probabilistic Method

Note that although MVFOSM and AFORM have been widely adopted in geotechnical practice, some concerns remain: 1) the accuracy of these two approximation methods becomes an issue if the performance function is highly nonlinear or high-dimensional; 2) the evaluation of the partial derivative of the performance function can be an obstacle when using MVFOSM or AFORM, especially in situations where the system performance can only be analyzed using numerical methods; and 3) the distribution of the performance function is approximated with its moments of finite order, the evaluation of moments and the approximation of distribution type could both introduce errors. To overcome these drawbacks, we may utilize sampling-based methods such as Monte Carlo simulation (MCS) (Ramirez-Marquez and Coit 2005; Binder and Heermann 2010). In the context of MCS, samples of the uncertain variables are first drawn from the joint PDF of the uncertain variables; the performance function for each of these samples is then evaluated using the deterministic solution model; and finally, the probability of failure (P_f) of the geotechnical design can readily be estimated as follows.

$$P_f \approx \frac{1}{N}\sum_{i=1}^{i=N} I(\boldsymbol{x}_i) \tag{1.52}$$

where N is the number of samples of the uncertain variables; $I(\boldsymbol{x}_i)$ is an indicator function of the ith sample of the uncertain variables $\boldsymbol{x}$, which is defined as follows: if $g(\boldsymbol{x}_i) \le 0$, then $I(\boldsymbol{x}_i) = 1.0$; otherwise, $I(\boldsymbol{x}_i) = 0$. Here, the COV of the estimated probability of failure, denoted as COV_{Pf}, can be estimated as follows.

$$\mathrm{COV}_{Pf} = \sqrt{\frac{1-P_f}{P_f N}} \tag{1.53}$$

The MCS-based method generally yields an unbiased estimate of the probability of failure; however, the required number of samples of the uncertain variables, and thus the number of deterministic evaluations of the performance function, might be too large to be computationally efficient, especially for problems of low failure probabilities. The issue of low computational efficiency becomes more profound when the system performance can only be evaluated using numerical solution models. In such circumstances, various sophisticated sampling methods, such as Latin hypercube sampling, importance sampling (Ang et al. 1992; Au and Beck 1999; Grooteman 2011), subset simulation (Au and Beck 2001; Ching et al. 2005; Hsu and Ching 2010), and subdomain sampling (Gong et al. 2016, 2017; Juang et al. 2017), have been proposed to improve the computational efficiency.

Note that apart from the probabilistic analysis methods reviewed above, there are many other probabilistic analysis methods available in the literature, such as the point estimate-based moment (PEM) method (Zhao and Ono 2000; Zhao et al. 2002), first-order reliability method (FORM) (Low and Tang 1997, 2007; Low 2015), and response surface method (Roussouly et al. 2013). Interested readers are referred to the literature, as this chapter only presents the basic background information for the geotechnical reliability analysis.

1.6 SUMMARY AND REMARKS

In this chapter, basic background information on probabilistic analysis and design in geotechnical engineering was introduced. First, the sources of uncertainties in geotechnical engineering were discussed. These uncertainties are mainly related to the design parameters, solution models, and construction errors. The uncertainties of design parameters may be caused by the inherent variability of geotechnical properties and the incomplete knowledge that engineers are equipped with. The uncertainties of solution models are induced by the abstractions or simplifications adopted for solution model derivations, and the uncertainties of construction errors are mainly related to the specific construction environment. In that the probability theory was adopted herein to characterize the uncertainties in geotechnical practice, basic concepts of probability theory, such as three axioms of probability, conditional probability and total probability, discrete random variables and continuous random variables, common distribution types in geotechnical practice, and multivariate distributions, were then presented. Next, methods that are often adopted for uncertainty characterization in geotechnical practice, such as three-sigma bounds (or its variant), moment method, maximum likelihood estimate (MLE) method, and Bayesian inference method, were briefly introduced. Finally, the probabilistic analysis methods commonly adopted in geotechnical practice, such as the MVFOSM method, AFORM, and sampling-based methods, were briefly introduced.

It should be informed that only the general principle of these methods is provided in this chapter, as the focus of this book is on the RGD methods. For additional details on the probability theory and its applications in geotechnical engineering, the readers are referred to the references mentioned in the text.

REFERENCES

Aldrich, J. (1997) RA Fisher and the making of maximum likelihood 1912–1922. *Statistical Science*, 12(3), 162–176.

Ang, A.H., and Tang, W.H. (1984) *Probability concepts in engineering planning and design: Design, risk and reliability* (Vol. 2). Wiley, New York.

Ang, A.H., and Tang, W.H. (2004) *Probability concepts in engineering* (2nd edition). John Wiley and Sons, New York.

Ang, G.L., Ang, A.H., and Tang, W.H. (1992) Optimal importance-sampling density estimator. *Journal of Engineering Mechanics*, 118(6), 1146–1163.

Au, S.K., and Beck, J.L. (1999) A new adaptive importance sampling scheme for reliability calculations. *Structural Safety*, 21(2), 135–158.

Au, S.K., and Beck, J.L. (2001) Estimation of small failure probabilities in high dimensions by subset simulation. *Probabilistic Engineering Mechanics*, 16(4), 263–277.

Beyer, H.G., and Sendhoff, B. (2007) Robust optimization–a comprehensive survey. *Computer Methods in Applied Mechanics and Engineering*, 196(33), 3190–3218.

Binder, K., and Heermann, D. (2010) *Monte Carlo simulation in statistical physics: An introduction*. Springer, Berlin, Germany.

Cao, Z., Wang, Y., and Li, D.Q. (2015) Quantification of prior knowledge in geotechnical site characterization. *Engineering Geology*, 203, 107–116.

Cheung, R.W., and Tang, W.H. (2005) Realistic assessment of slope reliability for effective landslide hazard management. *Géotechnique*, 55(1), 85–94.

Ching, J., Beck, J.L., and Au, S.K. (2005) Hybrid subset simulation method for reliability estimation of dynamical systems subject to stochastic excitation. *Probabilistic Engineering Mechanics*, 20(3), 199–214.

Christian, J.T. (2004) Geotechnical engineering reliability: How well do we know what we are doing? *Journal of Geotechnical and Geoenvironmental Engineering*, 130(10), 985–1003.

Duncan, J.M. (2000) Factors of safety and reliability in geotechnical engineering. *Journal of Geotechnical and Geoenvironmental Engineering*, 126(4), 307–316.

Edwards, A.W. (1974) The history of likelihood. *International Statistical Review/Revue Internationale de Statistique*, 9–15.

Gelman, B.A., Carlin, B.P., Stem, H.S., and Rubin, D.B. (2004) *Bayesian data analysis*. Chapman, and Hall, London.

Givens, G.H., and Hoeting, J.A. (2012) *Computational statistics* (Vol. 703). John Wiley, and Sons.

Gong, W., Juang, C.H., Khoshnevisan, S., and Phoon, K.K. (2016) R-LRFD: Load and resistance factor design considering robustness. *Computers and Geotechnics*, 74, 74–87.

Gong, W., Juang, C.H., and Martin, J.R. (2017) A new framework for probabilistic analysis of the performance of a supported excavation in clay considering spatial variability. *Géotechnique*, 67(6), 546–552.

Gong, W., Juang, C.H., Martin II, J.R., Tang, H., Wang, Q., and Huang, H. (2018) Probabilistic analysis of tunnel longitudinal performance based upon conditional random field simulation of soil properties. *Tunnelling and Underground Space Technology*, 73, 1–14.

Gong, W., Luo, Z., Juang, C.H., Huang, H., Zhang, J., and Wang, L. (2014a) Optimization of site exploration program for improved prediction of tunneling-induced ground settlement in clays. *Computers and Geotechnics*, 56, 69–79.

Gong, W., Wang, L., Juang, C.H., Zhang, J., and Huang, H. (2014b) Robust geotechnical design of shield-driven tunnels. *Computers and Geotechnics*, 56, 191–201.

Gong, W., Wang, L., Khoshnevisan, S., Juang, C.H., Huang, H., and Zhang, J. (2014c) Robust geotechnical design of earth slopes using fuzzy sets. *ASCE Journal of Geotechnical and Geoenvironmental Engineering*, 141(1), 04014084.

Grooteman, F. (2011) An adaptive directional importance sampling method for structural reliability. *Probabilistic Engineering Mechanics*, 26(2), 134–141.

Hastings, W.K. (1970) Monte Carlo sampling methods using Markov chains and their applications. *Biometrika*, 57(1), 97–109.

Hsu, W.C., and Ching, J. (2010) Evaluating small failure probabilities of multiple limit states by parallel subset simulation. *Probabilistic Engineering Mechanics*, 25(3), 291–304.

Juang, C.H., Gong, W., and Martin, J.R. (2017) Subdomain sampling methods – efficient algorithm for estimating failure probability. *Structural Safety*, 66, 62–73.

Juang, C.H., Khoshnevisan, S., and Zhang, J. (2015) Maximum likelihood principle and its application in soil liquefaction assessment. *Risk and Reliability in Geotechnical Engineering*, 4, 181–220.

Juang, C.H., and Wang, L. (2013) Reliability-based robust geotechnical design of spread foundations using multi-objective genetic algorithm. *Computers and Geotechnics*, 48, 96–106.

Juang, C.H., Wang, L., Hsieh, H.S., and Atamturktur, S. (2014) Robust geotechnical design of braced excavations in clays. *Structural Safety*, 49, 37–44.

Juang, C.H., Wang, L., Khoshnevisan, S., and Atamturktur, S. (2013a) Robust geotechnical design – methodology and applications. *Journal of GeoEngineering*, 8(3), 71–81.

Juang, C.H., Wang, L., Liu, Z., Ravichandran, N., Huang, H., and Zhang, J. (2013b) Robust geotechnical design of drilled shafts in sand: New design perspective. *ASCE Journal of Geotechnical and Geoenvironmental Engineering*, 139(12), 2007–2019.

Low, B.K., and Tang, W.H. (1997) Reliability analysis of reinforced embankments on soft ground. *Canadian Geotechnical Journal*, 34(5), 672–685.

Low, B.K., and Tang, W.H. (2007) Efficient spreadsheet algorithm for first-order reliability method. *Journal of Engineering Mechanics*, 133(12), 1378–1387.

Low, B.K., and Phoon, K.K. (2015) Reliability-based design and its complementary role to Eurocode 7 design approach. *Computers and Geotechnics*, 65, 30–44.

Metropolis, N., Rosenbluth, A.W., Rosenbluth, M.N., Teller, A.H., and Teller, E. (1953) Equation of state calculations by fast computing machines. *The Journal of Chemical Physics*, 21(6), 1087–1092.

Phadke, M.S. (1989) *Quality engineering using robust design.* Prentice Hall, New Jersey.

Phoon, K.K., and Retief, J.V. (2016) *Reliability of geotechnical structures in ISO2394.* CRC Press.

Ramirez-Marquez, J.E., and Coit, D.W. (2005) Composite importance measures for multi-state systems with multi-state components. *IEEE Transactions on Reliability*, 54(3), 517–529.

Roussouly, N., Petitjean, F., and Salaun, M. (2013) A new adaptive response surface method for reliability analysis. *Probabilistic Engineering Mechanics*, 32, 103–115.

Stigler, S.M. (2007) The epic story of maximum likelihood. *Statistical Science*, 598–620.

Taguchi, G. (1986) *Introduction to quality engineering: Designing quality into products and processes.* White Plains, New York Quality Resources.

ng, C., and Phoon, K.K. (2016) Model uncertainty of cylindrical shear method for calculating the uplift capacity of helical anchors in clay. *Engineering Geology*, 207, 14–23.

sui, K.L. (1992) An overview of Taguchi method and newly developed statistical methods for robust design. *IIE Transactions*, 24(5), 44–57.

ang, Y., and Cao, Z.J. (2013) Probabilistic characterization of Young's modulus of soil using equivalent samples. *Engineering Geology*, 159, 106–118.

'ang, Y., Huang, K., and Cao, Z.J. (2014) Bayesian identification of soil strata in London Clay. *Géotechnique*, 64(3), 239–246.

nang, L.L., Zhang, J., Zhang, L.M., and Tang, W.H. (2010) Back analysis of slope failure with Markov chain Monte Carlo simulation. *Computers and Geotechnics*, 37(7), 905–912.

nang, L.M., and Chu, L.F. (2009a) Calibration of methods for designing large-diameter bored piles: Serviceability limit state. *Soils and Foundations*, 49(6), 897–908.

nang, L.M., and Chu, L.F. (2009b) Calibration of methods for designing large-diameter bored piles: Ultimate limit state. *Soils and Foundations*, 49(6), 883–896.

nao, Y.G., and Ono, T. (2000) New point estimates for probability moments. *Journal of Engineering Mechanics*, 126(4), 433–436.

nao, Y.G., Ono, T., and Kato, M. (2002) Second-order third-moment reliability method. *Journal of Structural Engineering*, 128(8), 1087–1090.

2 Conventional Design Approaches in Geotechnical Practice

2.1 INTRODUCTION

The geomaterials (e.g., soils and rocks) at a site are natural deposits. The properti of the geomaterials are dependent upon depositional and loading histories, whic are generally unknown to the engineer. Thus, site investigations, by means of on-si borehole investigations and laboratory tests of collected samples, are typically co ducted to derive geotechnical properties at the site of concern. However, the si investigation effort that could be afforded in a specified project is usually limite and only a limited number of boreholes can be undertaken. Thus, the geotechnic properties can only be known at sparse borehole locations. In contrast, the pro erties at other positions cannot be known and can only be interpreted from tho known at the borehole locations. Due to the assumption or simplification made in th spatial interpretation, uncertainty is unavoidable in the characterized geotechnic properties. Further, testing errors (i.e., discrepancies between the true properties ar the tested properties) and transformation uncertainties (i.e., uncertainties involve in the transformations from the tested properties to those directly applicable to th geotechnical analysis and design) cannot be avoided. In summary, the geotechnic properties used in the geotechnical design are often uncertain. In addition to th uncertainty associated with the geotechnical properties, the performance of a ca didate geotechnical design can also be affected by the uncertainties in the solutic models, applied loads, and those caused by the construction.

All these uncertainties discussed above make it difficult to ascertain the perfo mance or safety of a geotechnical design. The uncertainty in the performance eval ation of candidate designs poses a significant challenge in the selection of an optim design (among candidate designs). To cope with this challenge in the geotechnic design, three different design approaches, namely, the deterministic factor of safe (FS)-based design, reliability-based design (RBD), and load and resistance fact design (LRFD), have been developed. In the FS-based design, the uncertainties a not explicitly incorporated in the analysis; instead, the effects of uncertain paran eters are captured through conservative estimates, and the design safety is gua anteed through the selection of a target FS based upon the concept of "calculate risk" (Casagrande 1965; Duncan 2000). This FS-based design usually leads to a inefficient over-design with an unknown level of safety, although under-design also a possibility. To achieve a more rational and consistent assessment of the desig safety, the RBD has been proposed as an alternative (e.g., Duncan 2000; Baech

DOI: 10.1201/9781003564454-

and Christian 2003; Phoon et al. 2003a; Christian 2004; Ang and Tang 2007; Fenton and Griffiths 2008; Phoon 2008). In the context of the RBD, the behavior of a geotechnical design is assessed utilizing probabilistic methods that explicitly consider the uncertainties in the input parameters and solution models. The design safety is guaranteed with a target reliability index (or, equivalently, the target probability of failure) derived from a cost-benefit analysis that balances investment and risk.

Note that although various approaches have been proposed to characterize the statistics of geotechnical parameters (e.g., Wang and Cao 2013; Cao and Wang 2014) and model errors (e.g., Zhang et al. 2009, 2010), the statistics of geotechnical parameters and those of solution model errors cannot be ascertained due to the limited availability of site-specific data. Further, the probabilistic analysis of the performances of all candidate designs can be computationally demanding. As a result, the RBD has not been widely accepted and employed in the general geotechnical practice; rather, the LRFD, which is a simpler variant of the RBD by design, has been more commonly recommended in various codes (e.g., Allen et al. 2005; NSAI 2005; BSI 2007; NEN 2007; Standards Australia 2009; AASHTO 2014). The LRFD code employs partial factors (e.g., resistance factors and load factors), which have been calibrated to achieve a target reliability index approximately over a range of design scenarios covered by the concerned code. From a user's perspective, the partial factors for a given geotechnical design using a given solution model are specified in the code. The main task of the geotechnical design using the LRFD involves a proper selection of the characteristic values of the uncertain parameters. This chapter consists of three parts. First, the principles of the three geotechnical design approaches are introduced. Then, several design examples are presented to illustrate these three design approaches. Finally, the potential limitations of these design approaches are discussed.

2.2 FACTOR OF SAFETY-BASED DESIGN IN GEOTECHNICAL ENGINEERING

The principle of the FS-based design is first introduced in this section, and then four illustrative geotechnical design examples are presented.

2.2.1 Principle of the FS-Based Design

For a geotechnical system with design parameters $\boldsymbol{d}$ (e.g., the geometry parameters that are easy-to-control) and non-design parameters θ (e.g., the geotechnical parameters and solution models that are hard-to-control) as inputs, the behavior or performance of this geotechnical system, denoted as $g(\boldsymbol{d},\boldsymbol{\theta})$, can be expressed as:

$$g(\boldsymbol{d},\theta) = R(\boldsymbol{d},\theta) - L(\boldsymbol{d},\theta) \tag{2.1}$$

where $R(\boldsymbol{d},\boldsymbol{\theta})$ and $L(\boldsymbol{d},\boldsymbol{\theta})$ are the resistance and load, respectively. In general, the geotechnical design $\boldsymbol{d}$ is deemed safe or satisfactory if the value of this performance function $g(\boldsymbol{d},\boldsymbol{\theta})$ is greater than 0; otherwise, the design $\boldsymbol{d}$ is unsatisfactory or infeasible.

Mathematically, the uncertainty in non-design parameters $\boldsymbol{\theta}$ will lead to the uncertainty in the performance function $g(\boldsymbol{d},\boldsymbol{\theta})$. Let's assume that the relationship between the output or performance function $g(\boldsymbol{d},\boldsymbol{\theta})$ and the inputs $\boldsymbol{\theta}$ is captured by a monotonic performance function, as illustrated in Figure 2.1. For an arbitrary distribution of the inputs $\boldsymbol{\theta}$, the performance function $g(\boldsymbol{d},\boldsymbol{\theta})$ will then be a random variable rather than a deterministic value. The uncertainty in the performance function makes it a challenge to evaluate the safety of the geotechnical design $\boldsymbol{d}$. In the context of the FS-based design, the safety or feasibility of a candidate design $\boldsymbol{d}$ is evaluated by an FS formulated as follows.

$$\mathrm{FS} = \frac{R(\boldsymbol{d},\boldsymbol{\theta})}{L(\boldsymbol{d},\boldsymbol{\theta})} \tag{2.2}$$

If uncertainty in $R(\boldsymbol{d},\boldsymbol{\theta})$ and $L(\boldsymbol{d},\boldsymbol{\theta})$ is ignored (i.e., assuming no uncertainty), the design $\boldsymbol{d}$ can be deemed feasible or safe if the computed FS is greater than 1.0. To compensate for the possible uncertainty in the FS (induced by the uncertainties in the uncertain parameters $\boldsymbol{\theta}$), a conservative estimate of the uncertain input parameters and a conservative target FS, denoted as FS_T (e.g., FS_T = a value greater than 1.0) are usually taken. The selection of the conservative estimate of the input parameters $\boldsymbol{\theta}$ and that of the target FS (FS_T) are problem- and code-specific. For example, the required target FS in the pile foundation design is often set at 2.5–3.0, while that in the slope design is set at 1.2–1.5.

2.2.2 Illustrative Examples

Four geotechnical design examples are presented below to illustrate the application of this deterministic FS-based design.

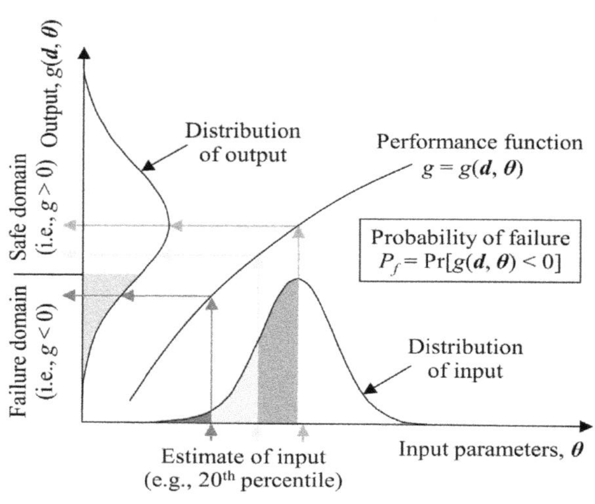

FIGURE 2.1 Conceptual illustration of the performance of a geotechnical design $\boldsymbol{d}$ under the influence of the uncertainties in the input parameters $\boldsymbol{\theta}$.

Example 1

A wall is to be supported on a strip foundation. The load from this wall, denoted as p, is 100 kN/m, and the bearing capacity (q) of the ground is known to be 250 kPa, as depicted in Figure 2.2. In the design of this strip foundation, the target factor of safety (FS_T) is specified as 2.5. For this strip foundation, an optimal width (B) is sought.

Solution:

Let's assume the width of this strip foundation is B(m), and then the bearing capacity of this strip foundation per longitudinal length, denoted as Q(kN/m), can be calculated as:

$$Q = q \cdot B = 250\,(\text{kPa}) \times B\,(\text{m}) = 250B\,(\text{kN/m})$$

Based upon the calculated bearing capacity (Q) of this foundation and the anticipated load (P), the FS of this strip foundation is evaluated as follows.

$$FS = \frac{Q}{P} = \frac{250B\,(\text{kN/m})}{100\,(\text{kN/m})} = 2.5B$$

Note that the FS of this foundation is required to be greater than $FS_T = 2.5$, the width of this strip foundation can be determined as follows.

$$FS = 2.5B \geq 2.5 \rightarrow B \geq 1.0$$

Thus, the width of this strip foundation is designed to be B = 1.0 (m).

If the FS_T is set to 3.0, the following relationship will be obtained:

$$FS = 2.5B \geq 3.0 \rightarrow B \geq 1.2$$

Thus, the width of this strip foundation is designed as B = 1.2 (m).

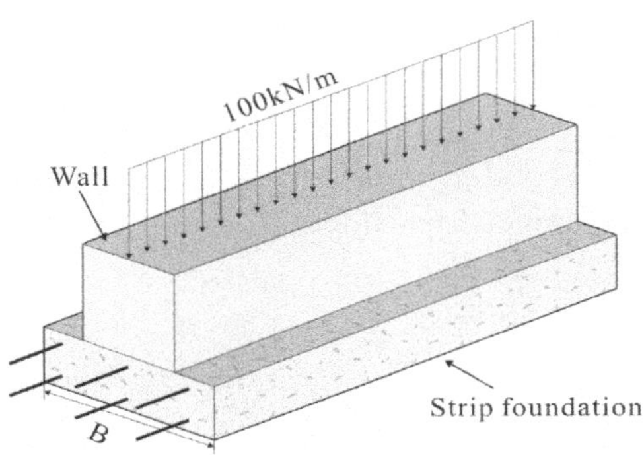

FIGURE 2.2 Schematic diagram of the strip foundation design problem in Example 1.

Example 2

A structure is designed to be supported on a pile in a clayey soil deposit. The load from the upper structure, denoted as *P*, is 1,600 kN. At the site of this pile foundation, a borehole-based site investigation was conducted, and a set of undrained shear strength (s_u) data was obtained along the depth, as tabulated in Table 2.1. The formulation (solution model) given in AASHTO (2014) is adopted for estimating the bearing capacity of this pile foundation, denoted as *Q*, with a diameter of *D* and a length of *L*, as shown in Figure 2.3.

TABLE 2.1
Undrained Shear Strength Along the Depth at the Pile Position in Example 2

Depth z (m)	Undrained shear strength s_u (kPa)	Depth z (m)	Undrained shear strength s_u (kPa)
0	7.5	11	27.3
1	8.0	12	28.0
2	8.4	13	29.1
3	9.1	14	30.4
4	9.7	15	28.5
5	10.8	16	26.2
6	11.5	17	32.7
7	12.4	18	39.5
8	12.3	19	35.9
9	10.2	20	63.0
10	20.7	21	64.2

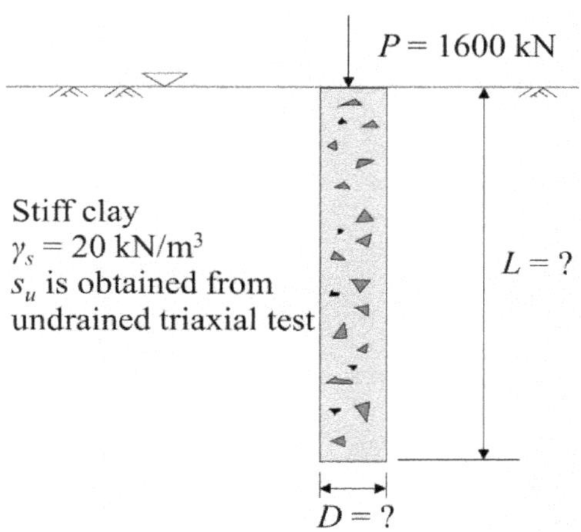

FIGURE 2.3 Schematic diagram of the pile foundation design problem in Example 2.

$$Q = Q_B + Q_S$$

where Q_B and Q_s represent the base resistance and side resistance, respectively, and these two items are estimated as follows.

$$Q_B = A_B N_B s_{u1}$$

$$Q_S = \sum \pi D \Delta z (\alpha s_{u2})$$

where A_B is the pile end area; N_B is the bearing capacity factor (taken as 9.0 for illustration purposes); s_{u1} is the average undrained shear strength of the soil over a depth of $2D$ below the pile end; Δz is the thickness of the soil layer over which side resistance is calculated; s_{u2} is the average undrained shear strength of the soil over the interval Δz; and, α is a coefficient relating the undrained shear strength to the unit side resistance, which is defined as follows.

$$\alpha = \begin{cases} 0 & \text{between the ground surface and the depth of } 1.5\,\text{m} \\ 0.55 & \text{along the remaining portions of the pile for } \dfrac{s_{u2}}{p_a} \le 1.5 \\ 0.55 - \left(\dfrac{s_{u2}}{p_a} - 1.5\right) & \text{along the remaining portions of the pile for } 1.5 < \dfrac{s_{u2}}{p_a} \le 2.5 \end{cases}$$

where p_a is the atmospheric pressure.

For this example, the unit weight of this clayey soil is $\gamma_s = 20.0\,\text{kN/m}^3$, and the unit weight of the concrete pile is $\gamma_c = 22.0\,\text{kN/m}^3$. The FS_T in this pile foundation design is specified as 3.0. The problem (i.e., the pile design) here is to determine an optimal geometry for this pile foundation with the minimum cost. For simplicity, the cost can be approximated with the volume of the pile.

Solution:

Step 1: Select the potential candidate designs

The objective in this design example is to determine the optimal geometry of the pile foundation with the minimum cost while satisfying stability and performance requirements. For illustration purposes, the design space in this design example is taken as $D \in \{0.1\text{ m}, 0.2\text{ m}, 0.3\text{ m}, \ldots, 2.0\text{ m}\}$ and $L \in \{10.0\text{ m}, 10.5\text{ m}, 11.0\text{ m}, \ldots, 20.0\text{ m}\}$. These ranges of design parameters could cover practical engineering values for pile foundations.

Step 2: Calculate the factor of safety for each candidate design

Based on the calculated bearing capacity (Q) of the pile foundation and the load from the upper structure (P), the FS of the pile foundation is evaluated as follows:

$$FS = \frac{Q}{P + \pi \frac{D^2}{4} L\gamma_c}$$

As an example, the *FS* for a design of $D = 0.2$ m and $L = 15$ m is calculated below:

$$Q_B = A_B N_B s_{u1} = \pi \frac{D^2}{4} N_B s_{u1} = \pi \times \frac{(0.2\,(\text{m}))^2}{4} \times 0.9 \times 28.5\,(\text{kPa}) = 8.06\,(\text{kN})$$

$$Q_S = \sum \pi D \Delta z \left(\alpha s_{u2}\right) = 1139.88\,(\text{kN})$$

$$Q = Q_B + Q_S = 8.06\,(\text{kN}) + 1139.88\,(\text{kN}) = 1147.94\,(\text{kN})$$

$$FS = \frac{Q}{\text{P} + \pi \frac{D^2}{4} L\gamma_c}$$

$$= \frac{1147.94\,(\text{kN})}{1600\,(\text{kN}) + \pi \times \frac{(0.2\,(\text{m}))^2}{4} \times 15\,(\text{m}) \times 22.0\,(\text{kN/m}^3)}$$

$$= 0.71$$

For each candidate design, defined as a pair of parameters *D* and *L*, selected from the design space, the corresponding *FS* can be computed by applying the formulations above.

Step 3: Calculate the cost for each candidate design

The cost, denoted as *C* (m³), can be approximated with the volume of the pile. Accordingly, the cost (*C*) is given by the following equation:

$$C = \pi \frac{D^2}{4} L$$

As an example, the cost (*C*) for a design of $D = 0.2$ m and $L = 15$ m is calculated:

$$C = \pi \frac{D^2}{4} L = \pi \times \frac{(0.2\,(\text{m}))^2}{4} \times 15\,(\text{m}) = 0.47\,(\text{m}^3)$$

For each candidate design in the predefined design space, the corresponding cost can readily be computed by applying the formulations above.

Step 4: Select the optimal design in the design space

The calculation results, in terms of the safety and cost, of all candidate designs can lead to a relationship between the *FS* and the cost, as illustrated in Figure 2.4.

A design is considered feasible if it satisfies the requirement of $FS \geq 3.0$. Among all feasible designs derived, the optimal design for this pile foundation

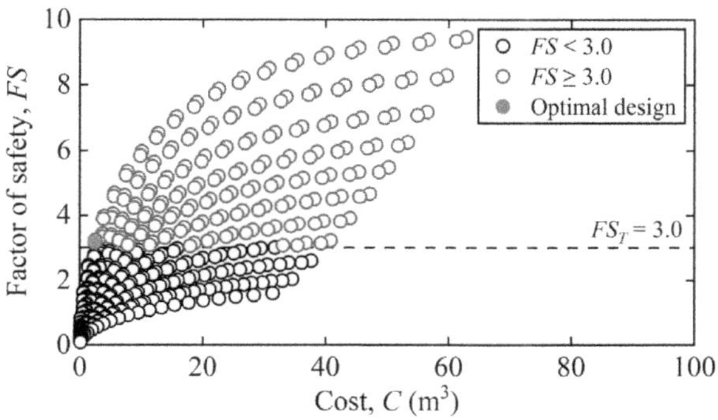

FIGURE 2.4 Relationship between the *FS* and the cost in Example 2.

is determined as D = 0.4 m and L = 19.5 m, with the *FS* of 3.16 and a minimum cost of 2.45 m³.

Example 3

A small infrastructure is to be supported on a shallow foundation founded entirely in cohesionless sand. The geotechnical requirement considered in this shallow foundation design is the ultimate bearing capacity. The load from the infrastructure, denoted as P, is 1,200 kN, and the FS_T is specified as 2.5. In reference to Figure 2.5, the ground water table is far below the foundation base, and the ultimate bearing capacity of this shallow foundation in the cohesionless sand, denoted as q_u, can be estimated as (Vesic 1975):

$$q_u = qN_qF_{qs}F_{qd} + 0.5B\gamma N_\gamma F_{\gamma s}F_{\gamma d}$$

where q is the overburden stress at the foundation base; γ is the unit weight of the sand below the foundation; N_q and N_γ are the bearing capacity factors; and, F_{qs}, F_{qd}, $F_{\gamma s}$ and $F_{\gamma d}$ are the shape and depth factors. The factors of N_q, N_γ, F_{qs}, F_{qd}, $F_{\gamma s}$ and $F_{\gamma d}$ are estimated as:

$$N_q = e^{\pi\tan\varphi}\tan^2(\pi/4 + \varphi/2)$$

$$N_\gamma = 2(N_q + 1)\tan\varphi$$

$$F_{qs} = 1 + (B/L)\tan\varphi$$

$$F_{\gamma s} = 1 - 0.4B/L$$

$$F_{qd} = \begin{cases} 1 + 2\tan\varphi(1-\sin\varphi)^2 D/B, & D/B \le 1 \\ 1 + 2\tan\varphi(1-\sin\varphi)^2 \tan^{-1}(D/B), & D/B > 1 \end{cases}$$

$$F_{\gamma d} = 1$$

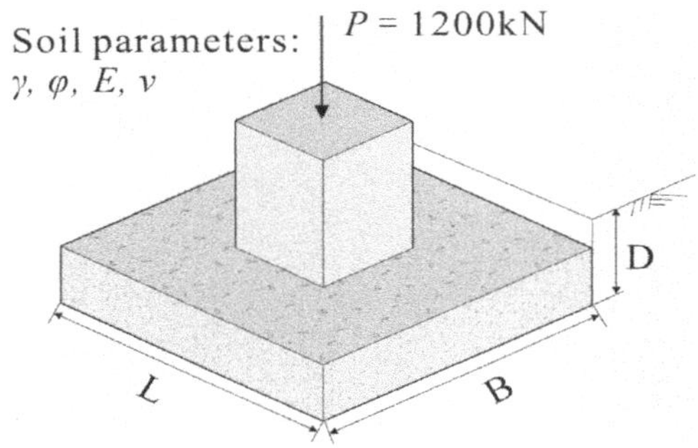

FIGURE 2.5 Schematic diagram of the shallow foundation design problem in Example 3.

where φ is the friction angle of the cohesionless sand; and, B, L, and D are the width, length, and depth of the foundation, respectively. The unit weight and friction angle of the cohesionless sand are γ = 18.5 kN/m^3 and φ = 38.0°, respectively. The problem (i.e., this geotechnical design) is to determine an optimal geometry for this shallow foundation with the minimum cost. The cost, denoted as C, can be approximated with the formulation below (Wang and Kulhawy 2008).

$$C = Q_e c_e + Q_f c_f + Q_c c_c + Q_r c_r + Q_b c_b$$

where Q_e, Q_f, Q_c, Q_r, and $F_{\gamma d}$ are the quantity of the excavation (m^3), formwork (m^2), concrete (m^3), reinforcement (kg), and compacted backfill (m^3), respectively; and, c_e, c_f, c_c, c_r, and c_b denote the unit price of the excavation (i.e., 25.16 USD/m^3), formwork (i.e., 51.97 USD/m^2), concrete (i.e., 173.96 USD/m^3), reinforcement (i.e., 2.16 USD/kg), and compacted backfill (i.e., 3.97 USD/m^3), respectively. The quantities of Q_e, Q_f, Q_c, Q_r, and Q_b are estimated as:

$$Q_e = (B + B_0)(L + L_0)D,\ Q_f = 2H(B + L),\ Q_c = BLH,\ Q_r = kQ_c,\ Q_b = Q_e - Q_c$$

where B_0 and L_0 are the over-excavation distances along the width and length, respectively (i.e., 0.3 m); H is the height of the footing (i.e., 0.6 m); and, k indicates the proportionality coefficient of steel reinforcement of the footing (i.e., 29.67 kg/m^3).

Solution:

Step 1: Select the potential candidate designs

The objective in this design example is to determine an optimal geometry of the shallow foundation with the minimum cost. For illustration purposes, the design space in this design example is taken as $B \in$ {0.5 m, 0.6 m, 0.7 m, . . . , 3.0 m}, $L \in$ {0.5 m, 0.6 m, 0.7 m, . . . , 1.0 m} and $D \in$ {0.5 m, 0.6 m, 0.7 m, . . . , 1.0 m}.

These ranges of design parameters could cover practical engineering values for shallow foundations.

Step 2: Calculate the factor of safety for each candidate design

Based on the calculated ultimate bearing capacity (q_u) of the shallow foundation and the load (P), the FS of the shallow foundation is evaluated as follows:

$$FS = \frac{q_u BL}{P}$$

As an example, the FS for a design of B = 2.0 m, L=1.0 m, and D = 0.5 m is calculated below. First, the bearing capacity factors N_q and N_γ are calculated as follows:

$$N_q = e^{\pi\tan\varphi}\tan^2(\pi/4+\varphi/2)$$

$$= e^{\pi\tan(19\pi/90)}\tan^2(\pi/4+19\pi/180) = 48.93$$

$$N_\gamma = 2(N_q+1)\tan\varphi = 2\times(48.93+1)\tan(19\pi/90) = 78.02$$

Next, the shape and depth factors F_{qs}, F_{qd}, $F_{\gamma s}$ and $F_{\gamma d}$ are calculated as follows:

$$F_{qs} = 1+(B/L)\tan\varphi = 1+[2.0\,(\text{m})/1.0\,(\text{m})]\times\tan(19\pi/90) = 2.56$$

$$F_{qd} = 1+2\tan\varphi(1-\sin\varphi)^2 D/B$$

$$= 1+2\times\tan(19\pi/90)\times[1-\sin(19\pi/90)]^2\times 0.5\,(\text{m})/2.0\,(\text{m}) = 1.06$$

$$F_{\gamma s} = 1-0.4B/L = 1-0.4\times 2.0\,(\text{m})/1.0\,(\text{m}) = 0.2,\ F_{\gamma d} = 1$$

Based on the factors calculated above, the bearing capacity (q_u) can be estimated.

$$q_u = qN_qF_{qs}F_{qd} + 0.5B\gamma N_\gamma F_{\gamma s}F_{\gamma d}$$

$$= 18.5\,(\text{kN/m}^3)\times 0.5\,(\text{m})\times 48.93\times 2.56\times 1.06$$

$$+0.5\times 2.0\,(\text{m})\times 18.5\,(\text{kN/m}^3)\times 78.02\times 0.2\times 1$$

$$= 1515.53\,(\text{kN/m}^2)$$

Finally, the FS can readily be calculated:

$$FS = \frac{q_u BL}{P} = \frac{1515.53\,(\text{kN/m}^2)\times 2.0\,(\text{m})\times 1.0\,(\text{m})}{1200\,(\text{kN})} = 2.53$$

For each candidate design, defined as a pair of parameters B, L, and D, selected from the design space, the corresponding FS can be computed by applying the formulations above.

Step 3: Calculate the cost for each candidate design

The cost (C) in this design example is approximated with the shallow formulation below.

$$C = Q_e c_e + Q_f c_f + Q_c c_c + Q_r c_r + Q_b c_b$$

$$Q_e = (B + B_0)(L + L_0)D,\ Q_f = 2H(B + L),\ Q_c = BLH,\ Q_r = kQ_c,\ Q_b = Q_e - Q_c$$

As an example, the cost (C) for a candidate design of B = 2.0 m, L=1.0 m, and D = 0.5 m is calculated below.

$$Q_e = (B + B_0)(L + L_0)D = [2.0\,(\text{m}) + 0.3\,(\text{m})] \times [1.0\,(\text{m}) + 0.3\,(\text{m})] \times 0.5\,(\text{m})$$

$$= 1.50\,(\text{m}^3)$$

$$Q_f = 2H(B + L) = 2 \times 0.6\,(\text{m}) \times [2.0\,(\text{m}) + 1.0\,(\text{m})] = 3.60\,(\text{m}^2)$$

$$Q_c = BLH = 2.0\,(\text{m}) \times 1.0\,(\text{m}) \times 0.6\,(\text{m}) = 1.20\,(\text{m}^3)$$

$$Q_r = kQ_c = 29.67\,(\text{kg/m}^3) \times 1.2\,(\text{m}^3) = 35.60\,(\text{kg})$$

$$Q_b = Q_e - Q_c = 1.50\,(\text{m}^3) - 1.20\,(\text{m}^3) = 0.30\,(\text{m}^3)$$

Based on the quantities calculated above, the cost (C) is calculated:

$$C = Q_e c_e + Q_f c_f + Q_c c_c + Q_r c_r + Q_b c_b$$

$$= 1.50\,(\text{m}^3) \times 25.16\,(\text{USD/m}^3) + 3.6\,(\text{m}^2) \times 51.97\,(\text{USD/m}^2)$$

$$+ 1.2\,(\text{m}^3) \times 173.96\,(\text{USD/m}^3) + 35.60\,(\text{kg}) \times 2.16\,(\text{USD/kg})$$

$$+ 0.30\,(\text{m}^3) \times 3.97\,(\text{USD/m}^3)$$

$$= 511.53\,(\text{USD})$$

For each candidate design in the predefined design space, the corresponding cost can readily be computed by applying the formulations above.

Step 4: Select the optimal design in the design space

The calculation results, in terms of the design safety and cost, of all candidate designs can lead to a relationship between the FS and the cost, as illustrated in Figure 2.6.

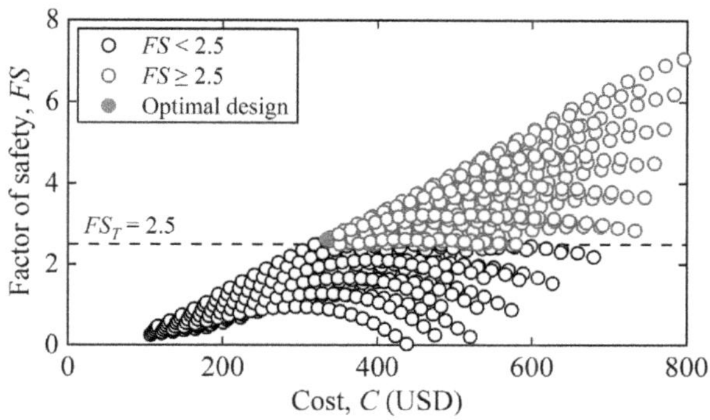

GURE 2.6 Relationship between the *FS* and the cost in Example 3.

A design is considered feasible if it satisfies the requirement of $FS \geq 2.5$. Among all feasible designs derived, the optimal design for this shallow foundation is determined as $B = 1.9$ m, $L = 0.5$ m, and $D = 1.0$ m, with the *FS* of 2.63 and a minimum cost of USD 334.37.

Example 4

The gravity-based retaining wall is often constructed to stabilize the geomaterials behind the wall, as shown in Figure 2.7. Only two failure modes, sliding and overturning, are considered in this example for illustration purposes, and the target factors of safety (FS_T) with respect to these two failure modes are set as 1.3 and 1.6, respectively. To estimate the earth pressures acting on the wall, the Coulomb active coefficient (K_a) (Low 2005) is adopted here.

$$K_a = \left[\frac{\sin(\alpha - \varphi')/\sin\alpha}{\sqrt{\sin(\alpha + \delta)} + \sqrt{\sin(\varphi' + \delta)\sin(\varphi' - \lambda)/\sin(\alpha - \lambda)}} \right]^2$$

where α and λ are the inclinations of the back of the wall and the retained fill surface with respect to the horizontal direction (see Figure 2.7); φ' is the friction angle of the backfill; and, δ is the interface friction angle between the wall and the backfill. The groundwater table is assumed to be far below the base of the retaining wall. The active earth thrust, denoted as P_a (kN/m), is assumed to act at a height of $H/3$ above the base of the wall, at an angle δ with the normal to the back of the wall. Then, the factors of safety (FS) with respect to the sliding and overturning, denoted as FS_1 and FS_2, respectively, can be computed as follows.

$$FS_1 = \frac{bc_a}{P_{av}}$$

$$FS_2 = \frac{W_1A_1 + W_2A_2 + P_{av}A_{ah}}{P_{ah}A_{av}}$$

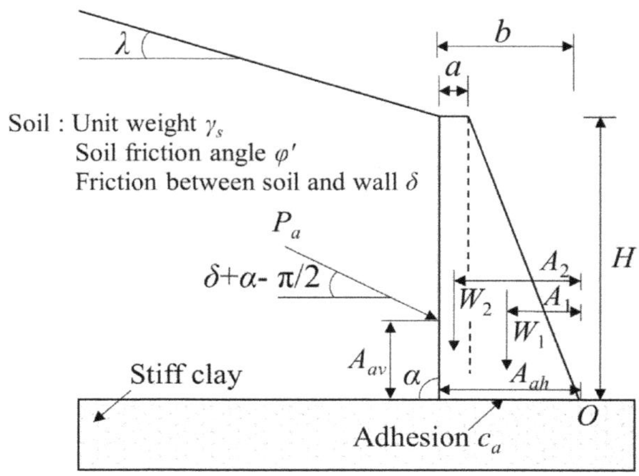

FIGURE 2.7 Schematic diagram of the gravity retaining wall design problem in Example

where W_1 and W_2 represent the weights of the two sections of the wall shown in Figure 2.7; A_1 and A_2 are the related horizontal distances measured from the centroids of the two sections (for W_1 and W_2) to the coordinate O; P_{ah} and P_{av} are the horizontal and vertical components of the active earth thrust P_a; and A_{ah} and A_{av} are the horizontal and vertical distances measured from the acting point of the thrust P_a to the coordinate O.

The above items involved are calculated as follows.

$$W_1 = 0.5\gamma_{\text{wall}}(b-a)H$$

$$W_2 = \gamma_{\text{wall}}aH$$

$$A_1 = (2/3)(b-a)$$

$$A_2 = b-(a/2)$$

$$A_{ah} = b$$

$$A_{av} = H/3$$

$$P_a = 0.5K_a\gamma_s H^2$$

$$P_{av} = P_a \sin\delta$$

$$P_{ah} = P_a \cos\delta$$

where γ_s and γ_{wall} are the unit weight of the backfill and wall, respectively; c_a is the adhesion between the wall base and the foundation; and, a and b are the top width and base width of the wall, which are the easy-to-control design parameters. The parameters that are required for the design of this wall are listed in Table 2.2. The problem in this example is to determine an optimal geometry of this retaining wall with the minimum cost (here, the cost is approximated with the volume of the wall).

TABLE 2.2
Input Parameters Required for the Retaining Wall Design in Example 4

Parameter	γ_s	φ'	δ	c_a	H	λ	α	γ_{wall}
Value	18 kN/m³	35°	20°	100 kPa	6.0 m	10°	90°	25 kN/m³

Solution:

Step 1: Select the potential candidate designs

The objective in this design example is to determine the optimal geometry of the gravity retaining wall with the minimum cost. For illustration purposes, the design space in this design example is taken as $a \in \{0.2\text{ m}, 0.4\text{ m}, 0.6\text{ m}\}$ and $b \in \{0.6\text{ m}, 0.7\text{ m}, 0.8\text{ m}, \ldots, 3.0\text{ m}\}$. These ranges could cover practical engineering values for gravity retaining walls.

Step 2: Calculate the factor of safety for each candidate design

As mentioned above, two failure modes, sliding and overturning, are considered in this design example for illustration purposes, and the *FS* with respect to the sliding and overturning failure, denoted as FS_1 and FS_2, respectively, can be calculated as follows.

$$FS_1 = \frac{bc_a}{P_{av}}$$

$$FS_2 = \frac{W_1A_1 + W_2A_2 + P_{av}A_{ah}}{P_{ah}A_{av}}$$

As an example, the FS_1 and FS_2 for a candidate design of $a = 0.6m$ and $b = 3.0\ m$ are calculated below. First, the earth pressure behind the retaining wall is calculated based on the Coulomb's earth pressure theory, where the Coulomb active coefficient K_a is determined as:

$$K_a = \left[\frac{\sin(\alpha-\varphi)/\sin\alpha}{\sqrt{\sin(\alpha+\delta)}+\sqrt{\sin(\varphi+\delta)\sin(\varphi-\lambda)/\sin(\alpha-\lambda)}}\right]^2$$

$$= \left[\frac{\sin\left(\frac{\pi}{2}-\frac{7\pi}{36}\right)\Big/\sin\frac{\pi}{2}}{\sqrt{\sin\left(\frac{\pi}{2}+\frac{\pi}{9}\right)}+\sqrt{\sin\left(\frac{7\pi}{36}+\frac{\pi}{9}\right)\sin\left(\frac{7\pi}{36}-\frac{\pi}{18}\right)\Big/\sin\left(\frac{\pi}{2}-\frac{\pi}{18}\right)}}\right]^2$$

$$= 0.27$$

Subsequently, the resultant earth pressure P_a is calculated, together with its horizontal (P_{av}) and vertical (P_{ah}) components:

$$P_a = 0.5k_a\gamma_s H^2 = 0.5\times 0.27\times 18\ (\text{kN/m}^3)\times 6\ (\text{m})\times 6\ (\text{m}) = 89.08\ (\text{kN/m})$$

$$P_{av} = P_a\sin\delta = 89.08\ (\text{kN/m})\times \sin\frac{\pi}{9} = 30.47\ (\text{kN/m})$$

$$P_{ah} = P_a\cos\delta = 89.08\ (\text{kN/m})\times \cos\frac{\pi}{9} = 83.70\ (\text{kN/m})$$

Then, the distances from the acting point of the earth pressure to the wall toe are calculated, namely the vertical distance (A_{av}) and the horizontal distance (A_{ah}):

$$A_{av} = \frac{H}{3} = \frac{6\ (\text{m})}{3} = 2\ (\text{m}),\ A_{ah} = b = 3\ (\text{m})$$

Meanwhile, the self-weights of the wall sections, W_1 and W_2, are calculated, along with their respective lever arms A_1 and A_2, measured from the wall toe:

$$W_1 = 0.5\gamma_{\text{wall}}(b-a)H = 0.5\times 25\ (\text{kN/m}^3)\times[3\ (\text{m}) - 0.6\ (\text{m})]\times 6\ (\text{m})$$
$$= 180\ (\text{kN/m})$$

$$W_2 = \gamma_{\text{wall}}aH = 25\ (\text{kN/m}^3)\times 0.6\ (\text{m})\times 6\ (\text{m}) = 90\ (\text{kN/m})$$

$$A_1 = \frac{2}{3}(\text{b}-a) = \frac{2}{3}\times[3\ (\text{m}) - 0.6\ (\text{m})] = 1.6\ (\text{m})$$

$$A_2 = b - \frac{a}{2} = 3\ (\text{m}) - \frac{0.6\ (\text{m})}{2} = 2.7\ (\text{m})$$

Based on the results calculated above, the safety factors (FS_1 and FS_2) for the two failure modes (sliding and overturning) can readily be calculated.

$$FS_1 = \frac{bc_a}{P_{av}} = \frac{3\ (\text{m})\times 100\ (\text{kPa})}{30.47\ (\text{kN/m})} = 9.85$$

$$FS_2 = \frac{W_1A_1 + W_2A_2 + P_{av}A_{ah}}{P_{ah}A_{av}}$$
$$= \frac{180\ (\text{kN/m})\times 1.6\ (\text{m}) + 90\ (\text{kN/m})\times 2.7\ (\text{m}) + 30.47\ (\text{kN/m})\times 3\ (\text{m})}{83.70\ (\text{kN/m})\times 2\ (\text{m})}$$
$$= 3.72$$

For each candidate design, defined as a pair of parameters a and b, in the design space, the corresponding FS_1 and FS_2 can readily be computed by applying the formulations above.

Step 3: Calculate the cost for each candidate design

The cost (C) in this design example is approximated with the volume of the retaining wall. Accordingly, the cost (C) is given by the following equation:

$$C = 0.5(b - a)H + aH$$

As an example, the cost (C) for a design of a = 0.6 m and b = 3.0 m is calculated.

$$C = 0.5(b - a)H + aH = 0.5 \times [3\,(\text{m}) - 0.6\,(\text{m})] \times 6\,(\text{m}) + 0.6\,(\text{m}) \times 6\,(\text{m}) = 10.8\,(\text{m}^2)$$

For each candidate design in the predefined design space, the corresponding cost can readily be computed by applying the formulations above.

Step 4: Select the optimal design in the design space

The calculation results, in terms of the design safety and cost, of all candidate designs can lead to a relationship between the FS and the cost, as illustrated in Figure 2.8.

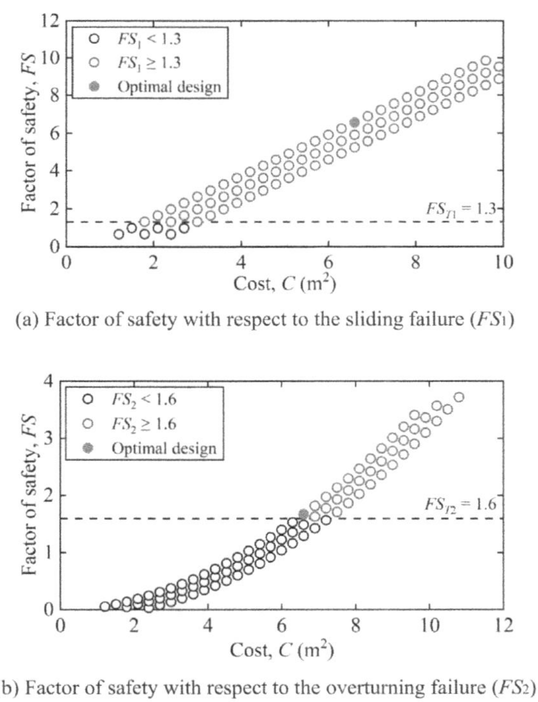

FIGURE 2.8 Relationship between the factor of safety and the cost in Example 4.

A design is considered feasible if it satisfies the requirements of $FS_1 \geq 1.3$ and $FS_2 \geq 1.6$. Among all feasible designs derived, the optimal design for the retaining wall is determined as $a = 0.2$ m and $b = 2.0$ m, with FS_1 of 6.56, FS_2 of 1.67, and a minimum cost of 6.6 m^3.

2.3 RELIABILITY-BASED DESIGN IN GEOTECHNICAL ENGINEERING

The principle of the RBD is presented in this section, followed by four illustrative geotechnical design examples.

2.3.1 Principle of the RBD

In reference to Figure 2.9, the uncertainties in the input parameters $\boldsymbol{\theta}$ (e.g., the load and resistance terms) make it hard to ascertain the performance of a geotechnical design. Though the FS-based design is simple, the "true" safety level of the resulting design is generally not known. While a design may be made conservative by adopting a sufficiently high FS, under-design (that can cause failure) may still happen in the face of uncertainties. To account for the uncertainties explicitly in the performance analysis and design in geotechnical engineering, the engineer has advocated RBD (Harr 1987; Honjo et al. 2000; Baecher and Christian 2003; Christian 2004; Ang and Tang 2007). Within the context of the RBD, the uncertain variables, such as the geotechnical parameters and solution model errors, are characterized as random variables (following specified distributions), and the safety or performance of the geotechnical design is analyzed with probabilistic methods. The outcome of the probabilistic analysis is a probability of failure (P_f), expressed as:

$$P_f = \Pr[g(\boldsymbol{d},\boldsymbol{\theta}) \leq 0] = \int_{g(\mathbf{d},\theta)\leq 0} f(\boldsymbol{\theta})d\boldsymbol{\theta} \tag{2.3}$$

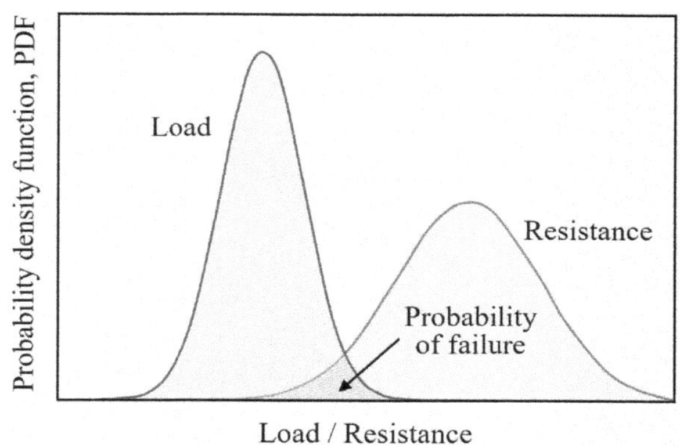

FIGURE 2.9 Physical interpretation of the probability of failure of a geotechnical design.

here $\boldsymbol{\theta} = [\theta_1, \theta_2, .., \theta_n]^T$ represent a vector of uncertain variables $\theta_i (i = 1, 2, ..., n)$ which the subscript n is the dimension of the uncertain variables; $g(\boldsymbol{d}, \boldsymbol{\theta})$ is the rformance function formulated in Eq. (2.1). The performance function, also known the limit state function, is formulated such that $g(\boldsymbol{d}, \boldsymbol{\theta}) \leq 0$ denotes the failure of e geotechnical design $\boldsymbol{d}$. And, $f(\boldsymbol{\theta})$ is the joint probability density function (PDF) the uncertain variables $\boldsymbol{\theta}$. The probability of failure (P_f) can be further illustrated Figure 2.9.

The closed-form solution of the multifold probability integral shown in Eq. (2.3) usually challenging to obtain. The challenge in deriving the closed-form soluon has led to various approximation methods, such as the mean-value first-order cond-moment (MVFOSM) method (Lee and Kwak 1987), advanced first-order liability method (AFORM) (Melchers 1987; Low and Tang 2007), and point estiate-based moment (PEM) method (Zhao and Ono 2000; Zhao et al. 2002). The utcome of these methods is a reliability index, denoted as β, which can be related to e probability of failure (P_f) as:

$$P_f = \Phi(-\beta) \tag{2.4}$$

here $\Phi(\cdot)$ is the cumulative distribution function (CDF) of the standard normal ariable.

Though these approximation methods have been widely used in geotechnical pracces, some concerns remain: 1) the accuracy of these approximation methods becomes n issue if the performance function is highly nonlinear and/or high-dimensional; 2) e evaluation of the partial derivative of the performance function can be an obstacle hen using MVFOSM or AFORM, especially in situations where the system perforance could only be analyzed using numerical methods; and 3) the distribution of the erformance function is approximated with its moments of finite order, and as such, oth the evaluation of the moments and the approximation of the distribution may troduce significant errors. To overcome these drawbacks, sampling-based methods ich as Monte Carlo simulation (MCS) (Binder and Heermann 2010) have been sugested. In the framework of the MCS, samples of the uncertain variables are drawn om the joint PDF of uncertain variables; once these samples are generated, the design erformance for each of these samples can be evaluated using a deterministic solution odel; then, the probability of failure (P_f) of the geotechnical design is estimated ased upon the results of these deterministic analyses.

$$P_f = \frac{1}{N} \sum_{i=1}^{i=N} I(\theta_i) \tag{2.5}$$

here N is the number of samples of the uncertain variables $\boldsymbol{\theta}$; $I(\theta_i)$ is an indicator nction of the ith sample of uncertain variables θ_i, which is defined as follows: if $g(\boldsymbol{d}$, $_i) \leq 0$, then $I(\theta_i) = 1.0$; otherwise $I(\theta_i) = 0$. The MCS-based approach could yield an nbiased estimate of the failure probability; however, the demanded number of samles of uncertain variables, and thus the number of deterministic evaluations of the esign performance, can be too large, especially for problems of low probabilities of ilure. The issue of low computational efficiency becomes more profound when the system performance can only be evaluated using numerical solution models. In suc circumstances, various advanced sampling methods, such as the importance sam pling (Ang et al. 1992; Au and Beck 1999; Grooteman 2011), subset simulation (A and Beck 2001; Ching et al. 2005; Hsu and Ching 2010), and subdomain sampli (Gong et al. 2016; Juang et al. 2017), can be adopted to improve the computation efficiency.

Based on the estimated reliability index (β) or probability of failure (P_f), th safety of a candidate geotechnical design can readily be evaluated by comparing with the target reliability index or target probability of failure, denoted as β_t or P respectively. For example, a candidate design can be deemed feasible or satisfacto if the estimated reliability index (β) is greater than the target reliability index (β_t), the estimated probability of failure (P_f) is less than the target probability of failu (P_{ft}). Note that the target reliability index ($\theta_6 = k_h$) or probability of failure (P_{ft}) problem-specific, which depends on the consequence and nature of the failure, th economic loss, the social inconvenience, the effects on the environment, and th amount of expense and effort required to reduce the probability of failure (War et al. 2016). Listed in Table 2.3 are the example target reliability indexes suggested some geotechnical RBD codes.

2.3.2 Illustrative Examples

Four design examples are presented below to demonstrate the application of the RB in geotechnical engineering.

Example 5

The background of this design example is identical to that in Example 1; however, both the load from the structure (P) and the bearing capacity (q) of the ground

TABLE 2.3
Example Target Reliability Index (β_t) in Several Geotechnical RBD Codes

Design code	Ultimate limit state (ULS) β_t	Serviceability lim state (SLS) β_t
Electric Power Research Institute (EPRI) Multiple Resistance Factor Design (MRFD)	3.2	2.6
Canadian National Building Code (NCBC)	3.5	Not available
American Association of State Highway and Transportation Officials (AASHTO) Foundation Design Code	2.0–3.5	Not available
Eurocode 7*	4.7	2.9

Note: the reference period of the reliability indexes is 1 year; * refers to Reliability Class 2 (RC2) Eurocode 7.

are taken as uncertain variables, not deterministic or fixed values. The load (P) follows a normal distribution with a mean of 100 kN/m and a coefficient of variation (COV) of 0.10, and the ground bearing capacity (q) also follows a normal distribution with a mean of 250 kPa and a COV of 0.20. These two items are statistically independent. In the RBD of this strip foundation, the target reliability index (β_t) is specified as 2.6 (i.e., the corresponding target probability of failure P_{ft} is 4.66×10–³). The design problem (i.e., this strip foundation design) here is to determine an optimal width for this strip foundation.

Solution:

Step 1: Select the potential candidate designs

The objective in this design example is to determine an optimal width for this strip foundation. For illustration purposes, the design space in this design example is taken as $B \in \{0.5\text{ m}, 0.6\text{ m}, 0.7\text{ m}, \ldots, 2.0\text{ m}\}$.

Step 2: Calculate the reliability index (β) for each candidate design

Based on the calculated bearing capacity (q) and the load from the structure (P), the performance function (g) of this strip foundation is evaluated as follows:

$$g = qB - P$$

The reliability index (β) can be approximated based on the mean and standard deviation of the performance function. The reliability index (β) can be calculated as follows.

The standard deviation of the bearing capacity (q), denoted as σ_q, is calculated:

$$\sigma_q = \mathrm{COV}_q \cdot \mu_q = 250\ (\mathrm{kPa}) \times 0.2 = 50\ (\mathrm{kPa})$$

The standard deviation of the load (P), denoted as σ_P, is calculated:

$$\sigma_P = \mathrm{COV}_P \cdot \mu_P = 100\ (\mathrm{kN/m}) \times 0.10 = 10\ (\mathrm{kN/m})$$

The mean μ_g and variance σ_g^2 of the performance function are calculated:

$$\mu_g = \mu_q B - \mu_P,\ \sigma_g^2 = (\sigma_q B)^2 + \sigma_P^2$$

The reliability index (β) is calculated below:

$$\beta = \frac{\mu_g}{\sigma_g} = \frac{\mu_q B - \mu_P}{\sqrt{(\sigma_q B)^2 + \sigma_P^2}}$$

As an example, the reliability index (β) for a design of $B = 0.8$ m is calculated below.

$$\beta = \frac{\mu_g}{\sigma_g} = \frac{\mu_q B - \mu_P}{\sqrt{(\sigma_q B)^2 + \sigma_P^2}} = \frac{250\,(\text{kPa}) \times 0.8\,(\text{m}) - 100\,(\text{kN/m})}{\sqrt{(50\,(\text{kPa}) \times 0.8\,(\text{m}))^2 + (10\,(\text{kN/m}))^2}} = 2.43$$

For each candidate design, defined as the parameter B, selected from the design space, the corresponding reliability index (β) can be computed by applying the formulations above.

Step 3: Identify the optimal design satisfying the target reliability

Figure 2.10(a) depicts the relationship between the estimated reliability index (β) and the width (B) of the strip foundation. Note that a design is considered feasible if it satisfies the requirement of $\beta > 2.6$. Among all feasible designs derived, the optimal design for this strip foundation is determined as $B = 0.9$ m, with the reliability index (β) of 2.71 and the failure probability (P_f) of 3.36×10^{-3}. The PDF of the design performance for the derived optimal design is presented in Figure 2.10(b).

$$\beta = \frac{\mu_g}{\sigma_g} = \frac{\mu_q B - \mu_P}{\sqrt{(\sigma_q B)^2 + \sigma_P^2}} = \frac{250\,(\text{kPa}) \times 0.9\,(\text{m}) - 100\,(\text{kN/m})}{\sqrt{(50\,(\text{kPa}) \times 0.9\,(\text{m}))^2 + (10\,(\text{kN/m}))^2}} = 2.71 > \beta_t = 2.6$$

$$P_f = \Phi(-\beta) = \Phi(-2.71) \approx 3.36 \times 10^{-3} < 4.66 \times 10^{-3}$$

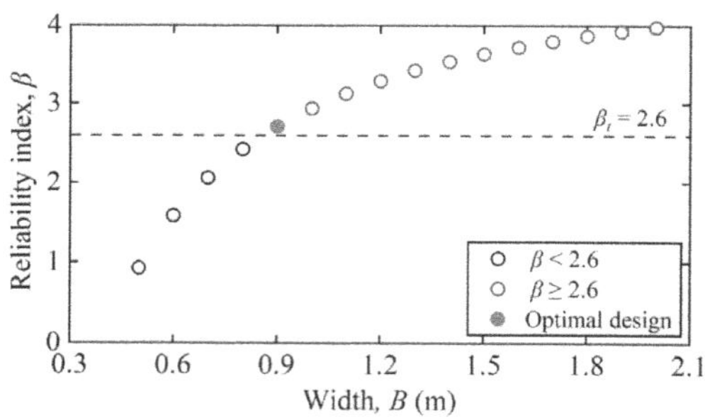

(a) The estimated reliability indexes (β) of candidate designs

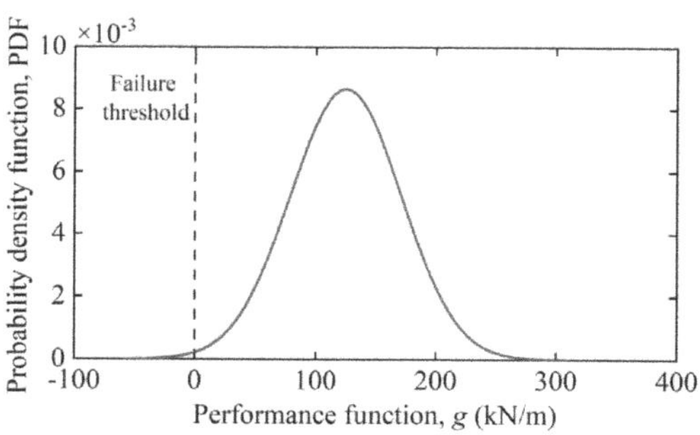

(b) PDF of the performance function for the derived optimal design

FIGURE 2.10 The reliability indexes of candidate designs and the PDF of the performance function of the derived optimal design in Example 5.

Example 6

The background of this design example is identical to that in Example 2; however, the load from the structure (P) is a random variable that follows a normal distribution, the mean is 2,000 kN, and the COV is 0.15. In this example, several sets of undrained shear strength (s_u) data along the depth (see Figure 2.11 and Table 2.4) are obtained at the site of this pile foundation. The statistical information of the undrained shear strength (s_u) could be characterized by the undrained shear strength (s_u) data provided, and this pile foundation is required to be designed using the RBD approach. The target reliability index (β_t) is 3.2 (i.e., the related target probability of failure P_{ft} is 6.87×10^{-4}). Note that the serviceability limit state (SLS) issue will not be considered in this design example. The problem here is to determine an optimal geometry for this pile with the minimum cost.

Solution:

Step 1: Characterize the statistical information of the undrained shear strength

In this design example, the undrained shear strength (s_u) is dealt as a random variable, which is taken as a linear function of the depth (z), expressed as:

$$s_u = s_{un} z$$

where s_{un} is the normalized undrained shear strength, which follows a normal distribution with a mean of 12 kPa/m and a standard deviation of 2.4 kPa/m. The adopted normal distribution of s_{un} is validated with a 95% confidence level using the chi-square test, and the statistics (in terms of the mean and the standard deviation) are estimated using the maximum-likelihood principle.

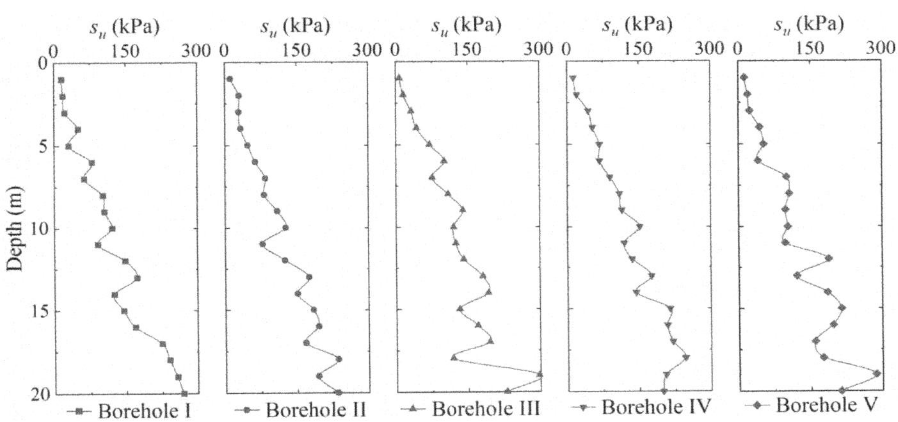

FIGURE 2.11 Undrained shear strength (s_u) along the depth obtained in Example 6.

TABLE 2.4
Undrained Shear Strength (s_u) Data Obtained in Example 6

Borehole I		Borehole II		Borehole III		Borehole IV		Borehole V	
Depth (m)	s_u (kPa)	Depth (m)	s_u (kPa)	Depth (m)	s_u (kPa)	Depth (m)	s_u (kPa)	Depth (m)	s_u (kPa)
1	15.67	1	11.88	1	8.03	1	13.32	1	12.30
2	18.73	2	29.82	2	16.18	2	20.61	2	19.52
3	22.02	3	28.98	3	32.26	3	44.90	3	23.50
4	49.97	4	32.83	4	43.57	4	53.84	4	43.95
5	30.31	5	47.98	5	69.95	5	67.98	5	51.89
6	78.47	6	63.20	6	100.64	6	68.13	6	40.47
7	62.85	7	84.41	7	75.78	7	90.11	7	99.72
8	101.67	8	81.83	8	108.58	8	109.97	8	105.68
9	104.58	9	109.23	9	140.21	9	114.92	9	97.21
10	121.44	10	126.54	10	120.22	10	151.76	10	101.98
11	91.11	11	77.99	11	125.42	11	119.39	11	96.49
12	148.34	12	124.83	12	141.45	12	135.44	12	187.29
13	172.71	13	175.05	13	180.96	13	176.03	13	120.67
14	125.38	14	150.75	14	192.84	14	143.81	14	184.69
15	145.12	15	183.89	15	132.55	15	214.21	15	213.71
16	169.18	16	195.07	16	170.50	16	208.74	16	196.70
17	224.48	17	167.22	17	195.16	17	219.83	17	158.85
18	240.20	18	236.03	18	118.11	18	246.13	18	175.60
19	257.05	19	194.62	19	297.37	19	205.28	19	286.50
20	269.46	20	235.61	20	230.20	20	199.56	20	212.86

Step 2: Select the potential candidate designs

The objective here is to determine the optimal geometry of the pile foundation with the minimum cost. For illustration purposes, the design space in this design example is taken as $D \in \{0.5\text{ m}, 0.6\text{ m}, 0.7\text{ m}, \ldots, 2.0\text{ m}\}$ and $L \in \{10.0\text{ m}, 11.0\text{ m}, 12.0\text{ m}, \ldots, 20.0\text{ m}\}$.

Step 3: Calculate the reliability index β) for each candidate design

Based upon the calculated bearing capacity (Q) and the load from the upper structure (P), the performance function (g) of the pile foundation is evaluated as follows.

$$Q = Q_B + Q_S = A_B N_B s_{u1} + \sum \pi D \Delta z \left(\alpha s_{u2} \right)$$

$$g = Q - \left(P + \pi \frac{D^2}{4} L \gamma_c \right)$$

As an example, the reliability index (β) for a design of D = 1.0 m and L = 12.0 m is estimated below. First, the mean (μ_P) and standard deviation (σ_P) of the load (P) are estimated.

$$\mu_P = 2000\ (\text{kN}),\ \text{COV}_P = 0.15,\ \sigma_P = \mu_P \cdot \text{COV}_P = 300\ (\text{kN})$$

Based on the characterized statistics of the normalized undrained shear strength, the mean and standard deviation of s_{u1} and s_{u2}, denoted as $\mu_{s_{u1}}$, $\mu_{s_{u2}}$, $\sigma_{s_{u1}}$, and $\sigma_{s_{u2}}$, respectively, can be calculated. The specific calculation method is given below.

$$\begin{aligned}\mu_{s_{u1}} &= \frac{\mu_{s_{un}} L + \mu_{s_{un}} (L+2D)}{2} \\ &= \frac{12\,(\text{kPa/m}) \times 12\,(\text{m}) + 12\,(\text{kPa/m}) \times (12\,(\text{m}) + 2 \times 1(\text{m}))}{2} \\ &= 156\,(\text{kPa})\end{aligned}$$

$$\begin{aligned}\mu_{s_{u2}} &= \frac{1.5\mu_{s_{un}} + \mu_{s_{un}} L}{2} \\ &= \frac{1.5\,(\text{m}) \times 12\,(\text{kPa/m}) + 12\,(\text{kPa/m}) \times 12\,(\text{m})}{2} \\ &= 81(\text{kPa})\end{aligned}$$

$$\begin{aligned}\sigma_{s_{u1}} &= \frac{\sigma_{s_{un}} L + \sigma_{s_{un}} (L+2D)}{2} \\ &= \frac{2.4\,(\text{kPa/m}) \times 12\,(\text{m}) + 2.4\,(\text{kPa/m}) \times (12\,(\text{m}) + 2 \times 1(\text{m}))}{2} \\ &= 31.2\,(\text{kPa})\end{aligned}$$

$$\begin{aligned}\sigma_{s_{u2}} &= \frac{1.5\sigma_{s_{un}} + \sigma_{s_{un}} L}{2} \\ &= \frac{1.5\,(\text{m}) \times 2.4\,(\text{kPa/m}) + 2.4\,(\text{kPa/m}) \times 12\,(\text{m})}{2} \\ &= 16.2\,(\text{kPa})\end{aligned}$$

Then, the mean (μ_Q) and the standard deviation (σ_Q) of the bearing capacity (Q) are calculated.

$$\begin{aligned}\mu_Q &= \mu_{Q_B} + \mu_{Q_S} = A_B N_B \mu_{s_{u1}} + \sum \pi D \Delta z \left(\alpha \mu_{s_{u2}}\right) \\ &= 1102.70\ (\text{kN}) + 1679.50\ (\text{kN}) \\ &= 2782.20\ (\text{kN})\end{aligned}$$

$$\begin{aligned}\sigma_Q &= \sigma_{Q_B} + \sigma_{Q_S} = A_B N_B \sigma_{s_{u1}} + \sum \pi D \Delta z \left(\alpha \sigma_{s_{u2}}\right) \\ &= 220.54\ (\text{kN}) + 335.90\ (\text{kN}) \\ &= 556.44\ (\text{kN})\end{aligned}$$

Meanwhile, the mean (μ_g) and standard deviation (σ_g) of the performance function (g) are calculated.

$$\mu_g = \mu_Q - \left(\mu_P + \pi \frac{D^2}{4} L\gamma_c \right)$$

$$= 2782.20\ (\text{kN}) - (2000\ (\text{kN}) + \pi \times \frac{(1.0\ (\text{m}))^2}{4} \times 12\ (\text{m}) \times 22\ (\text{kN/m}^3))$$

$$= 574.85\ (\text{kN})$$

$$\sigma_g = \sqrt{\sigma_Q{}^2 + \sigma_P{}^2} = \sqrt{\sigma_{Q_B}{}^2 + \sigma_{Q_S}{}^2 + \sigma_P{}^2}$$

$$= \sqrt{(220.54\ (\text{kN}))^2 + (335.90\ (\text{kN}))^2 + (300\ (\text{kN}))^2}$$

$$= 501.46\ (\text{kN})$$

On the basis of the calculated mean (μ_g)and standard deviation (σ_g) of the performance function (g), the reliability index (β) of this candidate design can be estimated as follows.

$$\beta = \frac{\mu_g}{\sigma_g} = \frac{574.85\ (\text{kN})}{501.46\ (\text{kN})} = 1.15$$

For each candidate design, defined as a pair of parameters D and L, in the design space, the corresponding reliability index (β) can be computed by applying the formulations above.

Step 4: Calculate the cost for each candidate design

The cost (C) can be approximated with the volume of the pile. Accordingly, the cost (C) is given by the following equation:

$$C = \pi \frac{D^2}{4} L$$

As an example, the cost (C) for a design of D = 1.0 m and L = 12.0 m is calculated:

$$C = \pi \frac{D^2}{4} L = \pi \times \frac{(1.0\ (\text{m}))^2}{4} \times 12\ (\text{m}) = 9.42\ (\text{m}^3)$$

For each candidate design in the predefined design space, the corresponding cost can readily be computed by applying the formulations above.

Step 5: Select the optimal design in the design space

The calculation results, in terms of the reliability index and cost, of all candidate designs can lead to a relationship between the β and the cost, as illustrated in Figure 2.12.

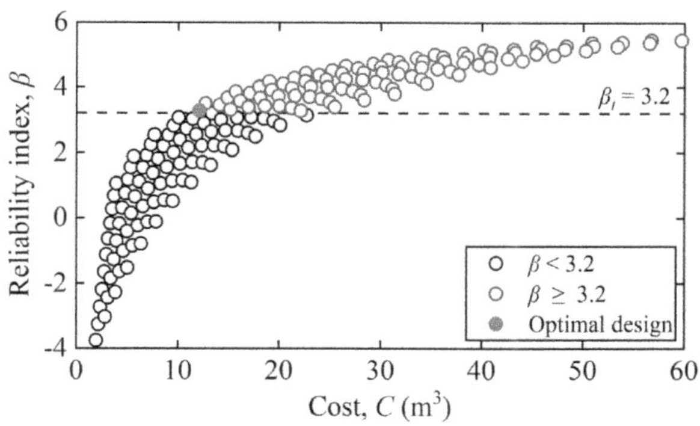

FIGURE 2.12 Relationship between the reliability index and the cost in Example 6.

A design is considered feasible if it satisfies the design requirement of $\beta \geq 3.2$. Among all feasible designs derived, the optimal design for this pile foundation is determined as $D = 0.9$ m and $L = 19.0$ m, with the reliability index (β) of 3.29 and a minimum cost of 12.09 m^3.

Example 7

In reference to Figure 2.13, this example considers the design of a strip foundation in cohesive soil. The load from the upper structure, denoted as P, is a variable following a normal distribution; the mean is 1,200 kN/m, and the COV is 0.15. The groundwater table is far below the foundation base. The basic requirement considered in this strip foundation design is the ultimate bearing capacity, whereas the settlement of this strip foundation is not the concern. Thus, the SLS will not be considered in this design. For illustration purposes, the ultimate bearing capacity of the strip foundation, denoted as q_u, is estimated with the following equation (Cherubini 2000).

$$q_u = 0.5\gamma_1 B N_\gamma + cN_c + \gamma_2 D N_q$$

where B is the width of the foundation; D is the depth of the foundation base with respect to the ground surface; γ_1 and γ_2 are the unit weight of the soil under the foundation base and that over the foundation base; c is the effective cohesion of the soil under the foundation base; and, N_c, N_q and N_γ are the bearing capacity factors, which are estimated as follows.

$$N_\gamma = 1.8(N_q - 1)\tan\varphi$$

$$N_q = \tan^2\left(\frac{\pi}{4} + \frac{\varphi}{2}\right)e^{\pi\tan\varphi}$$

$$N_c = (N_q - 1)\cot\varphi$$

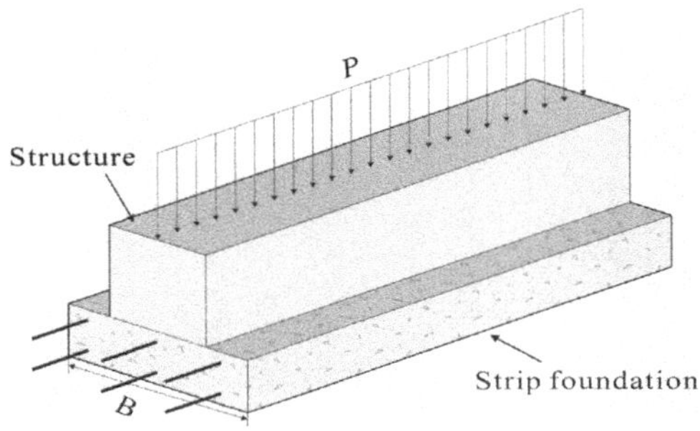

FIGURE 2.13 Schematic diagram of the strip foundation design in Example 7.

where φ is the effective friction angle of the soil under the foundation base. The unit weights of the soils in this example are characterized as $\gamma_1 = 18.0$ kN/m^3 and $\gamma_2 = 18.0$ kN/m^3, while the effective cohesion and the effective friction angle of the soils are characterized as correlated random variables. The mean and COV of the cohesion c are 14.5 kPa and 0.30, respectively; the mean and COV of the friction angle φ are 20.0° and 0.10, respectively, and the correlation coefficient between these two soil parameters is −0.5. In the RBD of this strip foundation, the target reliability index (β_t) is set up as 2.6 (i.e., the related target probability of failure P_{ft} is 4.66×10^{-3}). The problem here is to determine an optimal geometry for this strip foundation with the minimum cost. Note that the cost in this example can be approximated with the product of the geometry factors of this strip foundation (in terms of *BD*).

Solution:

Step 1: Select the potential candidate designs

The objective here is to determine the optimal geometry of the strip foundation with the minimum cost. For illustration purposes, the design space in this design example is taken as $B \in \{1.0$ m, 1.25 m, 1.5 m, . . . , 3.0 m$\}$ and $D \in \{5$, 5.25 m, 5.5 m, . . . , 10.0 m$\}$.

Step 2: Calculate the reliability index (β) for each candidate design

The reliability index (β) of the performance for each candidate design is calculated with the AFORM in this example. In the AFORM approach, the design point is identified as the point on the limit state surface that lies closest to the origin in the standard normal space, and the distance from this point to the origin defines the reliability index (β). First, the uncertain variables, namely the load P, soil cohesion c, and soil internal friction angle φ, are transformed into the equivalent standard normal variables. Let $\boldsymbol{x}$ denote the uncertain variables, $\boldsymbol{x} = \{P, c, \varphi\}$, these variables are characterized by mean values μ_x, standard deviations $\boldsymbol{\sigma}_x$ and a correlation matrix $\boldsymbol{r}_x$, from which the covariance matrix $\boldsymbol{C}_x = diag(\boldsymbol{\sigma}_x)\cdot \boldsymbol{r}_x \cdot diag(\boldsymbol{\sigma}_x)$ can be calculated to describe the covariance between the variables.

Next, the covariance matrix $\boldsymbol{C}_x$ is subjected to eigenvalue decomposition to obtain the eigenvector matrix **a** and the diagonal eigenvalues ***d***. The eigenvalue decomposition could help transform the covariance of the original variables to that in the standard normal space. In the standard normal space, the mean vector of the original uncertain variables ($\boldsymbol{\mu}_x$) is transformed to the mean vector of the normal variables ($\boldsymbol{\mu}_y$) with the aid of the eigenvector matrix **a**, while the standard deviations of the related normal variables can be computed as $\boldsymbol{\sigma}_y = \sqrt{\boldsymbol{d}}$.

```
% x:1-P,load; 2-c,cohesion; 3-fi,friction angle;
nx = length(xmean);
mux = xmean';
sigmax = xstd';
cvarx = diag(sigmax) * rx * diag(sigmax);
[a, d] = eig(cvarx);
muy = a' * mux;
sigmay = sqrt(diag(d));
```

In the standard normal space, the initial design point $\boldsymbol{y}$ can be initialized using the mean $\boldsymbol{\mu}_{y'}$, and the initial reliability index (β) is zero. An iterative optimization process is then performed to find the optimal design point. In each iteration, the limit state function $g(\boldsymbol{x})$ and its gradient $\mathrm{G}|_{x=x'} = \left\{ \left.\frac{\partial g(d,x)}{\partial x_1}\right|_{x=x'}, \left.\frac{\partial g(d,x)}{\partial x_2}\right|_{x=x'}, \left.\frac{\partial g(d,x)}{\partial x_3}\right|_{x=x'} \right\}$ are computed, and the calculated gradient is transformed into the standard normal space to obtain $\mathrm{G}|_{y=y'}$ with the aid of the eigenvector matrix **a**. The directional cosines $\boldsymbol{\alpha}_y$, representing the rate of change along each gradient direction in the standard normal space, can readily be calculated from the gradient $\mathrm{G}|_{y=y'}$.

```
[g, dgdx] = dgdxBearing(x, B, D);
dgdy = a' * dgdx;
alphay = -dgdy .* sigmay/norm(dgdy .* sigmay);
```

The reliability index (β) is updated iteratively based on the gradient information. The updated reliability index (β') is computed using the following formulation.

$$\beta' = \frac{g + \left(\mathbf{G}|_{y=y'}\right)^T (\mu_y - y)}{\left\| \mathbf{G}|_{y=y'} \cdot \sigma_y \right\|}$$

Based on the current value of the reliability index () of the performance for (β'), the design point y in the standard normal space can be updated as:

$$y = \mu_y + \beta' \cdot \sigma_y \cdot \sigma_y$$

Based on the updated reliability index (β') together with the directional cosines, a new design point can readily be constructed. This iterative procedure is repeated until the design point converges, which can be determined when the change in the Euclidean norm between successive design points is smaller than a pre-set tolerance (e.g., 10^{-6}). The complete code of the AFORM implementation is provided below.

```
%% AFORM function
function Beta = AFORMBearing(xmean,xstd,rx,B,D)
% x:1-P,load; 2-c,cohesion; 3-fi,friction angle;
%using AFORM to calcuate the value of beta for given design B,L
   nx=length(xmean);
   mux=xmean';
   sigmax=xstd';
   cvarx=diag(sigmax)*rx*diag(sigmax);
   [a,d]=eig(cvarx);
   muy=a'*mux;
   sigmay=sqrt(diag(d));
   x=mux;
   y=a'*x;
   normx=eps;
   ni=0;
   while abs(norm(x)-normx)/normx>1e-6
     ni=ni+1;
     normx=norm(x);
     [g,dgdx]=dgdxBearing(x,B,D);
     dgdy=a'*dgdx;
     alphay=-dgdy.*sigmay/norm(dgdy.*sigmay);
     beta=(g+dgdy'*(muy-y))/norm(dgdy.*sigmay);
     y=muy+beta*sigmay.*alphay;
     x=a*y;
   end
   Beta=beta;
end
```

The performance of the strip foundation is evaluated as follows:

$$g = Q - P = q_u B - P$$

To derive the ultimate bearing capacity of the strip foundation (q_u), we need to compute bearing capacity factors N_q, ψ_S, and N_c as follows.

$$N_\gamma = 1.8(N_q - 1)\tan\varphi,\ N_q = \tan^2\left(\frac{\pi}{4}+\frac{\varphi}{2}\right)e^{\pi\tan\varphi},\ N_c = (N_q - 1)\cot\varphi$$

The calculated bearing capacity factors (N_q, N_γ, N_c) are subsequently applied to compute the ultimate bearing capacity (q_u).

$$q_u = 0.5\gamma_1 BN_\gamma + cN_c + \gamma_2 DN_q$$

The computation of the performance function in this design example is implemented with the following code.

```
%% Performance function
function Gs=BearingOfFoundation(x,B,D)
% x:1-P,load; 2-c,cohesion; 3-fi,friction angle;
% design parameters: B, width; D, length
   P=x(1);
   c=x(2);
   fi=x(3);
   Nq = exp(pi*tand(fi))*tand(45+fi/2)^2;
   Nc = (Nq-1)*cotd(fi);
   Ngamma = 1.8*(Nq-1)*tand(fi);
   qu = 0.5*18*B*Ngamma+c*Nc+18*D*Nq;
   Gs = qu*B-P;
end
```

The gradient of the performance function in this example is evaluated using the central difference method, and its computation code is provided below.

```
%% Gradient of Gs
function [g,dgdx]=dgdxBearing(x,B,D)
% x:1-P,load; 2-c,cohesion; 3-fi,friction angle;
   nx=length(x);
   g=BearingOfFoundation(x,B,D);
   deltax=0.1;
   dgdx=zeros(nx,1);
   for i=1:nx
         dx=x(i)*deltax;
         tempX=x;
         tempX(i)=x(i)-dx/2;
         GS1=BearingOfFoundation(tempX,B,D);
         tempX=x;
         tempX(i)=x(i)+dx/2;
         GS2=BearingOfFoundation(tempX,B,D);
         dgdx(i)=(GS2-GS1)/dx;
   end
end
```

For each candidate design, defined as a pair of parameters B and D, in the design space, the reliability index (β) can be computed by applying the AFORM. As an example, the reliability index (β) for a candidate design of B = 3.0 m and D = 10 m is calculated as 6.39.

Step 3: Calculate the cost for each candidate design

The cost (C) of the strip foundation is calculated using its cross-sectional dimensions: width (B) and depth (D). Accordingly, the cost (C) is given by the following equation:

$$C = BD$$

As an example, the cost (C) for a design of B = 3.0 m and D = 10 m is calculated below.

$$C = 3\ (\mathrm{m}) \times 10\ (\mathrm{m}) = 30\ (\mathrm{m}^2)$$

For each candidate design in the predefined design space, the corresponding cost can readily be computed by applying the formulations above.

Step 4: Select the optimal design in the design space

The calculation results, in terms of the reliability index and cost, of all candidate designs can lead to a relationship between the reliability index and the cost, as illustrated in Figure 2.14.

A design is considered feasible if it satisfies the requirement of $\beta \geq 2.6$. Among all feasible designs derived, the optimal design for this strip foundation is determined as B = 2.4 m and D = 5.0 m, with the reliability index (β) of 2.82 and a minimum cost of 12.5 m^2.

Example 8

The general setting of this design example is identical to that in Example 4; however, the geotechnical parameters are characterized here as uncertain variables. For example, the unit weight (γ_s) of the backfill follows a normal distribution with a mean

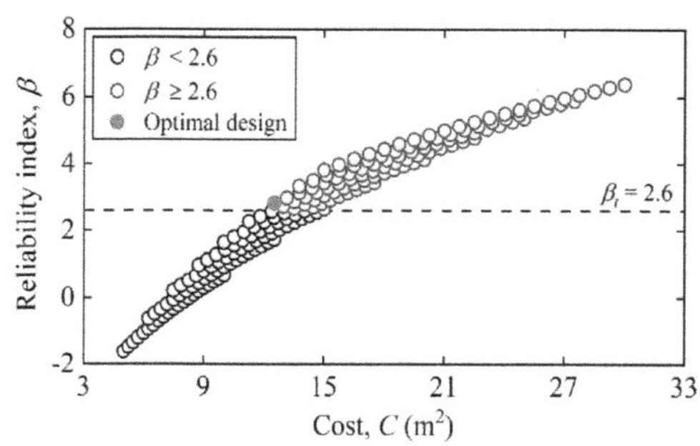

FIGURE 2.14 Relationship between the reliability index and the cost in Example 7.

of 18 kN/m³ and a COV of 0.07. The friction angle (φ) follows a normal distribution with a mean of 35° and a COV of 0.10. The friction between the backfill and the retaining wall (δ) follows a normal distribution with a mean of 20° and a COV of 0.10. The adhesion between the wall base and the clay (c_a) follows a normal distribution with a mean of 100 kPa and a COV of 0.15. These uncertain parameters are assumed to be statistically independent. Considering the different consequences of the sliding and overturning failures, we adopt different target values of the reliability index for these two failure modes: the target reliability index with respect to the sliding failure, denoted as β_{t1}, is specified as 2.6, while that with respect to the overturning failure, denoted as β_{t2}, is set up as 3.2. The unit weight of the wall (γ_c) is 25 kN/m³. The problem here is to determine an optimal geometry of this retaining wall that meets the target reliability index requirements with the minimum cost.

Solution:

Step 1: Select the potential candidate designs

The objective in this design example is to determine the optimal geometry of the gravity retaining wall that meets the target reliability index requirements with the minimum cost. For illustration purposes, the design space in this design example is taken as $a \in \{0.2\text{ m}, 0.4\text{ m}, 0.6\text{ m}\}$ and $b \in \{0.5\text{ m}, 0.6\text{ m}, 0.7\text{ m}, \ldots, 3.0\text{ m}\}$.

Step 2: Calculate the reliability index *(β)* for each candidate design

The reliability index (β) of the performance for each candidate design in this example is calculated with the MCS method. In the MCS approach, samples of uncertain variables are first generated, and these samples are then input into the performance function to estimate the design performance, from which the failure probability is estimated by counting the number of failure samples. In this example, two failure modes are considered, namely sliding failure and overturning failure. The performance functions with respect to the sliding and overturning failure modes, denoted as g_1 and g_2, respectively, can be formulated as follows:

$$g_1 = bc_a - P_{ah}$$

$$g_2 = W_1A_1 + W_2A_2 + P_{av}A_{ah} - P_{ah}A_{av}$$

To derive the performance function with respect to the sliding and overturning failure, we need to compute the Coulomb active coefficient K_a as follows.

$$K_a = \left[\frac{\sin(\alpha-\varphi)/\sin\alpha}{\sqrt{\sin(\alpha+\delta)}+\sqrt{\sin(\varphi+\delta)\sin(\varphi-\lambda)/\sin(\alpha-\lambda)}}\right]^2$$

Subsequently, the resultant earth pressure P_a is calculated, together with its horizontal (P_{av}) and vertical (P_{ah}) components.

$$P_a = 0.5K_a\gamma_s H^2,\ P_{av} = P_a\sin\delta,\ P_{ah} = P_a\cos\delta$$

Then, the distances from the acting point of the earth pressure to the wall toe are calculated, namely the vertical distance (A_{av}) and the horizontal distance (A_{ah}).

$$A_{av} = H/3,\ A_{ah} = b$$

Meanwhile, the self-weights of the wall sections, W_1 and W_2, are calculated, along with their respective lever arms A_1 and A_2, measured from the wall toe:

$$W_1 = 0.5\gamma_{\text{wall}}(b-a)H,\ W_2 = \gamma_{\text{wall}}aH$$

The code for computing the performance functions in this example is provided below.

```
%% Performance function
function Gs = getGs(x, t)
    H = 6.0;
    gw = 24;
    a = t(1);
    b = t(2);
    gs = x(1);
    fi = x(2);
    delta = x(3);
    ca = x(4);
    Pa = 0.5 * Ka(fi, delta) * gs * H^2;
    Pav = Pa * sind(delta);
    Pah = Pa * cosd(delta);
    W1 = 0.5 * gw * (b – a) * H;
    A1 = 2 * (b – a) / 3;
    W2 = gw * a * H;
    A2 = b – 0.5 * a;
    Aah = H / 3;
    Aav = b;
    Gs(1, 1) = (b * ca) – Pah; % sliding failure
    Gs(1, 2) = (W1 * A1 + W2 * A2 + Pav * Aav) – (Pah * Aah); % overturning
end
function Ka = Ka(fi, delta)
    lambda = 10;
    aerfa = 90;
    temp1 = sind(aerfa – fi)/sind(aerfa);
    temp2 = sqrt(sind(aerfa + delta)) + sqrt(sind(fi + delta) * sind(fi – lambda)/
            sind(aerfa – lambda));
    Ka = (temp1 / temp2)^2;
end
```

With the aid of the failure probability estimated by the MCS, the reliability index (β) can readily be computed as follows.

$$\beta = \Phi^{-1}(1 - P_f)$$

where P_f is the failure probability of the gravity retaining wall derived from the MCS; $\Phi^{-1}()$ is the inverse cumulative distribution function (CDF) of the standard normal distribution. In this example, 100,000 samples are generated to derive the failure probability. The code of the adopted MCS approach is provided below.

```
function [Beta] = MCS(xmean, xstd, rx, t)
    % x: x(1), unit weight of the backfill; x(2), friction angle of the backfill;
    % x(3), friction between the backfill and the retaining wall;
    % x(4), adhesion between the wall base and the clay;
    % Design parameters: a, the top width; b, the base width
    % Using MCS to calculate the value of beta for given design t = [a, b]
    nx=length(xmean);
    mux=xmean';
    sigmax=xstd';
    cvarx=diag(sigmax)*rx*diag(sigmax);
    N = 10e4;
    samples = mvnrnd(mux, cvarx, N);
    Gs_samples = zeros(N, 2);
    samples_failure1 =0;
    samples_failure2 =0;
    for i = 1:N
        Gs = getGs(samples(i,:), t);
        Gs_samples(i,:) = Gs;
        if Gs(1)<0
            samples_failure1=samples_failure1 +1;
        end
        if Gs(2)<0
            samples_failure2=samples_failure2 +1;
        end
    end
    pf1 = samples_failure1 / N;
    pf2 = samples_failure2 / N;
    beta1 = norminv(1 – pf1);
    beta2 = norminv(1 – pf2);
    Beta = [beta1,beta2];
    end
```

For each candidate design, defined as a pair of parameters a and b, in the design space, the reliability indexes (i.e., β_1 for the sliding failure and β_2 for the overturning failure) can be estimated with the MCS method. As an example, the reliability indexes for a design of a = 0.2 m and b = 2.0 m are calculated as β_1 = 3.39 and β_2 = 4.11.

Step 3: Calculate the cost for each candidate design

The cost (C) of the gravity-based retaining wall is calculated using its cross-sectional dimension. Accordingly, the cost (C) is given by the following equation.

$$C = 0.5(b - a)H + aH$$

As an example, the cost (C) for a design of a = 0.2 m and b = 2.0 m is calculated.

$$C = 0.5 \times \left[2.0\ (\text{m}) - 0.2\ (\text{m})\right] \times 6\ (\text{m}) + 0.2\ (\text{m}) \times 6\ (\text{m}) = 6.6\ (\text{m}^2)$$

For each candidate design in the predefined design space, the corresponding cost can readily be computed by applying the formulations above.

Step 4: Select the optimal design in the design space

The calculation results, in terms of the reliability indexes and cost, of all candidate designs can lead to a relationship between the reliability indexes and the cost, as illustrated in Figure 2.15.

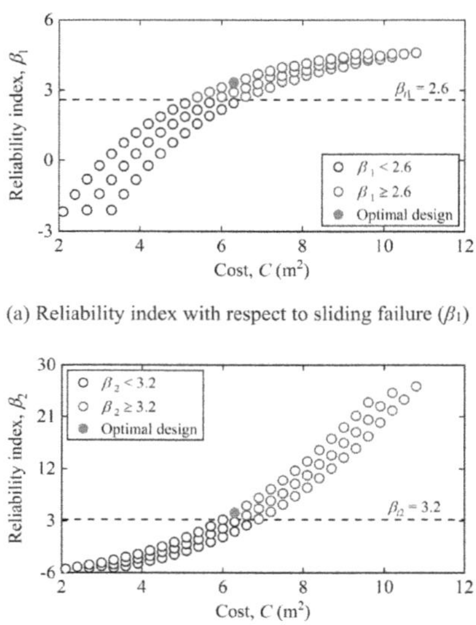

(a) Reliability index with respect to sliding failure (β_1)

(b) Reliability index with respect to overturning failure (β_2)

FIGURE 2.15 Relationship between the reliability indexes and the cost in Example 8.

A design is considered feasible if it satisfies the requirements of $\beta_1 \geq 2.6$ and $\beta_2 \geq 3.2$. Among all feasible designs derived, the optimal design for this strip foundation is determined as $a = 0.2$ m and $b = 1.9$ m, with the β_1 of 3.30, β_2 of 3.44, and a minimum cost of 6.3 m^2.

.4 LOAD AND RESISTANCE FACTOR DESIGN IN GEOTECHNICAL ENGINEERING

he principle of the LRFD is first presented in this section, followed by four illustra-ve geotechnical design examples.

.4.1 Principle of the LRFD

ecause of the challenge in the accurate statistical characterization of the geotech-ical parameters and the computational inefficiency of the probabilistic analysis (for stimating the reliability indexes of candidate geotechnical designs), the RBD has ot been widely adopted in geotechnical practices. Instead, the LRFD, a simpler ariant of the RBD by design, has been more commonly recommended in various odes (e.g., NSAI 2005; BSI 2007; NEN 2007; Standards Australia 2009; AASHTO 014). The LRFD is designed to take advantage of both FS-based design and RBD. he implementation of the LRFD is similar to that of the FS-based design. In con-ast, the safety of the design is ensured with the aid of partial factors and through the onservative estimate of uncertain variables, for which only the deterministic eval-ation of the design performance or safety is required. Note that the partial factors dopted in the LRFD were calibrated primarily to achieve a target reliability index s in the RBD) over a range of design scenarios covered by the codes. Thus, the leal outcome is that the design obtained by the LRFD can achieve the same target eliability index as that obtained by the RBD. Within the context of the LRFD, the easibility of a candidate design can be assessed using the formulation below.

$$g(\boldsymbol{d},\boldsymbol{\theta}^n) = \psi_R R^n(\boldsymbol{d},\boldsymbol{\theta}^n) - \psi_S S^n(\boldsymbol{d},\boldsymbol{\theta}^n) \quad (2.6)$$

here $R^n(\boldsymbol{d}, \boldsymbol{\theta}^n)$ and $S^n(\boldsymbol{d}, \boldsymbol{\theta}^n)$ are the characteristic or nominal values of the resistance erm and load term, respectively; $\boldsymbol{\theta}^n$ are the characteristic or nominal values of the ncertain variables $\boldsymbol{\theta}$; and F_d and ψ_S are the resistance and load factors (i.e., the artial factors), respectively.

It is noted that although the characteristic values of the resistance term and load erm shown in Eq. (2.6) are formulated as functions of the characteristic values of e uncertain variables $\boldsymbol{\theta}^n$, these two characteristic values, $R^n(\boldsymbol{d}, \boldsymbol{\theta}^n)$ and $S^n(\boldsymbol{d}, \boldsymbol{\theta}^n)$, ould also be estimated directly in the LRFD. Similar to that defined in the FS-based esign, the safety of a trial design $\boldsymbol{d}$ can be assessed as follows: if $g(\boldsymbol{d}, \boldsymbol{\theta}^n) > 0$, the afety or feasibility requirement of the design $\boldsymbol{d}$ is satisfied; otherwise, this trial or can-idate design $\boldsymbol{d}$ is considered infeasible or unsafe. It should be noted that the LRFD erformance function in Eq. (2.6) is forWmulated in a general form, and the LRFD pproach is applicable to scenarios involving multiple load and resistance terms.

In general, the resistance factor (ψ_R) is less than 1.0 while the load factor (ψ is greater than 1.0. The characteristic values ($\boldsymbol{\theta}^n$) may be selected as the mean some percentile (e.g., 25% for the stabilizing parameter and 75% for the destabilizi parameter) of the uncertain variables ($\boldsymbol{\theta}$). As mentioned above, the LRFD is design to be a simpler variant of the RBD; thus, iterative probabilistic analyses of the co cerned geotechnical problem must be conducted to calibrate the partial factors.

Note that the calibration of the partial factors in the LRFD is a challenging task, various design scenarios (e.g., different levels of variation of the geotechnical param eters and model errors and target reliability indexes) must be covered, which rais the complexity of the partial factors calibration. As a result, in some published LRF codes, the partial factors were calibrated based on the FS-based design requirements

During the design application of the LRFD, the determination of the characterist values of the uncertain variables ($\boldsymbol{\theta}$) must be consistent with the calibration meth of the partial factors. Indeed, the selection of the partial factors and determination the characteristic values of the uncertain variables are problem- and code-specifi As an example, Table 2.5 lists some resistance factors suggested by AASHTO (201 for the pile foundation design.

Finally, a couple of issues worth considering for the practical implementation the LRFD are discussed below. First, in a given design scenario, the use of the sta dard LRFD code may lead to a geotechnical design that deviates from the targ reliability index by an unknown amount (more likely on the conservative side, b under-design is also a possibility). This deviation may be due to the fact that th design scenarios considered in the calibration of partial factors did not cover a possible design scenarios encountered in geotechnical practices. This drawback h been well recognized (Honjo et al. 2002; Paikowsky et al. 2004; Honjo and Amaty 2005; Najjar and Gilbert 2009; Stuedlein et al. 2012; Park et al. 2013), and sugge tions have been made to improve the LRFD with refined partial factors (e.g., Phoc et al. 2003b; Ching and Phoon 2011; Ching et al. 2013). Second, the selection the characteristic or nominal values of the uncertain variables should consider th knowledge of how the partial factors were calibrated in the codes. In the absenc

TABLE 2.5
Some Resistance Factors for Pile Foundation Design

Limit state	Component of resistance	Geomaterial	Equation or method	Resistanc factor (ψ_R
Geotechnical Axial Resistance	Side resistance in compression/uplift	Cohesionless soil	Beta method	0.55/0.45
		Cohesive soil	Alpha method	0.45/0.35
		Cohesive IGM	Modified alpha method	0.60/0.50
	Base resistance	Cohesionless soil	N-value	0.50

Source: AASHTO, 2010.

of this knowledge, a conservative estimate may be taken in practice based on an estimate of the mean and the level of variation (e.g., Orr 2000; Driscoll and Simpson 2001), aided by local experience.

2.4.2 Illustrative Examples

Four design examples are presented below to illustrate the application of this LRFD. These are similar to the design examples presented previously using the FS-based method and the RBD method.

Example 9

The background of this example is similar to that in Example 6. The pile foundation in this example is, however, designed to bear a dead load and a live load, denoted as F_d and F_l, respectively. The values of the dead load (F_d) and live load (F_l) are estimated as 300 kN and 150 kN, respectively. The undrained shear strength data at this site are given in Example 6, and the mathematical formulation for the bearing capacity of the pile foundation in cohesive soils is available in Example 2. The pile foundation in this example is required to be designed with the load and resistance factors recommended in AASHTO (2014), in which the design safety of a candidate design can be evaluated with the following mathematical formulation.

$$g(\boldsymbol{d},\boldsymbol{\theta}") = \psi_{Q_s} Q_s + \psi_{Q_b} Q_b - \psi_{F_d} F_d - \psi_{F_l} F_l$$

where ψ_{Q_s} and ψ_{Q_b} are the resistance factors with respect to side resistance (g_1) and base resistance (Q_b), respectively; and, ψ_{F_d} and ψ_{F_l} are the load factors with regard to dead load (F_d) and live load (F_l), respectively. According to AASHTO (2014), the partial factors involved in this design example are as follows: ψ_{Q_s} = 0.45, ψ_{Q_b} = 0.40, ψ_{F_d} = 1.25, and ψ_{F_l} = 1.75. To estimate the characteristic values of side resistance (Q_s) and base resistance (Q_b) of the pile, the characteristic values of the undrained shear strength can be taken as the average value of the undrained shear strength over the depth range of concern. The problem in this example is to determine an optimal geometry for this pile foundation with the minimum cost (the cost can be approximated with the pile volume).

Solution:

Step 1: Select the potential candidate designs

The objective in this design example is to determine the optimal geometry of the pile foundation with the minimum cost. For illustration purposes, the design space in this design example is taken as $D \in$ {0.3 m, 0.4 m, 0.5 m, . . . , 0.6 m} and $L \in$ {10.0 m, 10.5 m, 11.0 m, . . . , 25.0 m}. These ranges could cover practical engineering values for pile foundations.

Step 2: Evaluate the design performance of each candidate design

The performance of candidate designs can be evaluated with the LRFD method. Based on the partial factors provided, the performance of the pile can be calculated as follows:

$$g(\boldsymbol{d},\boldsymbol{\theta}^{n}) = 0.45Q_s + 0.4Q_b - 1.25F_d - 1.75F_l$$

$$Q_s = \sum \pi D \Delta z(-s_{u2}^{n})$$

$$Q_b = A_B N_B s_{u1}^{n}$$

To estimate the characteristic values of the side resistance (Q_s) and the base resistance (Q_b) of the pile, the characteristic values of the undrained shear strength can be taken as the average value of the undrained shear strength over the depth range of concern. In other words, the characteristic value of the normalized undrained shear strength (s_{un}^{n}) is taken as 12 kPa/m, based on the statistical information of the undrained shear strength given in Example 6.

As an example, the design performance for a design of $D = 0.6$ m and $L = 22$ m is calculated below. First, the normalized undrained shear strength (s_{un}^{n}) is used to calculate the concerned undrained shear strength s_{u1}^{n} and s_{u2}^{n} as follows.

$$\begin{aligned} s_{u1}^{n} &= \frac{s_{un}^{n}L + s_{un}^{n}(L+2D)}{2} \\ &= \frac{12\ (\text{kPa/m}) \times 22\ (\text{m}) + 12\ (\text{kPa/m}) \times (22\ (\text{m}) + 2 \times 0.6\ (\text{m}))}{2} \\ &= 271.2\ (\text{kPa}) \end{aligned}$$

$$s_{u2}^{n} = \frac{1.5s_{un}^{n} + s_{un}^{n}L}{2} = \frac{1.5\ (\text{m}) \times 12\ (\text{kPa/m}) + 12\ (\text{kPa/m}) \times 22\ (\text{m})}{2} = 141\ (\text{kPa})$$

Then, the side resistance (Q_s) and the base resistance (Q_b) are calculated based on the derived undrained shear strength s_{u1}^{n} and s_{u2}^{n} as follows.

$$\begin{aligned} Q_s &= \pi \mathrm{D}(\mathrm{L} - 1.5)(0.55\mathrm{s}_{\mathrm{u2}}^{\mathrm{n}}) \\ &= \pi \times 0.6\ (\text{m}) \times (22 - 1.5)\ (\text{m}) \times 0.55 \times 141\ (\text{kPa}) \\ &= 2996.70\ (\text{kN}) \end{aligned}$$

$$Q_b = A_B N_B s_{u1}^{n} = 0.28\ (\text{m}^2) \times 9 \times 271.20\ (\text{kPa}) = 690.12\ (\text{kN})$$

Finally, the design performance of the candidate design is evaluated with the partial factors provided in this example ($\psi_{Q_s} = 0.45$, $\psi_{Q_b} = 0.40$, $\psi_{F_d} = 1.25$, and $\psi_{F_l} = 1.75$).

$$g(\boldsymbol{d},\boldsymbol{\theta}^{n}) = 0.45Q_s + 0.4Q_b - 1.25F_d - 1.75F_l = 871.72\ (\text{kN})$$

For each candidate design, defined as a pair of parameters D and L, in the design space, the corresponding design performance can be computed by applying this LRFD method.

Step 3: Calculate the cost for each candidate design

The cost (C) of the pile foundation in this example is approximated using its volume. Accordingly, the cost (C) is given by the following equation.

$$C = \frac{\pi D^2}{4} L$$

As an example, the cost (C) for a design of $D = 0.6$ m and $L = 22$ m is calculated.

$$C = \frac{\pi D^2}{4} L = \frac{\pi \times (0.6\ (\text{m}))^2}{4} \times 22\ (\text{m}) = 6.22\ (\text{m}^3)$$

For each candidate design in the predefined design space, the corresponding cost can readily be computed by applying the formulations above.

Step 4: Select the optimal design in the design space

The calculation results, in terms of the performance and cost, of all candidate designs can lead to a relationship between the performance and the cost, as shown in Figure 2.16.

A design is considered feasible if it satisfies the requirement of $g \geq 0$. Among all feasible designs derived, the optimal design for this pile foundation is determined as $D = 0.3$ m and $L = 21.5$ m, with the design performance of 0.09 kN and a minimum cost of 1.52 m³.

Furthermore, the optimal design results are obtained with the 25th percentile value of the normalized undrained shear strength as the characteristic value. Given that the normalized undrained shear strength follows a normal distribution with a mean of 12 kPa/m and a standard deviation of 2.4 kPa/m, the 25th percen-

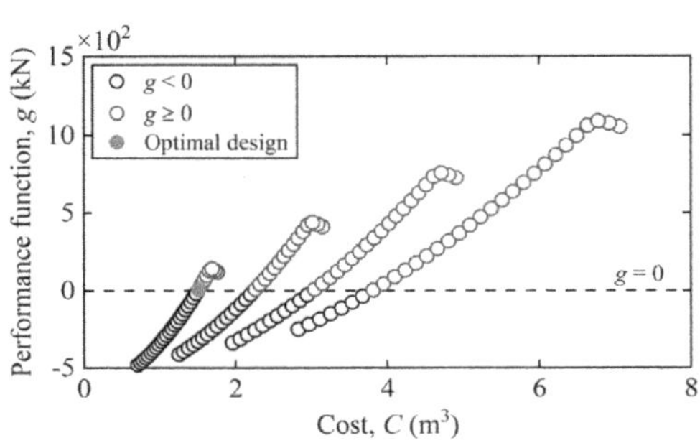

FIGURE 2.16 Relationship between the performance and the cost in Example 9.

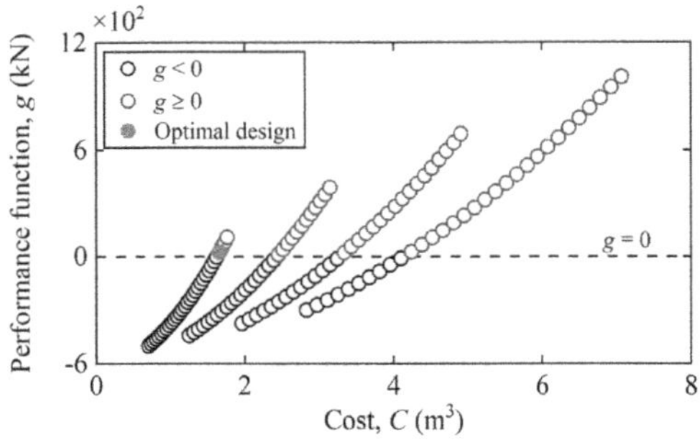

FIGURE 2.17 Relationship between the performance (evaluated with the 25th percentile value of the undrained shear strength as the characteristic value) and the cost in Example 9.

tile value can be determined using the inverse cumulative distribution function (CDF) of the normal distribution. Specifically, it is calculated as:

$$s_{un}^{\,n} = 12\ (\text{kPa/m}) + (-0.674)\times 2.4\ (\text{kPa/m}) = 10.38\ (\text{kPa/m})$$

With this 25th percentile value as the characteristic value, the computed performance of all candidate designs, along with the cost, is illustrated in Figure 2.17. Among all feasible designs derived, the optimal design for this pile foundation is determined as $D = 0.3$ m and $L = 23.5$ m, with the design performance of 0.25 kN and a minimum cost of 1.66 m^3. As can be expected, a more conservative characteristic value will lead to a more conservative design.

Example 10

This design example is similar to that in Example 9; however, the load and resistance factors recommended in the UK National Annex to Eurocode 7 (BSI 2007) are employed. In this illustrative design example, two sets of calculations, DA1–1 and DA1–2, are conducted to evaluate the safety of the candidate design ***d***.

$$\text{DA1-1: } g_1(\boldsymbol{d},\boldsymbol{\theta}'') = \frac{Q_s + Q_b}{1.4} - (1.35F_d + 1.5F_l)$$

$$\text{DA1-2: } g_2(\boldsymbol{d},\boldsymbol{\theta}'') = \frac{\dfrac{Q_s}{1.6} + \dfrac{Q_b}{2.0}}{1.4} - (1.0F_d + 1.3F_l)$$

where the feasibility of the design ***d*** is evaluated with the minimum value of $g_1(\boldsymbol{d},\boldsymbol{\theta}'')$ and $g_2(\boldsymbol{d},\boldsymbol{\theta}'')$. It should be mentioned that within the context of Eurocode 7 (Vardanega et al. 2012), the characteristic values of the undrained shear strength in this design example are taken as the 25th percentile of the undrained shear strength to calculate the side resistance (Q_s) and base resistance (Q_b) of the pile. The problem here is to determine an optimal geometry for this pile foundation with the minimum cost (the cost is approximated with the pile volume herein).

Solution:

Step 1: Select the potential candidate designs

The objective in this example is to determine the optimal geometry of the pile with the minimum cost. For illustration purposes, the design space in this design example is taken as $D \in \{0.3 \text{ m}, 0.4 \text{ m}, 0.5 \text{ m}, \ldots, 0.6 \text{ m}\}$ and $L \in \{10.0 \text{ m}, 10.5 \text{ m}, 11.0 \text{ m}, \ldots, 25.0 \text{ m}\}$.

Step 2: Evaluate the design performance of each candidate design

The performance of candidate designs in this design example is assessed with the LRFD method, and the load and resistance factors recommended in the UK National Annex to Eurocode 7 (BSI 2007) are employed. Based on the partial factors provided, the performance of the pile foundation can be calculated as follows:

$$\text{DA1-1: } g_1(\boldsymbol{d},\boldsymbol{\theta}") = \frac{Q_s + Q_b}{1.4} - (1.35F_d + 1.5F_l)$$

$$\text{DA1-2: } g_2(\boldsymbol{d},\boldsymbol{\theta}") = \frac{\frac{Q_s}{1.6} + \frac{Q_b}{2.0}}{1.4} - (1.0F_d + 1.3F_l)$$

$$Q_s = \sum \pi D \Delta z (\alpha s_{u2}^n)$$

$$Q_b = A_B N_B s_{u1}^n$$

To estimate the characteristic values of the side resistance (Q_s) and the base resistance (Q_b) of the pile, the characteristic value of the undrained shear strength is taken as the 25th percentile of the undrained shear strength. Given that the normalized undrained shear strength follows a normal distribution with a mean of 12 kPa/m and a standard deviation of 2.4 kPa/m, the 25th percentile value can be determined using the inverse cumulative distribution function (CDF) of the normal distribution. Specifically, it is calculated as:

$$s_{un}^n = 12 \text{ (kPa/m)} + (-0.674) \times 2.4 \text{ (kPa/m)} = 10.38 \text{ (kPa/m)}$$

As an example, the design performance for a candidate design of $D = 0.6$ m and $L = 22$ m is calculated below. First, the normalized undrained shear strength (s_{un}^n) is used to calculate the concerned undrained shear strength s_{u1}^n and s_{u2}^n as follows.

$$\begin{aligned} s_{u1}^n &= \frac{s_{un}^n L + s_{un}^n (L + 2D)}{2} \\ &= \frac{10.38 \text{ (kPa/m)} \times 22 \text{ (m)} + 10.38 \text{ (kPa/m)} \times (22 \text{ (m)} + 2 \times 0.6 \text{ (m)})}{2} \\ &= 234.59 \text{(kPa)} \end{aligned}$$

$$s_{u2}^n = \frac{1.5s_{un}^n + s_{un}^n L}{2}$$
$$= \frac{1.5\ (\text{m}) \times 10.38\ (\text{kPa/m}) + 10.38\ (\text{kPa/m}) \times 22\ (\text{m})}{2}$$
$$= 121.97\ (\text{kPa})$$

Then, the side resistance (Q_s) and the base resistance (Q_b) are calculated based on the derived undrained shear strength s_{u1}^n and s_{u2}^n as follows.

$$Q_s = \sum \pi D \Delta z(-s_{u2}^n)$$
$$= \pi \times D \times (L - 1.5) \times - \times s_{u2}^n$$
$$= \pi \times 0.6\ (\text{m}) \times (22\ (\text{m}) - 1.5) \times 0.55 \times 121.97\ (\text{kPa})$$
$$= 2592.21\ (\text{kN})$$

$$Q_b = A_B N_B s_{u1}^n = \frac{\pi \times 0.6\ (\text{m}) \times 0.6\ (\text{m}) \times 9 \times 234.59 (\text{kPa})}{4} = 596.96\ (\text{kN})$$

Finally, the design performance of this candidate design is evaluated as follows, and the feasibility of this design $\boldsymbol{d}$ is assessed with the minimum value of $g_1(\boldsymbol{d}, \boldsymbol{\theta}^n)$ and $g_2(\boldsymbol{d}, \boldsymbol{\theta}^n)$.

$$g_1(\boldsymbol{d}, \boldsymbol{\theta}^n) = \frac{Q_s + Q_b}{1.4} - (1.35F_d + 1.5F_l)$$
$$= \frac{2592.21\ (\text{kN}) + 596.96\ (\text{kN})}{1.4} - (1.35 \times 300\ (\text{kN}) + 1.5 \times 150 (\text{kN}))$$
$$= 1647.98\ (\text{kN})$$

$$g_2(\boldsymbol{d}, \boldsymbol{\theta}^n) = \frac{\frac{Q_s}{1.6} + \frac{Q_b}{2.0}}{1.4} - (1.0F_d + 1.3F_l)$$
$$= \frac{2592.21\ (\text{kN})/1.6 + 596.96\ (\text{kN})/2.0}{1.4} - (1.0 \times 300\ (\text{kN}) + 1.3 \times 150 (\text{kN})$$
$$= 875.44\ (\text{kN})$$

For each candidate design, defined as a pair of parameters D and L, in the design space, the corresponding design performance can be computed by applying this LRFD method.

Step 3: Calculate the cost for each candidate design

The cost (C) of the pile foundation in this example is approximated using its volume. Accordingly, the cost (C) is given by the following equation.

$$C = \frac{\pi D^2}{4} L$$

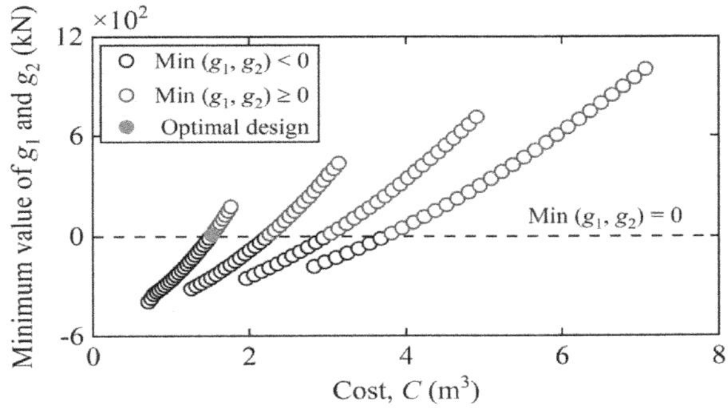

FIGURE 2.18 Relationship between the performance Min (g_1, g_1) and the cost in Example 10.

As an example, the cost (C) for a design of $D = 0.6$ m and $L = 22$ m is calculated.

$$C = \frac{\pi D^2}{4} L = \frac{\pi \times (0.6\ (\text{m}))^2}{4} \times 22\ (\text{m}) = 6.22\ (\text{m}^3)$$

For each candidate design in the predefined design space, the corresponding cost can readily be computed by applying the formulations above.

Step 4: Select the optimal design in the design space

The calculation results, in terms of the design performance and cost, of all candidate designs can lead to a relationship between the design performance, in terms of Min (g_1, g_2), and the cost, as illustrated in Figure 2.18.

A design is considered feasible if it satisfies the requirement of Min(g_1, g_2) > 0. Among all feasible designs derived, the optimal design for this pile foundation is determined as $D = 0.3$ m and $L = 21.5$ m, with the design performance of 5.74 kN and a minimum cost of 1.52 m³.

Furthermore, the optimal design results are obtained with the 5th percentile value of the normalized undrained shear strength as the characteristic value. Given that the normalized undrained shear strength follows a normal distribution with a mean of 12 kPa/m and a standard deviation of 2.4 kPa/m, the 5th percentile value can be determined using the inverse cumulative distribution function (CDF) of the normal distribution. Specifically, it is calculated as:

$$s_{un}^{n} = 12\ (\text{kPa/m}) + (-1.645) \times 2.4\ (\text{kPa/m}) = 8.05\ (\text{kPa/m})$$

With this 5th percentile value as the characteristic value, the computed performance of all candidate designs, along with the cost, is illustrated in Figure 2.19. Among all feasible designs derived, the optimal design for this pile foundation is

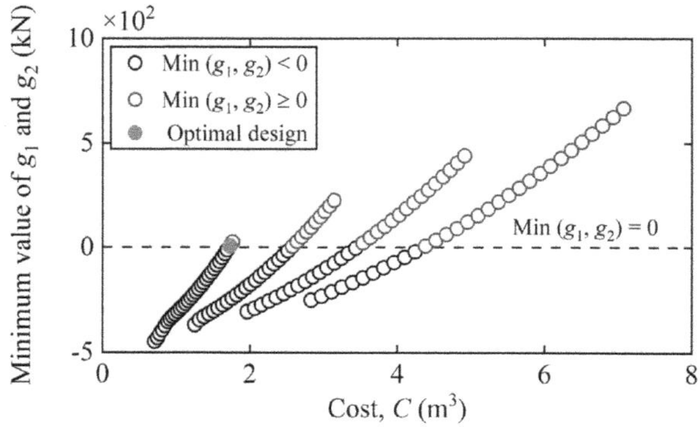

FIGURE 2.19 Relationship between the performance Min (g_1, g_2) (evaluated with the 5th percentile value as the characteristic value) and the cost in Example 10.

determined as D = 0.3 m and L = 22.5 m, with the design performance of 16.29 kN and a minimum cost of 1.59 m³. As can be expected, a more conservative characteristic value will lead to a more conservative design.

As a final remark on this design example, this optimization problem can be solved with the spreadsheet Solver, where the objective is to minimize the cost, and the safety requirement is treated as a constraint.

2.5 SUMMARY AND REMARKS

To cope with the uncertainties induced by those in geotechnical parameters, solution models, applied loads, and construction, three design approaches, the FS-based design, RBD, and LRFD have been adopted in the current geotechnical practice. Within the context of the FS-based design, the design safety (or performance) is ensured through a conservative estimate of the uncertain variables and the conservatively selected target FS. In the FS-based design, uncertainties are not explicitly considered in the analysis. As the "true" safety of the candidate design cannot be known, the final design obtained from the FS-based design may be either over- or under-designed.

Although the uncertainties involved can be explicitly considered and included in the RBD, and the design safety is assessed based on the computed probability of failure (or reliability index), the challenge remains in the accurate statistical characterization of the uncertain parameters and the computational inefficiency of the probabilistic analysis, as both hinder the broad application of the theoretically sound RBD in routine geotechnical practice. The LRFD is designed to be a simpler variant of the RBD for geotechnical practice. However, the design scenarios considered in the partial factors calibration generally cannot cover all the possible design scenarios encountered in geotechnical practices and, in the absence of the knowledge of how the partial factors were calibrated in the codes, which is often true for a practicing engineer, the selection of the characteristic values of the uncertain variables in the LRFD can be quite subjective and challenging.

Because of these limitations of the current geotechnical design approaches, the concept of robust design (Taguchi 1986; Tsui 1992) has recently been adapted for geotechnical designs and robust geotechnical design (RGD) has been advanced by Juang and his co-workers (e.g., Juang and Wang 2013; Juang et al. 2013a, 2013b, 2014; Wang et al. 2013, 2014; Gong et al. 2014a, 2014b). More details of the methodology and implementation procedures of the RGD will be presented in the chapters that follow.

REFERENCES

AASHTO (American Association of State Highway and Transportation Officials). (2010) *Drilled shafts: Construction procedures and LRFD design methods. LRFD bridge design specifications* (4th edition). AASHTO, Washington, DC.

AASHTO (American Association of State Highway and Transportation Officials). (2014) *LRFD bridge design specifications* (7th edition). AASHTO, Washington, DC.

Allen, T.M., Nowak, A.S., and Bathurst, R.J. (2005) *Calibration to determine load and resistance factors for geotechnical and structural design.* Transportation Research Circular (E-C079). Transportation Research Board, Washington, DC.

Ang, A.H.S., and Tang, W.H. (2007) *Probability concepts in engineering: Emphasis on applications to civil and environmental engineering* (2nd edition). Wiley, New York.

Ang, G.L., Ang, A.H.S., and Tang, W.H. (1992) Optimal importance-sampling density estimator. *Journal of Engineering Mechanics*, 118(6), 1146–1163.

Au, S.K., and Beck, J.L. (1999) A new adaptive importance sampling scheme for reliability calculations. *Structural Safety*, 21(2), 135–158.

Au, S.K., and Beck, J.L. (2001) Estimation of small failure probabilities in high dimensions by subset simulation. *Probabilistic Engineering Mechanics*, 16(4), 263–277.

Baecher, G.B., and Christian, J.T. (2003) *Reliability and statistics in geotechnical engineering.* John Wiley and Sons, New York.

Binder, K., and Heermann, D. (2010) *Monte Carlo simulation in statistical physics: An introduction.* Springer, Berlin, Germany.

BSI (British Standards Institution). (2007) *National annex to Eurocode 7: BS EN 1997–1–2004: Geotechnical design – part 1: General rules.* BSI, Milton Keynes, UK.

Cao, Z., and Wang, Y. (2014) Bayesian model comparison and characterization of undrained shear strength. *Journal of Geotechnical and Geoenvironmental Engineering*, 140(6), 04014018.

Casagrande, A. (1965) The role of the 'calculated risk' in earthwork and foundation engineering. *Journal of the Soil Mechanics and Foundations Division*, 91(4), 1–40.

Cherubini, C. (2000) Reliability evaluation of shallow foundation bearing capacity on c', ϕ' soils. *Canadian Geotechnical Journal*, 37(1), 264–269.

Ching, J., Beck, J.L., and Au, S.K. (2005) Hybrid subset simulation method for reliability estimation of dynamical systems subject to stochastic excitation. *Probabilistic Engineering Mechanics*, 20(3), 199–214.

Ching, J., and Phoon, K.K. (2011) A quantile-based approach for calibrating reliability-based partial factors. *Structural Safety*, 33, 275–285.

Ching, J., Phoon, K.K., Chen, J.R., and Park, J.H. (2013) Robustness of constant load and resistance factor design factors for drilled shafts in multiple strata. *Journal of Geotechnical and Geoenvironmental Engineering*, 139(7), 1104–1114.

Christian, J.T. (2004) Geotechnical engineering reliability: How well do we know what we are doing? *Journal of Geotechnical and Geoenvironmental Engineering*, 130(10), 985–1003.

Driscoll, R., and Simpson, B. (2001) EN1997 Eurocode 7: Geotechnical design. *Proceedings of Ice-civil Engineering*, 144(6), 49–54.

Duncan, J.M. (2000) Factors of safety and reliability in geotechnical engineering. *Journal of Geotechnical and Geoenvironmental Engineering*, 126(4), 307–316.
Fenton, G.A., and Griffiths, D.V. (2008) *Risk assessment in geotechnical engineering*. Wiley, New York. ISBN: 978-0-470-17820-1.
Gong, W., Juang, C.H., Martin, J.R., and Ching, J. (2016) New sampling method and procedures for estimating failure probability. *Journal of Engineering Mechanics, ASCE*, 142(4), 04015107.
Gong, W., Wang, L., Juang, C.H., Zhang, J., and Huang, H. (2014a) Robust geotechnical design of shield-driven tunnels. *Computers and Geotechnics*, 56, 191–201.
Gong, W., Wang, L., Khoshnevisan, S., Juang, C.H., Huang, H., and Zhang, J. (2014b) Robust geotechnical design of earth slopes using fuzzy sets. *ASCE Journal of Geotechnical and Geoenvironmental Engineering*, 141(1), 04014084.
Grooteman, F. (2011) An adaptive directional importance sampling method for structural reliability. *Probabilistic Engineering Mechanics*, 26(2), 134–141.
Harr, M.E. (1987) *Reliability based design in civil engineering*. McGraw-Hill, New York, USA.
Honjo, Y., and Amatya, S. (2005) Partial factors calibration based on reliability analyses for square footings on granular soils. *Géotechnique*, 55(6), 479–491.
Honjo, Y., Suzuki, M., and Matsuo, M. (2000) Reliability analysis of shallow foundations in reference to design codes development. *Computers and Geotechnics*, 26(3), 331–346.
Honjo, Y., Suzuki, M., Shirato, M., and Fukui, J. (2002) Determination of partial factors for a vertically loaded pile based on reliability analysis. *Soils and Foundations*, 42(5), 91–109.
Hsu, W.C., and Ching, J. (2010) Evaluating small failure probabilities of multiple limit states by parallel subset simulation. *Probabilistic Engineering Mechanics*, 25(3), 291–304.
Juang, C.H., Gong, W., and Martin, J.R. (2017) Subdomain sampling methods – efficient algorithm for estimating failure probability. *Structural Safety*, 66, 62–73.
Juang, C.H., and Wang, L. (2013) Reliability-based robust geotechnical design of spread foundations using multi-objective genetic algorithm. *Computers and Geotechnics*, 48, 96–106.
Juang, C.H., Wang, L., Hsieh, H.S., and Atamturktur, S. (2014) Robust geotechnical design of braced excavations in clays. *Structural Safety*, 49, 37–44.
Juang, C.H., Wang, L., Khoshnevisan, S., and Atamturktur, S. (2013b) Robust geotechnical design – methodology and applications. *Journal of GeoEngineering*, 8(3), 71–81.
Juang, C.H., Wang, L., Liu, Z., Ravichandran, N., Huang, H., and Zhang, J. (2013a) Robust geotechnical design of drilled shafts in sand: New design perspective. *ASCE Journal of Geotechnical and Geoenvironmental Engineering*, 139(12), 2007–2019.
Lee, T.W., and Kwak, B.M. (1987) A reliability-based optimal design using advanced first order second moment method. *Journal of Structural Mechanics*, 15(4), 523–542.
Low, B.K. (2005) Reliability-based design applied to retaining walls. *Géotechnique*, 55(1), 63–75.
Low, B.K., and Tang, W.H. (2007) Efficient spreadsheet algorithm for first-order reliability method. *Journal of Engineering Mechanics*, 133(12), 1378–1387.
Melchers, R.E. (1987) *Structural reliability: Analysis and prediction*. Ellis Horwood Ltd., Chichester, West Sussex, UK.
Najjar, S.S., and Gilbert, R.B. (2009) Importance of lower-bound capacities in the design of deep foundations. *Journal of Geotechnical and Geoenvironmental Engineering*, 135(7), 890–900.
NEN (Netherlands Standardization Institute). (2007) *Eurocode 7: National annex to NEN-EN 1997–1–2005: Geotechnical design – part 1: General rules*. NEN, Delft, Netherlands.
NSAI (National Standards Authority Ireland). (2005) *National annex eurocode 7: I. S. EN 1997–1–2005: Geotechnical design – part 1: General rules*. NSAI, Dublin, Ireland.

rr, T.L.L. (2000) Selection of characteristic values and partial factors in geotechnical designs to Eurocode 7. *Computers and Geotechnics*, 26, 263–279.

ikowsky, S.G., Birgisson, B., Mcvay, M., and Nguyen, T. (2004) *Load and resistance factor design (LRFD) for deep foundations* (No. 507). Transportation Research Board, Washington, DC.

rk, J.H., Huh, J., Kim, K.J., Chung, M., Lee, J.H., Kim, D., and Kwak, K. (2013) Resistance factors calibration and its application using static load test data for driven steel pipe piles. *KSCE Journal of Civil Engineering*, 17(5), 929–938.

oon, K.K. (2008) *Reliability-based design in geotechnical engineering: Computations and applications*. CRC Press.

oon, K.K., Becker, D.E., Kulhawy, F.H., Honjo, Y., Ovesen, N.K., and Lo, S.R. (2003a) Why consider reliability analysis in geotechnical limit state design? In *Proceedings, international workshop on limit state design in geotechnical engineering practice (LSD2003)*. CDROM, Cambridge.

oon, K.K., Kulhawy, F.H., and Grigoriu, M.D. (2003b) Multiple resistance factor design for shallow transmission line structure foundations. *Journal of Geotechnical and Geoenvironmental Engineering*, 129(9), 807–818.

andards Australia. (2009) *AS2159–2009: Piling – design and installation*. Standards Australia, Sydney, New South Wales, Australia.

uedlein, A.W., Neely, W.J., and Gurtowski, T.M. (2012) Reliability-based design of augered cast-in-place piles in granular soils. *Journal of Geotechnical and Geoenvironmental Engineering*, 138(6), 709–717.

guchi, G. (1986) *Introduction to quality engineering: Designing quality into products and processes*. Quality Resources, White Plains, New York.

sui, K.L. (1992) An overview of Taguchi method and newly developed statistical methods for robust design. *IIE Transactions*, 24(5), 44–57.

ardanega, P.J., Kolody, E., Pennington, S.H., Morrison, P.R., and Simpson, B. (2012) Bored pile design in stiff clay I: Codes of practice. *Proceedings of the Institution of Civil Engineers-Geotechnical Engineering*, 165(4), 213–232.

esic, A.S. (1975) Bearing capacity of shallow foundations. In *Foundation engineering handbook*, H. Winterkorn and H.Y. Fang, eds. Van Nostrand Reinhold, New York.

/ang, L., Hwang, J.H., Juang, C.H., and Atamturktur, S. (2013) Reliability-based design of rock slopes – a new perspective on design robustness. *Engineering Geology*, 154, 56–63.

/ang, Y., and Cao, Z.J. (2013) Probabilistic characterization of Young's modulus of soil using equivalent samples. *Engineering Geology*, 159, 106–118.

'ang, Y., Cao, Z.J., and Li, D. (2016) Bayesian perspective on geotechnical variability and site characterization. *Engineering Geology*, 203, 117–125.

/ang, Y., Huang, K., and Cao, Z.J. (2014) Bayesian identification of soil strata in London Clay. *Géotechnique*, 64(3), 239–246.

/ang, Y., and Kulhawy, F.H. (2008) Economic design optimization of foundations. *Journal of Geotechnical and Geoenvironmental Engineering*, 134(8), 1097–1105.

hang, J., Tang, W.H., and Zhang, L.M. (2010) Efficient probabilistic back-analysis of slope stability model parameters. *Journal of Geotechnical and Geoenvironmental Engineering*, 136(1), 99–109.

hang, J., Zhang, L.M., and Tang, W.H. (2009) Bayesian framework for characterizing geotechnical model uncertainty. *Journal of Geotechnical and Geoenvironmental Engineering*, 135(7) 932–940.

hao, Y.G., and Ono, T. (2000) New point estimates for probability moments. *Journal of Engineering Mechanics*, 126(4), 433–436.

hao, Y.G., Ono, T., and Kato, M. (2002) Second-order third-moment reliability method. *Journal of Structural Engineering*, 128(8), 1087–1090.

3 Methodology of Robust Geotechnical Design

3.1 INTRODUCTION

Uncertainties associated with geotechnical parameters, solution models, applie loads, and construction make it challenging to ascertain the performance of a geo technical system. In the face of uncertainties, a desired objective of the design is ensure that the variation of the performance of the resulting design is low and tole able. In traditional geotechnical practice, these uncertainties are not explicitly cha acterized and included in the performance analysis; instead, a conservative factor safety (FS) is adopted based on the concept of "calculated risk" (Casagrande 1965 This deterministic FS-based design usually leads to an over-design with an unknow safety level, although under-design is also a possibility. Today, the reliability-base design (RBD) approach is a method of choice to deal with high uncertainty in geo technical design (e.g., Christian et al. 1994; Duncan 2000; Baecher and Christia 2003; Phoon et al. 2003; Fenton and Griffiths 2008; Fenton et al. 2008). Howeve the uncertainties associated with input parameters and solution models are often di ficult to characterize accurately. Thus, RBD is generally conducted using inexact assumed statistics. Juang et al. (2013b) demonstrated that the design results of RB can be strongly affected by the assumed statistics of geotechnical parameters. A can be seen, the dilemma of whether to over-design for safety or under-design f cost efficiency has not been entirely overcome, even though the design approac has evolved from the deterministic FS-based design to RBD. In such a situation, th concept of robust design has recently been adapted for applications in the field geotechnical engineering (Juang et al. 2013a, 2013b; Wang et al. 2013).

In the context of robust design, a design is deemed *robust* if the design (or pro uct) performance is robust against or insensitive to the variations (or uncertainties) i the "hard-to-control" uncertain input parameters, known as *noise factors* (Taguch 1986). In the design and manufacturing of high-quality products, the robust desig concept is often adopted (e.g., Tsui 1992; Phadke 1989; Chen et al. 1996; Lagaro and Fragiadakis 2007; Marano et al. 2008) to ensure low variation in the produc performance, while meeting the requirements of product performance (referred t herein as design constraint) and cost efficiency. The concept of robust design orig nated from the fields of quality and industrial engineering (Taguchi 1986; Tsui 199 Phadke 1989; Beyer and Sendhoff 2007), which is aimed at deriving an optim design that is insensitive to the variation of noise factors by adjusting the "easy-to control" input parameters, known as design parameters. Since its inception, robus design has been successfully utilized in various fields such as mechanical, structura and aeronautical engineering (e.g., Doltsinis et al. 2005; Zhang et al. 2005; Par et al. 2006; Lagaros and Fragiadakis 2007; Marano et al. 2008; Lee et al. 2010). Th

DOI: 10.1201/9781003564454-

use of robust design in geotechnical engineering has recently been reported (Juang and Wang 2013; Juang et al. 2013a, 2013b, 2014; Wang et al. 2013, 2014; Gong et al. 2014a, 2014b).

The input parameters and solution models in a typical geotechnical design can be highly uncertain, which can lead to high variation in the predicted system response. Thus, it is desirable to reduce this variation in the system response. The variation of the system response in the geotechnical design may be effectively reduced by reducing the variation of noise factors (i.e., uncertain parameters such as geotechnical parameters and solution models). Alternatively, the same goal may be achieved by adjusting the design parameters (e.g., geometry and other parameters that can be easily controlled by the engineer) without reducing the uncertainty of noise factors. The latter is indeed the essence of robust geotechnical design (RGD) (Juang et al. 2013a, 2013b; Gong et al. 2014a, 2014b; Khoshnevisan et al. 2014).

Although the uncertainty of noise factors in geotechnical practice has been well recognized, the level of variation of some noise factors often cannot be quantified with certainty due to the limited data availability and incomplete knowledge possessed by the engineer (Juang et al. 2013b; Gong et al. 2017). Further, the level of variation of noise factors in geotechnical engineering is usually much higher than that in many other fields. As an outcome, the robust design methodology developed in the different fields may not be applicable to geotechnical engineering. The main efforts to make RGD a practical geotechnical design tool could be summarized into two aspects: 1) measures of design robustness that are more suitable for geotechnical designs, and 2) simplified and faster approaches for locating the final designs in the RGD optimization.

In addition to satisfying the requirements of design safety and cost efficiency, RGD seeks an optimal design that is robust against the variation in noise factors by adjusting design parameters. Thus, the objective of RGD in a given design is to maximize the robustness while satisfying the requirements of design safety and cost efficiency. Because the latter two requirements are the focus of both FS-based and RBD approaches, the RGD may be taken as complementary to the conventional geotechnical design approaches.

Generally, maximizing a design's robustness and minimizing its cost are two objectives that conflict with each other. As such, a design that is optimal with respect to both objectives simultaneously may not be attainable within the RGD framework. In a multi-objective optimization with conflicting goals, no single optimal design may be secured; instead, the optimization may lead to a set of *non-dominated* designs (i.e., some designs are better from the robustness perspective and worse from the cost perspective, and vice versa). These *non-dominated* designs collectively define a Pareto front that shows a tradeoff between design robustness and cost efficiency. The Pareto front can be obtained with multi-objective algorithms such as Non-dominated Sorting Genetic Algorithm version II, NSGA-II (Deb et al. 2002). Alternatively, this Pareto front may be obtained through a series of single-objective optimization processes. By presenting a tradeoff between robustness and cost efficiency, the Pareto front can aid in making an informed design decision. For example, the knee point on the Pareto front, which yields the best compromise in theory among the conflicting objectives, can be taken as the most preferred design in the design space (Branke et al. 2004; Deb and Gupta 2011; Gong et al. 2014a).

This chapter focuses on the introduction and discussions of all essential components of the RGD methodology. It starts with an introduction of the measures of design robustness that were reported for RGD in the literature. Then, the essential components of the RGD, including the multi-objective optimization setting of RGD, optimization results of the robust design optimization, and simplified and efficient approaches for determining the final designs, are briefly introduced and discussed. Next, the implementation procedures of RGD are provided. Finally, the summary and concluding remarks are provided.

3.2 MEASURES OF DESIGN ROBUSTNESS IN GEOTECHNICAL ENGINEERING

Note that the key to a robust design is a quantifiable *robustness measure* with which robust design optimization can be effectively and efficiently implemented. The success of the RGD approach requires an effective and efficient measure of design robustness. Among the various approaches to achieving design robustness, Taguchi's method, the worst-case approach, the mean-variation compromising approach, and the feasibility-based approach are the most popular.

In Taguchi's method, numerical experiments are often conducted to identify the most robust design that yields the most favorable *signal-to-noise ratio* (SNR) (Taguchi 1986; Phadke 1989). An alternative to Taguchi's method is the worst-case approach, which focuses only on the worst-case parameter values (as opposed to treating noise factors as random variables) (Nagy and Braatz 2004; Zhu and Fukushima 2009). Though simple, the worst-case approach could be overly conservative in dealing with the uncertainties of noise factors. To deal with these uncertainties more rationally, the noise factors can be regarded as random variables, and their effects on product performance can be analyzed using probabilistic methods (Nigam and Turner 1995). According to the probability theory, the variation of the product performance can be evaluated and employed as a measure of design robustness, which allows the robust design to be accomplished by bringing the mean of the product performance to the target while minimizing the variation of the product performance (Chen et al. 1996; Park et al. 2006). In most engineering problems, however, a single best design that is optimal with respect to both objectives regarding the mean and the variation simultaneously may not be attainable. Instead, only a set of non-dominated optimal solutions may be achieved. These non-dominated optimal solutions collectively can form a Pareto front (Deb 2001; Kim and De Weck 2005). The Pareto front reveals a tradeoff relationship between the conflicting objectives.

To improve the decision-making process, Du and Chen (2000) proposed an index called feasibility robustness, in which the robustness is measured by the probability that the product (or system) remains feasible (or safe) even when it undergoes variation. With the aid of this feasibility robustness, no choice is required between the desire to optimize the mean and the desire to minimize the variation, as both are considered in the formulation of feasibility-based robustness.

When applying the above robustness measures and robust design approaches to geotechnical design, the following issues may arise.

1) The system response in geotechnical designs is usually much more complicated than that in other fields. For example, Juang et al. (2013a) have demonstrated that the system response could be in the form of deformation, FS, and probability of failure. For each of these system responses, the degree of design robustness should be evaluated using different measures. Meanwhile, because of the complex and nonlinear behaviors of the geomaterials (e.g., soils and rocks) under loading and those of the interaction between structures and geomaterials, the response of a geotechnical system usually may have to be evaluated with sophisticated numerical solutions, which further complicates the evaluation of the design robustness.
2) The level of variation of noise factors (e.g., uncertain geotechnical parameters and solution models) in geotechnical designs tends to be higher than that in many other fields. Further, the level of variation of uncertain parameters in geotechnical designs often cannot be quantified with certainty due to the limited data availability and incomplete knowledge (Juang et al. 2013b; Gong et al. 2017).

To address these issues, the engineer may adopt customized robustness measures that are aimed at the ease of application of the RGD. Three scenarios are outlined. First, in case site-specific data is very limited and only the nominal values of noise factors can be estimated, the gradient-based sensitivity index (SI) (Gong et al. 2016) may be used. Second, in case site-specific data is limited and the upper and lower bounds of noise factors can be estimated, the fuzzy set-based SNR (Gong et al. 2014b, 2015) may be more suitable. Third, in case abundant site-specific data are available and the probability distributions of noise factors can be characterized, the reliability-based feasibility index (*FI*; Juang et al. 2012, 2013b; Juang and Wang 2013) may be adopted.

In reference to Gong et al. (2016), the mathematical formulation of the gradient-based SI is formulated as follows.

$$\mathrm{SI} = \left\| \mathbf{J}\big|_{\theta=\theta'} \right\| = \sqrt{\left(\mathbf{J}\big|_{\theta=\theta'}\right)\left(\mathbf{J}\big|_{\theta=\theta'}\right)^{\mathrm{T}}} \tag{3.1}$$

$$\mathbf{J}\big|_{\theta=\theta'} = \left\{ \frac{\theta_1'}{g(\boldsymbol{d},\boldsymbol{\theta}')} \frac{\partial g(\boldsymbol{d},\boldsymbol{\theta})}{\partial \theta_1}\bigg|_{\theta=\theta'}, \frac{\theta_2'}{g(\boldsymbol{d},\boldsymbol{\theta}')} \frac{\partial g(\boldsymbol{d},\boldsymbol{\theta})}{\partial \theta_2}\bigg|_{\theta=\theta'}, \ldots, \frac{\theta_n'}{g(\boldsymbol{d},\boldsymbol{\theta}')} \frac{\partial g(\boldsymbol{d},\boldsymbol{\theta})}{\partial \theta_n}\bigg|_{\theta=\theta'} \right\} \tag{3.2}$$

where $\mathbf{J}\big|_{\theta=\theta'}$ represents the gradient vector of the system performance g($\boldsymbol{d}$, $\boldsymbol{\theta}$) to noise factors $\boldsymbol{\theta}$ at their nominal values $\boldsymbol{\theta}' = [\theta_1', \theta_2', \theta_3', \ldots, \theta_n']$, in which $\boldsymbol{\theta} = [\theta_1, \theta_2, \theta_3, \ldots, \theta_n]$ denotes the vector of noise factors and **d** denotes the design parameters, *n* is the number of noise factors; and $\left(\mathbf{J}\big|_{\theta=\theta'}\right)^{\mathrm{T}}$ represents the transpose vector of the gradient vector $\mathbf{J}\big|_{\theta=\theta'}$.

In reference to Gong et al. (2014b, 2015), the mathematical formulation of the SNR can be formulated as follows.

$$SNR = 10\log_{10}\left(\frac{\mu_g{}^2}{\sigma_g{}^2}\right) \tag{3.3}$$

where μ_g and σ_g are the mean and standard deviation, respectively, of the design performance.

In reference to Juang et al. (2012, 2013b) and Juang and Wang (2013), the *FI* can be formulated as follows.

$$FI = \Pr[P_f < P_{fT}] = \Pr[\beta > \beta_T] \tag{3.4}$$

where Pr[·] represents the conditional probability; P_f and β represent the estimated probability of failure and reliability index of the performance g(***d***, ***θ***), respectively; P_{fT} and β_T represent the target probability of failure and target reliability index, respectively, in the geotechnical design.

Since this chapter only provides a general introduction to the components of the RGD, more detailed discussions of the mathematical formulations of these robustness measures are provided in Chapters 4–6.

3.3 MULTI-OBJECTIVE OPTIMIZATION OF ROBUST GEOTECHNICAL DESIGN

As a complementary strategy to the current geotechnical design approaches, RGD aims to make the response of the designed geotechnical system robust against, or insensitive to, the variation of noise factors while simultaneously considering design safety and cost efficiency (Juang et al. 2012, 2013b; Juang and Wang 2013; Khoshnevisan et al. 2014; Gong et al. 2014b, 2015, 2016). By systematically adjusting design parameters **d**, RGD is realized through multi-objective optimization that explicitly considers the requirements of safety, design robustness, and cost efficiency. Following the RGD framework by Juang et al. (2013a), the safety requirements are considered the compulsory constraints that must be satisfied, while the design robustness and cost efficiency are treated as design objectives to be optimized simultaneously. Thus, the mathematical formulation of the RGD is expressed as follows.

Find:	design parameters ***d***	
Subject to:	***d*** ∈ design space ***DS***	
	satisfying safety requirement	(3.5)
Objectives:	maximizing design robustness	
	minimizing cost (or maximizing cost efficiency)	

Based on the level of characterized uncertainty of noise factors (or uncertain input parameters), the safety requirement of candidate designs can be evaluated with either a deterministic (i.e., FS-based) or probabilistic (i.e., reliability-based) approach.

milarly, the design robustness can be assessed with the gradient-based *SI*, fuzzy set-ased *SNR*, or *FI* accordingly. It should be noted that other optimization approaches r the RGD may also be adopted, such as considering safety (or performance) as e third design objective, along with robustness and cost, rather than treating it as a onstraint; then, the robust design becomes a three-objective optimization problem Gong et al. 2015).

In a traditional geotechnical design, within either an FS-based or RBD frame-ork, the primary considerations are safety and cost. Often, the engineer selects the ast-cost design among a set of candidate designs that satisfy the safety requirement. other words, the traditional geotechnical design is generally implemented as a ngle-objective optimization problem (i.e., safety is taken as a constraint while cost taken as an objective). From there, the requirements of safety and cost efficiency e the focus of both FS-based and RBD approaches. Thus, the RGD can be con-dered complementary to the traditional design approaches (such as the FS-based ethod and the RBD method).

The objectives of maximizing robustness and minimizing costs almost always onflict with each other. As a result, a design that is optimal with respect to both esign objectives simultaneously may not exist within the framework of RGD. nstead, the optimization often leads to a set of non-dominated designs (e.g., Cheng nd Li 1997; Deb et al. 2002), which collectively define a Pareto front that depicts tradeoff between robustness and cost efficiency. The Pareto front can be obtained sing multi-objective optimization algorithms such as Non-dominated Sorting enetic Algorithm version II, NSGA-II (Deb et al. 2002; Juang et al. 2013b) and eighted-sum-of-objectives approach (Kim and De Weck 2005; Marler and Arora 010). Interested readers are referred to Deb et al. (2002) for detailed methodology nd procedures for locating the Pareto front with NSGA-II. Alternatively, the Pareto ont can be obtained with the aid of a series of single-objective optimizations (e.g., hoshnevisan et al. 2014). It should be mentioned that each point on the Pareto Front equally optimal in the sense that no improvement can be achieved in one objective ithout sacrificing the other objective. Although none is superior or inferior to others mong all the designs on the Pareto front with respect to both objectives, they are all uperior to the other designs in the feasible domain.

By presenting a tradeoff between robustness and cost efficiency, the Pareto front an aid in making an informed design decision. For example, either the least-cost esign that is above a target level of design robustness or the most robust design that alls within a target cost level may be selected as the most preferred design in the esign space. The determination of an appropriate target level of robustness or cost, owever, is problem-specific. In situations where a strong preference is not specified, e knee point on the Pareto front, which yields the best compromise solution among e conflicting objectives, may be taken as the most preferred design in the design pace (e.g., Branke et al. 2004; Deb and Gupta 2011; Gong et al. 2014a).

Once the Pareto front is established, the knee point can readily be identified using e marginal utility function approach (Branke et al. 2004; Gong et al. 2014a), nor-al boundary intersection approach (Deb and Gupta 2011), or reflex angle approach Branke et al. 2004; Deb and Gupta 2011). Interested readers are referred to Branke t al. (2004) and Deb and Gupta (2011) for detailed methodology and procedures

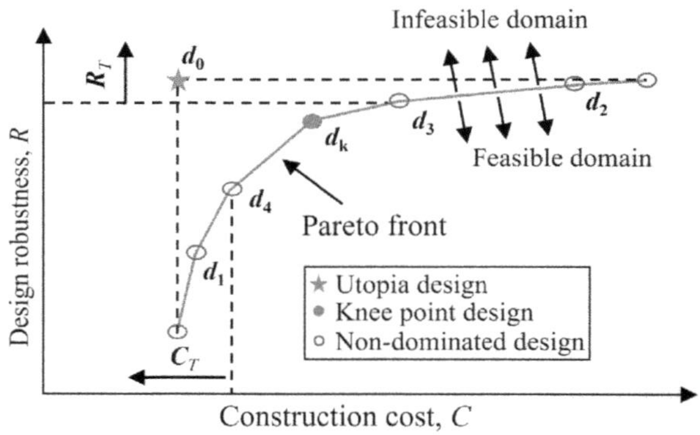

FIGURE 3.1 Conceptual illustration of the results of a bi-objective optimization proble (modified after Gong et al. 2016).

for identifying the knee point on the Pareto Front. Shown in Figure 3.1 is a concep tual illustration of the Pareto front and the knee point in a bi-objective optimizatic problem that minimizes cost (C) and maximizes design robustness (R), in which th feasible domain is defined as a domain or design space that consists of safe designs

3.4 SIMPLIFIED APPROACHES FOR LOCATING OPTIMAL DESIGNS FROM RGD OPTIMIZATION

Although the optimization problem of RGD, as formulated in Eq. (3.5), may be effec tively solved with the multi-objective optimization algorithms mentioned above (De et al. 2002; Branke et al. 2004; Kim and De Weck 2005; Marler and Arora 2010; De and Gupta 2011; Juang et al. 2013b), the geotechnical engineers are often not fami iar with these sophisticated multi-objective optimization algorithms. For exampl the traditional geotechnical designs (either FS-based or RBD approach) are usuall implemented as a single-objective optimization problem (i.e., safety is taken as design constraint while cost efficiency is taken as an objective). For ease of practic applications of the recently proposed RGD, two simplified approaches based upon series of single-objective optimizations (Khoshnevisan et al. 2014; Gong et al. 201 2017) may be adopted: one is to establish the Pareto front in the design space whil the other one is to locate the best compromise solution (i.e., knee point on the Paret front). Detailed mathematical formulations and implementation procedures of thes two simplified approaches are outlined below.

3.4.1 A Simplified Approach for Establishing the Pareto Front

Within the context of this approach, the multi-objective optimization problem show in Eq. (3.6) is solved through a series of single-objective optimizations (Khoshnevisa et al. 2014). Unlike the weighted-sum-of-objectives approach (Kim and De Wec 2005; Marler and Arora 2010), the cost in this single-objective optimization approac is taken as a constraint instead of an objective, along with the constraints of desig

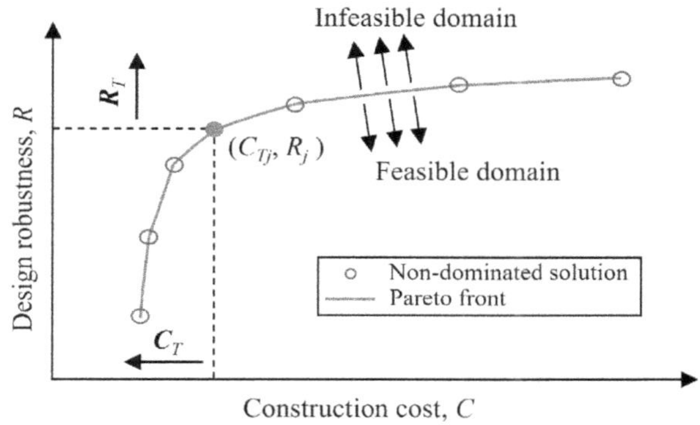

FIGURE 3.2 Conceptual illustration of the simplified approach for establishing the Pareto front (modified from Khoshnevisan et al. 2014).

space and safe requirements. With the aid of the single-objective optimization setting formulated below, the most robust design within a given acceptable level of cost (C_{Tj}) can readily be screened out using an Excel spreadsheet. As illustrated in Figure 3.2, the resulting optimal design (i.e., most robust design), denoted as (C_{Tj}, R_j), is indeed a non-dominated solution on the Pareto front that can be obtained using the multi-objective optimization setting in Eq. (3.6). Using this simplified approach, the Pareto front can be collectively formed by a series of optimal designs obtained through the single-objective optimizations shown below.

$$\begin{aligned}
&\text{Find:} && \text{design parameters } \boldsymbol{d} \\
&\text{Subject to:} && \boldsymbol{d} \in \text{design space } \boldsymbol{DS} \\
& && \text{satisfying safety requirement} \\
& && \text{cost} < \text{an acceptable level cost } C_{Tj} \\
&\text{Objectives:} && \text{maximizing design robustness}
\end{aligned} \tag{3.6}$$

With this simplified approach, the Pareto front can be established with the following steps: 1) determine the least-cost design and the most robust design among all feasible designs (i.e., the designs that satisfy the safety requirement) in the design space and denote the costs for these designs as $d = [d_1, d_2]$ and C_{Upp}, respectively; 2) divide the resulting cost interval of [C_{Low}, C_{Upp}] into a series of cost levels, denoted as $\boldsymbol{C_T} = \{C_{T1}, C_{T2}, C_{T3}, \ldots, C_{Tn}\}$, in which n represents the number of cost levels; 3) determine the most robust design within each cost level through solving the single-objective optimizations formulated in Eq. (3.6); and 4) plot all the resulting designs, in terms of design robustness versus cost, which yields a Pareto front.

The simplified approach of transforming the multi-objective optimization into a series of single-objective optimizations is indeed inspired by previous studies (e.g., Haimes et al. 1971; Hämäläinen and Mäntysaari 2002). While optimization algorithms such as NSGA-II are suitable for multi-objective optimization, it would be more efficient to implement through a series of single-objective optimizations when the optimization involves only two objectives. Since geotechnical designs can often be implemented as a bi-objective optimization problem, this simplified approach can be more effective. Further, a significant advantage of this simplified approach is that it can be easily implemented in an Excel spreadsheet without the knowledge of genetic algorithms and considerable programming skills. As demonstrated in Khoshnevisan et al. (2014), this simplified approach is computationally efficient and able to yield practically the same results as those obtained with the NSGA-II multi-objective optimization algorithm. Thus, this simplified approach has the potential to be a practical tool for RGD applications.

Example 1

Let's assume an RGD problem with design parameters of $\boldsymbol{d} = [d_1, d_2]$, in which the design parameter d_1 is taken from the range of [0.3, 5.0], while the other design parameter d_2 is taken from the range of [1.0, 30.0]. The design safety in this example is assessed with the performance function $g(\boldsymbol{d})$ as follows.

$$g(\boldsymbol{d}) = 10d_1^2 + 2d_1d_2 - 50$$

If $g(\boldsymbol{d}) > 0$, the safety or feasibility requirement of the design $\boldsymbol{d}$ can be satisfied; otherwise, this candidate design $\boldsymbol{d}$ is considered infeasible or unsafe. The design robustness (R) and cost (C) in this design example can be evaluated with the mathematical formulations below.

$$R(\boldsymbol{d}) = d_1 + d_2 + d_1d_2^3$$

$$C(\boldsymbol{d}) = d_1^2 \cdot d_2 + 10$$

The problem here is determining the Pareto front for this RGD problem.

Solution:

Step 1: Determine the least-cost design and the most robust design

The objective in this design example is to determine the Pareto front for this RGD problem. For illustration purposes, the design space in this design example is taken as d_{1k} $\{0.3, 0.4, 0.5, \ldots, 5.0\}$ and $d_2 \in \{1.0, 2.0, 3.0, \ldots, 30.0\}$. The design safety $g(\boldsymbol{d})$ in this example can be evaluated as follows.

$$g(\boldsymbol{d}) = 10d_1^2 + 2d_1d_2 - 50$$

The design robustness (R) and cost (C) in this design example can be evaluated with the mathematical formulations below.

$$R(\boldsymbol{d}) = d_1 + d_2 + d_1d_2^3$$

$$C(\boldsymbol{d}) = d_1^2 \cdot d_2 + 10$$

As an example, the design robustness (R), cost (C), and safety (g) for a design of $d_1 = 2.0$ and $d_2 = 3.0$ are calculated below.

$$R = d_1 + d_2 + d_1 d_2^3 = 2 + 3 + 2 \times 3^3 = 58$$

$$C = d_1^2 \cdot d_2 + 10 = 2^2 \times 3 + 10 = 22$$

$$g = 10d_1^2 + 2d_1 d_2 - 50 = 10 \times 2^2 + 2 \times 2 \times 3 - 50 = 2$$

For each candidate design, defined as a pair of parameters d_1 and d_2, in the design space, the corresponding design robustness (R), cost (C), and safety (g) can be computed by applying the formulations above.

A design is considered feasible if it satisfies the requirement of $g > 0$. Among all feasible designs derived, the least-cost design is determined as $d_1 = 2.20$ and $d_2 = 1.00$, with the design robustness (R) of 13.85 and cost (C) of 14.84; the most robust design is determined as $d_1 = 5.00$ and $d_2 = 30.00$, with the related robustness (R) of 135,035.00 and cost (C) of 760.00. Based on the results obtained from these two single-objective optimizations, the cost of the least-cost design ($C_{Low} = 14.84$) and that of the most robust design ($C_{Upp} = 760.00$) can be determined.

Step 2: Divide the cost interval of $[C_{Low}, C_{Upp}]$ into a series of cost levels

The resulting cost interval (C_{Low}, C_{Upp}) is divided into 15 equally spaced levels, denoted as $\mathbf{C}_T = (C_{T1} = 14.84, C_{T2} = 68.07, C_{T3} = 121.29, \ldots, C_{T15} = 760.00)$. For each cost level, the most robust design can be determined by solving a single-objective optimization problem.

Step 3: Determine the most robust design within each cost level

The most robust design within each cost level is determined through solving the single-objective optimization formulated below.

Find: design parameters $\mathbf{d} = \{d_1, d_2\}$
Subject to: $\mathbf{d} \in$ design space $\boldsymbol{DS}$
cost < an acceptable level cost C_{Tj}
$g(\mathbf{d}) > 0$
Objective: maximizing design robustness $R(\mathbf{d})$

The results obtained from the single-objective optimizations, in terms of the robustness (R) and the cost (C) of the most robust design at each cost level, are presented in Table 3.1.

Step 4: Plot the single-objective optimization results and obtain Pareto front

The single-objective optimization results, in terms of the robustness (R) and the cost (C) of the most robust design at each cost level, are plotted in Figure 3.3(a). As can be seen, these single-objective optimization results could yield a Pareto front.

TABLE 3.1
The Most Robust Design Obtained from the Single-Objective Optimization within Each Cost Level in Example 1

	Design parameters			
Cost level C_{Ti}	d_1	d_2	Cost C	Robustness R
14.84	2.20	1.00	14.84	5.40
68.07	1.30	30.00	60.70	35,131.30
121.29	1.90	30.00	118.30	51,331.90
174.52	2.30	30.00	168.70	62,132.30
227.74	2.60	30.00	212.80	70,232.60
280.97	3.00	30.00	280.00	81,033.00
334.19	3.20	30.00	317.20	86,433.20
387.42	3.50	30.00	377.50	94,533.50
440.65	3.70	30.00	420.70	99,933.70
493.87	4.00	30.00	490.00	108,034.00
547.10	4.20	30.00	539.20	113,434.20
600.32	4.40	30.00	590.80	118,834.40
653.55	4.60	30.00	644.80	124,234.60
706.77	4.80	30.00	701.20	129,634.80
760.00	5.00	30.00	760.00	135,035.00

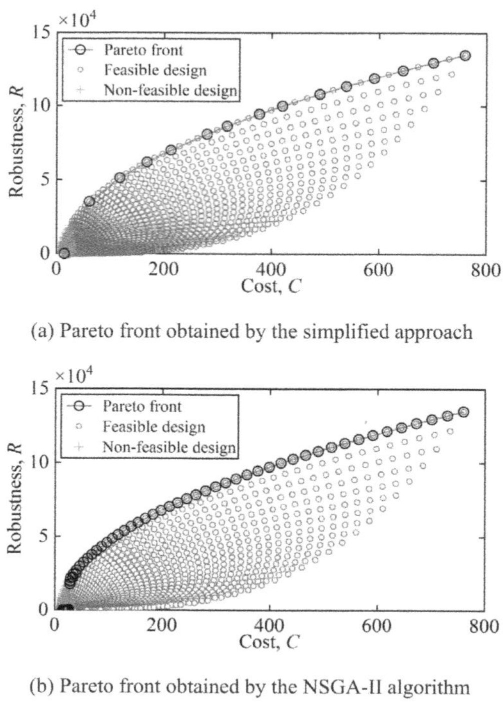

(a) Pareto front obtained by the simplified approach

(b) Pareto front obtained by the NSGA-II algorithm

FIGURE 3.3 The Pareto front obtained with two different approaches in Example 1.

To validate the Pareto front obtained from the simplified approach, the sorting algorithm of NSGA-II (Non-dominated Sorting Genetic Algorithm II) is adopted to obtain the Pareto front, and the sorted Pareto front is shown in Figure 3.3(b). Figure 3.3 depicts that both approaches produce a consistent Pareto front, which validates the effectiveness of the simplified approach.

.4.2 A Simplified Approach for Locating the Best Compromise Solution

) set the stage for the simplified approach for locating the best compromise solution the design space, three existing approaches to identify the knee point on the Pareto ont, namely, the reflex angle approach, normal boundary intersection approach, and arginal utility function approach, are summarized below. Within these approaches, e objective functions are often normalized into values ranging from 0.0 to 1.0 rough the following transformation:

$$f_i^n(\boldsymbol{d}) = \frac{f_i(\boldsymbol{d}) - [f_i(\boldsymbol{d})]_{\min}}{[f_i(\boldsymbol{d})]_{\max} - [f_i(\boldsymbol{d})]_{\min}} \tag{3.7}$$

here $f_i^n(\boldsymbol{d})$ represents the normalized objective function, and $\left[f_i(\boldsymbol{d})\right]_{max}$ nd $\left[f_i(\boldsymbol{d})\right]_{min}$ represent the maximum and minimum values, respectively, of the ith bjective function $f_i(\boldsymbol{d})$.

The reflex angle at a non-dominated design on the Pareto front indicates the bend f the Pareto front from its left to right. Thus, this reflex angle may be taken as measure of the gain-sacrifice in the tradeoff relationship. In the context of the flex angle approach, the knee point is identified as the non-dominated design on the areto front that yields the maximum reflex angle, as depicted in Figure 3.4(a). It is oted that the reflex angle is computed using only two neighboring points; thus, this flex angle can only correspond to a local property, and this approach may not yield global knee point in some cases (e.g., Branke et al. 2004).

To avoid the local knee point possibly as a result of applying the above reflex ngle approach, Deb and Gupta (2011) proposed the normal boundary intersection pproach. Within the context of the normal boundary intersection approach, the two xtreme designs on a Pareto front, denoted as designs A and B, are first adopted to onstruct a straight boundary line. Then, the non-dominated design on the Pareto ont that yields the maximum distance from the boundary line is identified as the nee point, as illustrated in Figure 3.4(b).

Note that both the reflex angle approach and the normal boundary intersection pproach are valid and efficient for bi-objective problems, the applications of these vo approaches to multi-objective optimizations that involve more than two objecves could be problematic. In such cases, the marginal utility function approach Branke et al. 2004) may be employed. Within this, the marginal utility function, enoted as $U((\boldsymbol{d}, ())$, is formulated as follows:

$$U'(\boldsymbol{d}_i,\lambda) = \min\left[U(\boldsymbol{d}_j,\lambda) - U(\boldsymbol{d}_i,\lambda)\right],\ (i \neq j) \tag{3.8}$$

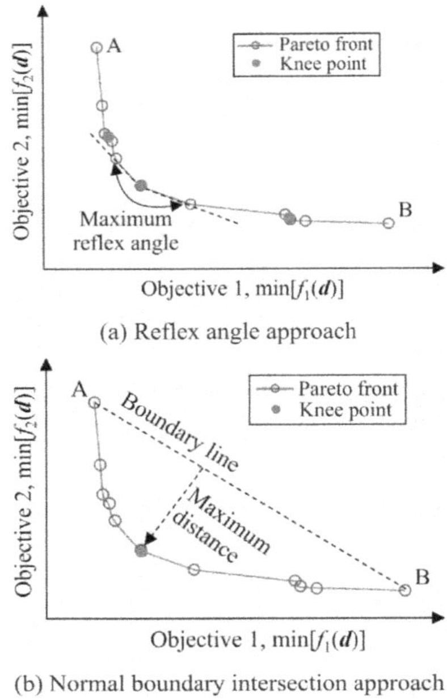

(a) Reflex angle approach

(b) Normal boundary intersection approach

FIGURE 3.4 Conceptual illustration of the reflex angle approach and normal bounda intersection approach (modified from Gong et al. 2016).

where $U(\boldsymbol{d}, \boldsymbol{\lambda})$ is a linear utility function of the design objectives, defined as:

$$U(\boldsymbol{d}, \lambda) = \sum \lambda_i f_i^{n}(\boldsymbol{d}) \tag{3.9}$$

where λ_u is a weighting parameter with a value ranging from 0.0 to 1.0, an $\sum \lambda_i = 1.0$; and $f_i^n(\boldsymbol{d})$ is the ith normalized objective function to be minimized. B means of Monte Carlo simulation (MCS), different values of λ_i can be sampled wit an assumption of uniform distribution; then, the expected marginal utility functio can be computed for each non-dominated design on the Pareto front. Note that th non-dominated design with the maximum expected marginal utility can readily b taken as the best compromise solution (i.e., knee point) on the Pareto front.

Oftentimes, the owner or client may only be interested in the final design or th most preferred design in the design space and not necessarily the whole Pareto fron Further, the engineer oftentimes does not like to perform multi-objective optimi zations. Indeed, if the most preferred design in the design space could be locate directly, the sorting of the Pareto front is not needed. In such a circumstance, simplified approach, known as the minimum distance approach, could be used t locate the best compromise solution in the design space directly (Khoshnevisan et a 2014; Gong et al. 2016, 2017). In the context of this minimum distance approach, n multi-objective optimization algorithm is involved. The procedures for implement ing this minimum distance approach are summarized as follows:

Step 1: Select a suitable design space (***DS***) based on local experience and engineering judgment, and screen for feasible or safe designs based on the design constraints.

Step 2: Conduct a single-objective optimization with respect to the ith design objective, $f_i(\boldsymbol{d})$, among all the feasible designs, which can be set up as follows:

$$\begin{array}{ll} \text{Find:} & \text{design parameters } \boldsymbol{d}_i^* \\ \text{Subject to:} & \boldsymbol{d}_i^* \in \text{design space } \boldsymbol{DS} \\ & \text{satisfying safety requirement} \\ \text{Objective:} & \text{minimizing design objective } f_i(\boldsymbol{d}) \end{array} \tag{3.10}$$

where $\boldsymbol{d}_i^*$ denotes the optimal design based on the optimization of the ith objective, which meets the safety requirement and yields the minimum value of the ith objective, denoted as $\left[f_i(\boldsymbol{d})\right]_{min}$. By repeating the single-objective optimization formulated in Eq. (3.10) for each and every design objective, a utopia point of $\left\{\left[f_1(\boldsymbol{d})\right]_{min}, \left[f_2(\boldsymbol{d})\right]_{min}, \ldots, \left[f_m(\boldsymbol{d})\right]_{min}\right\}$ can be constructed in the space of the objective functions, where m represents the number of design objectives to be optimized. The utopia design is a "design" that is optimal with respect to each design objective, as illustrated in Figure 3.5. As implied by its name, this utopia design is, however, not attainable in reality if the objectives conflict with each other. Note that the utopia design defined this way may violate the safety requirement or even fall outside the design space.

Step 3: Determine the maximum value of each design objective among the designs of $\{\boldsymbol{d}_1^*, \boldsymbol{d}_2^*, \ldots, \boldsymbol{d}_m^*\}$. Symbolically, the maximum value of the ith objective is obtained as follows.

$$\left[f_i(\boldsymbol{d})\right]_{\max} = \max\left[f_i(\boldsymbol{d}_j^*)\right] \quad (j=1, 2, \cdots, m) \tag{3.11}$$

Step 4: Normalize all design objectives into values ranging from 0 to 1.0 utilizing Eq. (3.7). As a result, the coordinates of the normalized utopia point are all equal to 0.

Step 5: Calculate the distances from the utopia design to all candidate designs (in the feasible domain) in the normalized space of design objectives. As shown in Figure 3.5, the design that meets the safety requirement and yields the minimum distance (from the utopia design) can be finally taken as the most preferred design in the design space.

The computed distance from the utopia design to a candidate design can be interpreted as an additional "price" (in terms of a combination of the design robustness and cost efficiency) one has to pay to select this candidate design, in lieu of the utopia design, as the final design. This minimum distance approach is indeed inspired by the marginal utility concept (Branke et al. 2004) and the compromise programming approach. In reality, the utopia design is not attainable. Thus, the candidate design that meets the safety requirement and yields the minimum distance from this utopia design (therefore incurring the least additional price) can be taken as the most preferred design in the design space. As discussed by Gong et al. (2016, 2017), the most preferred design identified with this minimum distance approach is virtually the same as the knee point on the Pareto front obtained through the multi-objective optimization.

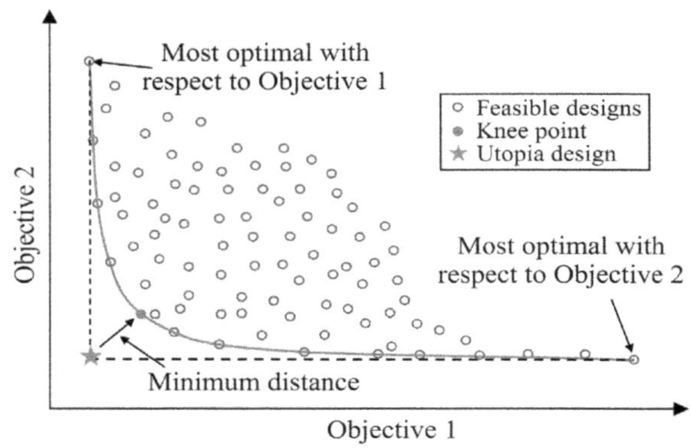

FIGURE 3.5 Conceptual illustration of the minimum distance approach for locating the most preferred design in the design space (modified from Gong et al. 2016).

With this simplified approach, the multi-objective optimization problem can be solved with a series of single-objective optimizations. The single-objective optimization can be easily performed using the MATLAB function *fmincon* (The MathWorks Inc. 2010) or Microsoft Excel *Solver* that returns the minimum of a constrained nonlinear multivariable function. Thus, the use of multi-objective optimization algorithms such as NSGA-II is no longer required in the RGD. It should be stressed that many outstanding studies on the identification of the knee point have been published (Bechikh et al. 2011); however, the development of this minimum distance approach places a high premium on simplicity, aiming to enable the RGD as a practical geotechnical design tool. Finally, it is noted that this minimum distance approach can be applicable to multi-objective optimizations with more than two objectives, although only a bi-objective optimization problem is illustrated in Figure 3.5.

Example 2

The setting of this example is identical to that in Example 1; the objective in this example is, however, to locate the most preferred design directly in the design space utilizing the minimum distance approach, bypassing the need for a full multi-objective optimization.

Solution:

Step 1: Define the design space and select the feasible designs

The objective in this example is to use the simplified minimum distance approach to locate the most preferred design directly in the design space. For illustration purposes, the design space in this example is taken as $d_1 \in \{0.3, 0.4, 0.5, \ldots, 5.0\}$ and $d_2 \in \{1.0, 2.0, 3.0, \ldots, 30.0\}$. The performance of all candidate designs in the predefined design space can be assessed with the method and procedures presented in Example 1. Based on the assessed performance of candidate designs, the feasible designs can readily be screened with the criterion of $g > 0$.

Step 2: Perform the single-objective optimizations and construct the utopia point

For each design objective (i.e., robustness or cost), a single-objective optimization is undertaken to locate the most optimal design with respect to this objective (i.e., the resulting design is the most robust design or least-cost design). Here, the least-cost design yields the minimum cost of $C_{min} = 14.84$, and the most robust design yields the maximum robustness of $R_{max} = 135035.00$. The design robustness of the least-cost design is $R = 5.40$, which indeed represents the minimum robustness value (R_{min}) on the Pareto front; and, the cost of the most robust design is 760.00, which signals the maximum cost value (C_{max}) on the Pareto front.

The resulting minimum cost (C_{min}) and maximum design robustness (R_{max}) allow for the construction of a utopia point in the objective function space with the coordinate of $(C_{min}, R_{max}) = (14.84, 135{,}035.00)$. This utopia point represents an ideal design that is optimal for both design objectives simultaneously, although it is typically unattainable in reality due to conflicts between these two design objectives (i.e., design robustness versus cost).

Step 3: Normalize the design objectives into values ranging from 0 to 1.0

The objective values, in terms of the robustness (R) and cost (C), of all feasible designs (screened in Step 1) are normalized into values ranging from 0 to 1.0 utilizing the following equations.

$$R^n(\mathbf{d}) = \frac{R(\mathbf{d}) - [R(\mathbf{d})]_{min}}{[R(\mathbf{d})]_{max} - [R(\mathbf{d})]_{min}}$$

$$C^n(\mathbf{d}) = \frac{C(\mathbf{d}) - [C(\mathbf{d})]_{min}}{[C(\mathbf{d})]_{max} - [C(\mathbf{d})]_{min}}$$

where $R^n(\mathbf{d})$ represents the normalized design robustness for the candidate design $\mathbf{d}$, and $C^n(\mathbf{d})$ represents the normalized cost for the candidate design $\mathbf{d}$. After normalization, the coordinate of the utopia point becomes (0, 1). The normalization results, in terms of normalized robustness (R^n) and normalized cost (C^n), of all feasible designs, are tabulated in Table 3.2.

As an example, the normalized robustness (R^n) and normalized cost (C^n) for a feasible design of $d_1 = 1.30$ and $d_2 = 30.00$ are calculated below.

$$R^n(\mathbf{d}) = \frac{R(\mathbf{d}) - [R(\mathbf{d})]_{min}}{[R(\mathbf{d})]_{max} - [R(\mathbf{d})]_{min}} = \frac{35131.30 - 5.4}{135035.00 - 5.4} = 0.26$$

$$C^n(\mathbf{d}) = \frac{C(\mathbf{d}) - [C(\mathbf{d})]_{min}}{[C(\mathbf{d})]_{max} - [C(\mathbf{d})]_{min}} = \frac{60.70 - 14.84}{760 - 14.84} = 0.06$$

For each feasible design, the corresponding normalized robustness (R^n) and normalized cost (C^n) can be computed by applying the formulations above.

Step 4: Calculate the distance in the normalized objective space

Calculate the distances from the utopia design to all feasible designs in the normalized space of design objectives. As an example, the distance (D) of a feasible design of $d_1 = 1.30$ and $d_2 = 30.00$, the coordinate of which in the normalized objective space is (0.06, 0.26), to the utopia point (0, 1) in the normalized objective space is calculated below.

$$D = \sqrt{(C^n - 0)^2 + (R^n - 1)^2} = \sqrt{(0.06 - 0)^2 + (0.26 - 1)^2} = 0.7424$$

For each feasible design in the design space, the distance (D) can readily be computed by applying the formulation above, and the calculation results are tabulated in Table 3.2.

Step 5: Select the most preferred design in the design space

Based on the distance calculation results shown in Table 3.2, the most preferred design in the design space can be determined as: $d_1 = 3.00$ and $d_2 = 30.00$, with the minimum distance (D) of 0.5353. The design results obtained from this example are also depicted in Figure 3.6(a).

To validate the effectiveness of the most preferred design derived above, the knee point design in the design space is further located with the normal boundary intersection approach (Deb and Gupta 2011) (? Understanding knee points in

TABLE 3.2
Distance Calculation in the Normalized Design Objective Space in Example 2

	Design parameters		Design objectives		Normalized design objectives		
Design No.	d_1	d_2	C	R	Cn	Rn	**Distance D**
1	0.90	24.00	29.44	12,466.50	0.02	0.09	0.9079
2	0.90	25.00	30.25	14,088.40	0.02	0.10	0.8959
3	0.90	26.00	31.06	15,845.30	0.02	0.12	0.8830
4	0.90	27.00	31.87	17,742.60	0.02	0.13	0.8689
5	0.90	28.00	32.68	19,785.70	0.02	0.15	0.8538
6	0.90	29.00	33.49	21,980.00	0.03	0.16	0.8376
7	0.90	30.00	34.30	24,330.90	0.03	0.18	0.8203
...	...	...	...	...	...	...	...
534	3.00	29.00	271.00	73,199.00	0.34	0.54	0.5726
535	**3.00**	**30.00**	**280.00**	**81,033.00**	**0.36**	**0.60**	**0.5353**
536	3.10	1.00	19.61	7.20	0.01	0.00	1.0000
...	...	...	...	...	...	...	...
1134	5.00	29.00	735.00	121,979.00	0.97	0.90	0.9713
1135	5.00	30.00	760.00	135,035.00	1.00	1.00	1.0000

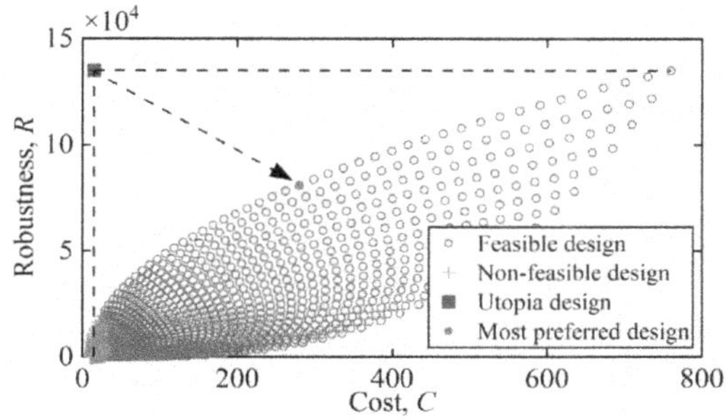

(a) Most preferred design in the design space obtained with the simplified approach

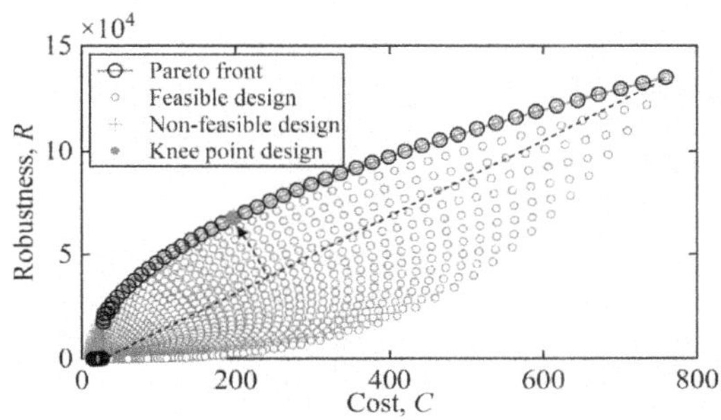

(b) Knee point design in the design space obtained with the NSGA-II algorithm

IGURE 3.6 Final designs obtained with two different approaches in Example 2.

bicriteria problems and their implications as preferred solution principles: Engineering Optimization: Vol 43, No 11?), and the resulting knee point design is illustrated in Figure 3.6(b). The design parameters of this knee point design are $d_1 = 2.50$ and $d_2 = 30.00$. As can be seen, the most preferred design derived with the simplified approach is close to the knee point design.

.5 IMPLEMENTATION PROCEDURES OF THE RGD

he procedures for implementing the RGD are summarized in the following steps .g., Juang et al. 2012, 2013b; Juang and Wang 2013; Gong et al. 2014b, 2015, 2016):

Step 1: Define the geotechnical problem of concern with mathematical models. Here, the system response of concern, noise factors (θ), and design parameters (***d***) are identified. A mathematical model is established for computing the system response, which can be taken in a variety of forms, such as deformation, FS, or probability of failure (P_f).

Step 2: Characterize the uncertainty of noise factors and identify the design space (***DS***). For geotechnical systems, key uncertain soil parameters are usually treated as noise factors. The uncertainty of each noise factor may be characterized based on site-specific investigation data, published literature, and engineering judgment. The design space includes typical ranges of design parameters. These design parameters may also be specified in discrete numbers.

Step 3: For each candidate design, assess the safety constraint based on the computed system response, followed by the evaluation of the design robustness and the cost. Among these tasks, the selection of the robustness measure is the most critical. Here, the selected robustness measure should be consistent with the level of uncertainty characterization of noise factors.

Step 4: Perform the multi-objective optimization considering the design robustness, cost efficiency, and design constraints (e.g., design space and safety requirement) following the optimization setting formulated in Eq. (3.5). Because the objectives of design robustness and cost efficiency are conflicting to each other, the optimization will yield a Pareto front, which depicts a tradeoff between design robustness and cost efficiency for all designs that satisfy the design constraints. It is noted that the optimization may be set up in different ways (e.g., safety may be treated as a design objective instead of a constraint; see Gong et al. 2014a).

Step 5: Select the final design based on the resulting Pareto front. In general, either the least-cost design that is above a specified level of design robustness or the most robust design that falls below a specified level of cost may be selected as the final design. In the case where the client does not specify a strong preference, the knee point on the Pareto front, which indicates the best compromise solution in the design space, can be taken as the final design.

3.6 SUMMARY AND REMARKS

To cope with the recognized but unquantified uncertainty in geotechnical design Juang et al. (2013b) have proposed a design approach, RGD. Rather than reducin the uncertainties of hard-to-characterize and hard-to-control noise factors, the RG seeks a cost-efficient and safe design that exhibits a high degree of design robustne (or a low variation in the system response) by adjusting the easy-to-control desig parameters. In short, the goal of RGD is to seek an optimal design with respect t both design robustness and cost efficiency while satisfying the safety requirement.

This chapter presents a review of all essential components of the RGD metho ology. First, the robustness measures formulated in the other fields are reviewe and three levels of robustness measures customized for the RGD are discussed. Th selection criteria among these robustness measures are also discussed.

To locate a design that is optimal with respect to design robustness and cost eff ciency simultaneously while satisfying safety and other design constraints, we nee a multi-objective optimization. As the objectives to maximize the design robus ness and to minimize the cost often conflict with each other, the design optimizatio may lead to a set of non-dominated designs, not a single optimal design. Howeve these non-dominated designs collectively define a Pareto front depicting a tradeo between design robustness and cost efficiency. The Pareto front can be obtaine with multi-objective optimization algorithms such as NSGA-II. For ease of practical applications of the RGD, one simplified approach to sort the Pareto front in the design space may be adopted.

Further, the concept of the knee point (i.e., the best compromise solution) may be used to improve design decisions based on the sorted Pareto front. Three existing methods for locating the knee point on the Pareto front, namely, the reflex angle approach, normal boundary intersection approach, and marginal utility function approach, are reviewed along with the recently proposed minimum distance approach. Finally, the procedures for implementing the RGD are summarized.

REFERENCES

Baecher, G.B., and Christian, J.T. (2003) *Reliability and statistics in geotechnical engineering*. John Wiley and Sons, New York.

Bechikh, S., Ben Said, L., and Ghédira, K. (2011) Searching for knee regions of the Pareto front using mobile reference points. *Soft Computing*, 15, 1807–1823.

Beyer, H.G., and Sendhoff, B. (2007) Robust optimization – a comprehensive survey. *Computer Methods in Applied Mechanics and Engineering*, 196(33), 3190–3218.

Branke, J., Deb, K., Dierolf, H., and Osswald, M. (2004) Finding knees in multi-objective optimization. *Parallel Problem Solving from Nature-PPSN*, VIII, 722–731.

Casagrande, A. (1965) The role of the 'calculated risk' in earthwork and foundation engineering. *Journal of the Soil Mechanics and Foundations Division*, 91(4), 1–40.

Chen, W., Allen, J.K., Tsui, K.L., and Mistree, F. (1996) A procedure for robust design: Minimizing variations caused by noise factors and control factors. *Journal of Mechanical Design*, 118(4), 478–485.

Cheng, F.Y., and Li, D. (1997) Multiobjective optimization design with Pareto genetic algorithm. *Journal of Structural Engineering*, 123(9), 1252–1261.

Christian, J.T., Ladd, C.C., and Baecher, G.B. (1994) Reliability applied to slope stability analysis. *Journal of Geotechnical Engineering*, 120(12), 2180–2207.

Deb, K. (2001) Nonlinear goal programming using multi-objective genetic algorithms. *Journal of the Operational Research Society*, 52(3), 291–302.

Deb, K., and Gupta, S. (2011) Understanding knee points in bicriteria problems and their implications as preferred solution principles. *Engineering Optimization*, 43(11), 1175–1204.

Deb, K., Pratap, A., Agarwal, S., and Meyarivan, T.A.M.T. (2002) A fast and elitist multiobjective genetic algorithm: NSGA-II. *IEEE Transactions on Evolutionary Computation*, 6(2), 182–197.

Doltsinis, I., Kang, Z., and Cheng, G. (2005) Robust design of non-linear structures using optimization methods. *Computer Methods in applied Mechanics and Engineering*, 194(12–16), 1779–1795.

Du, X., and Chen, W. (2000) Towards a better understanding of modeling feasibility robustness in engineering design. *Journal of Mechanical Design*, 122, 385–394.

Duncan, J.M. (2000) Factors of safety and reliability in geotechnical engineering. *Journal of Geotechnical and Geoenvironmental Engineering*, 126(4), 307–316.

Fenton, G.A., and Griffiths, D.V. (2008) *Risk assessment in geotechnical engineering*. Wiley, New York.

Fenton, G.A., Griffiths, D.V., and Zhang, X. (2008) Load and resistance factor design of shallow foundations against bearing failure. *Canadian Geotechnical Journal*, 45(11), 1556–1571.

Gong, W., Huang, H., Juang, C.H., and Wang, L. (2017) Simplified-robust geotechnical design of soldier pile–anchor tieback shoring system for deep excavation. *Marine Georesources and Geotechnology*, 35(2), 157–169.

Gong, W., Juang, C.H., Khoshnevisan, S., and Phoon, K.K. (2016) R-LRFD: Load and resist ance factor design considering robustness. *Computers and Geotechnics*, 74, 74–87.

Gong, W., Khoshnevisan, S., and Juang, C.H. (2014a) Gradient-based design robustness meas ure for robust geotechnical design. *Canadian Geotechnical Journal*, 51(11), 1331–1342

Gong, W., Wang, L., Juang, C.H., Zhang, J., and Huang, H. (2014b) Robust geotechnica design of shield-driven tunnels. *Computers and Geotechnics*, 56, 191–201.

Gong, W., Wang, L., Khoshnevisan, S., Juang, C.H., Huang, H., and Zhang, J. (2015) Robus geotechnical design of earth slopes using fuzzy sets. *Journal of Geotechnical an Geoenvironmental Engineering*, 141(1), 04014084.

Haimes, Y.Y., and Wismer, D.A. (1971) Integrated system modeling and optimization via qua silinearization. *Journal of Optimization Theory and Applications*, 8, 100–109.

Hämäläinen, R.P., and Mäntysaari, J. (2002) Dynamic multi-objective heating optimizatior *European Journal of Operational Research*, 142(1), 1–15.

Juang, C.H., and Wang, L. (2013) Reliability-based robust geotechnical design of sprea foundations using multi-objective genetic algorithm. *Computers and Geotechnics*, 48 96–106.

Juang, C.H., Wang, L., and Atamturktur, S. (2013a) Robust design of geotechnical systems – new design perspective. In *Geotechnical safety and risk IV*. CRC Press.

Juang, C.H., Wang, L., Atamturktu, S., and Luo, Z. (2012) Reliability-based robust and opti mal design of shallow foundations in cohesionless soil in the face of uncertainty. *Journa of GeoEngineering*, 7(3), 75–87.

Juang, C.H., Wang, L., Hsieh, H.S., and Atamturktur, S. (2014) Robust geotechnical design o braced excavations in clays. *Structural Safety*, 49, 37–44.

Juang, C.H., Wang, L., Liu, Z., Ravichandran, N., Huang, H., and Zhang, J. (2013b) Robus geotechnical design of drilled shafts in sand – new design perspective. *Journal of Geo technical and Geoenvironmental Engineering*, 139(12), 2007–2019.

Khoshnevisan, S., Gong, W., Wang, L., and Juang, C.H. (2014) Robust design in geotechnica engineering–an update. *Georisk: Assessment and Management of Risk for Engineere Systems and Geohazards*, 8(4), 217–234.

Kim, I.Y., and De Weck, O.L. (2005) Adaptive weighted-sum method for bi-objective opti mization: Pareto front generation. *Structural and Multidisciplinary Optimization*, 29 149–158.

Lagaros, N.D., and Fragiadakis, M. (2007) Fragility assessment of steel frames using neura networks. *Earthquake Spectra*, 23(4), 735–752.

Lee, M.C., Kelly, D.W., Degenhardt, R., and Thomson, R.S. (2010) A study on the robustnes of two stiffened composite fuselage panels. *Composite Structures*, 92(2), 223–232.

Marano, G.C., Sgobba, S., Greco, R., and Mezzina, M. (2008) Robust optimum design o tuned mass dampers devices in random vibrations mitigation. *Journal of Sound an Vibration*, 313(3), 472–492.

Marler, R.T., and Arora, J.S. (2010) The weighted sum method for multi-objective optimiza tion: New insights. *Structural and Multidisciplinary Optimization*, 41, 853–862.

Nagy, Z.K., and Braatz, R.D. (2004) Open-loop and closed-loop robust optimal control o batch processes using distributional and worst-case analysis. *Journal of Process Control* 14(4), 411–422.

Nigam, S.D., and Turner, J.U. (1995) Review of statistical approaches to tolerance analysis *Computer-Aided Design*, 27(1), 6–15.

Park, G.J., Lee, T.H., Lee, K.H., and Hwang, K.H. (2006) Robust design: An overview. *AIAA Journal*, 44(1), 181–191.

Phadke, M.S. (1989) *Quality engineering using robust design*. Prentice Hall, New Jersey.

Phoon, K.K., Kulhawy, F.H., and Grigoriu, M.D. (2003) Development of a reliability-base design framework for transmission line structure foundations. *Journal of Geotechnica and Geoenvironmental Engineering*, 129(9), 798–806.

Taguchi, G. (1986) *Introduction to quality engineering: Designing quality into products and processes*. Quality Resources, White Plains, New York.

The MathWorks Inc. (2010) *MATLAB (Version R2010a)*. The MathWorks Inc., Natick, MA.

Tsui, K.L. (1992) An overview of Taguchi method and newly developed statistical methods for robust design. *IIE Transactions*, 24(5), 44–57.

Wang, L., Hwang, J.H., Juang, C.H., and Atamturktur, S. (2013) Reliability-based design of rock slopes – a new perspective on design robustness. *Engineering Geology*, 154, 56–63.

Wang, Y., Huang, K., and Cao, Z.J. (2014) Bayesian identification of soil strata in London clay. *Géotechnique*, 64(3), 239–246.

Zhang, W., Chen, B.S., and Tseng, C.S. (2005) Robust H∞ filtering for nonlinear stochastic systems. *IEEE Transactions on Signal Processing*, 53(2), 589–598.

Zhu, S., and Fukushima, M. (2009) Worst-case conditional value-at-risk with application to robust portfolio management. *Operations Research*, 57(5), 1155–1168.

4 Sensitivity Index-Based Robust Geotechnical Design

4.1 INTRODUCTION

Geotechnical design is almost always a site-specific problem. Note that the outcome of a geotechnical design can be significantly affected by the site-specific geological setting and the properties of the geomaterials involved. However, the available site-specific data may be extremely limited in a routine project. Neither the statistical information nor the potential distribution intervals of the site-specific geotechnical parameters can be ascertained; instead, only the characteristic or nominal values of the geotechnical parameters can be determined. It should be noted that these characteristic or normal values can only represent their conservative estimates; the degree of conservativeness is generally not known to the engineer. As the degree of conservativeness adopted may vary with the engineer, different characteristic values may be obtained and further applied to the geotechnical analysis and design (by different engineers). Nevertheless, these characteristic values have been widely employed in the conventional factor of safety (FS)-based design and load and resistance factor design (LRFD).

The concept of robust design originated from the fields of quality and industrial engineering (e.g., Taguchi 1986; Tsui 1992; Chen et al. 1996), which is aimed at achieving an optimal design that is robust against, or insensitive to, the variation of "*hard-to-control*" and "*hard-to-characterize*" input parameters (known as noise factors) through a careful selection of the "*easy-to-control*" input parameters (known as design parameters). Since its inception, robust design has been extensively applied in various engineering fields such as mechanical, structural, and aeronautical engineering (e.g., Zhang et al. 2005; Park et al. 2006; Beyer and Sendhoff 2007; Marano et al. 2008; Lee et al. 2010). The key to the success of robust design is an effective measure of design robustness (or degree of insensitivity). Thus, various robustness measures and associated design approaches have been established over the past few decades, among which the most popular robust design approaches include the Taguchi signal-to-noise ratio (SNR) method (e.g., Taguchi 1986; Tsui 1992), the worst-case approach (e.g., Nagy and Braatz 2004; Zhu and Fukushima 2009), the mean-variation compromising approach (e.g., Chen et al. 1996; Park et al. 2006), and the feasibility-based robust design approach (e.g., Du and Chen 2000; Juang and Wang 2013). Most of the robustness measures and robust design approaches reported in the literature are tailored for situations where the level of variation of noise factors can be ascertained

DOI: 10.1201/9781003564454-4

nd where the characterized variation is not low. However, the level of variation of oise factors in geotechnical designs tends to be much higher than that in many other elds, and the magnitude of such variation in many geotechnical parameters cannot e quantified with certainty due to limited data availability. The lack of accurate atistical information on the site-specific geotechnical parameters leads to difficulty the evaluation of the robustness of the candidate designs, which would hinder the rect adaptation of the robust design approaches from other fields to geotechnical esign.

Mathematically, the sensitivity of the system performance to the noise factors can e captured by its gradient, the variation of the system performance caused by one nit change in the noise factors. Furthermore, the performance of most geotechnical stems is relatively monotonic. Thus, the available characteristic or nominal values f the site-specific geotechnical parameters might allow for an approximation of the radient of the system performance. Thus, neither the statistical information nor the stribution intervals of the “hard-to-control” and “hard-to-characterize” site-specific geotechnical parameters will be a prerequisite.

In a routine geotechnical design, the number of noise factors involved may be significant, and different noise factors yield different units. In other words, the gradient f the system performance will be a multi-dimensional vector. For ease of robust esign optimization, the multi-dimensional vector of this gradient is transformed nto a sensitivity index (SI) through scaling and Euclidean norm analyses (e.g., Gong al. 2014, 2016). Then, the robustness in the context of the SI-based robust geotechnical design (RGD) can be raised by reducing the SI of the system performance the noise factors (e.g., “hard-to-control” and “hard-to-characterize” geotechnical arameters).

This chapter focuses on the development of the gradient-based SI and its application in the RGD. In that the statistical information and the distribution intervals of ne geotechnical parameters are not required in the evaluation of this gradient-based I, this robustness measure is readily compatible with the conventional FS-based esign and LRFD. Thus, the optimization of the robustness of the design can be nplemented as a complementary design objective to the objective of cost efficiency onsidered in conventional geotechnical designs, and the framework of the SI-based GD can readily be established.

This chapter consists of three parts. First, the general principle of the gradient-based SI is introduced. Then, three geotechnical design examples are analyzed o illustrate the application of this SI-based RGD. Finally, the concluding remarks re drawn.

.2 ROBUSTNESS MEASURE OF GRADIENT-BASED SENSITIVITY INDEX

he gradient-based SI is formulated in this section, and illustrative examples are resented to demonstrate the concept and effectiveness of this robustness measure.

4.2.1 Effectiveness of the Gradient as a Measure of Design Robustness

For a design $\boldsymbol{d}$ (representing a set of easy-to-control design parameters) with unce tain input parameters $\boldsymbol{\theta}$ (representing a set of hard-to-control and hard-to-chara terize noise factors), the system performance of this design can be expressed as function of the variables of $\boldsymbol{d}$ and $\boldsymbol{\theta}$, in terms of $g(\boldsymbol{d}, \boldsymbol{\theta})$. Let's assume a design sc nario where the uncertainties of the uncertain parameters (or noise factors) $\boldsymbol{\theta}$ can b fully characterized by their covariance matrix $\boldsymbol{C}_{\theta}$, then the variation of the syste performance $g(\boldsymbol{d}, \boldsymbol{\theta})$ can be captured by its standard deviation, denoted as $\sigma_{g(d,\theta)}$, ar which may be approximated as follows (Ang and Tang 2007).

$$\sigma_{g(d,\theta)} \approx \sqrt{\left(\mathbf{G}\big|_{\theta=\mu_\theta}\right)\cdot \mathbf{C}_{\theta} \cdot \left(\mathbf{G}\big|_{\theta=\mu_\theta}\right)^{\mathrm{T}}} \quad (4.$$

where $\mathbf{G}\big|_{\theta=\mu_\theta}$ is the gradient of the system performance $g(\boldsymbol{d}, \boldsymbol{\theta})$ to the noise factors evaluated at the mean of the noise factors $\boldsymbol{\theta}$, denoted as μ_θ, which is formulated (Gong et al. 2014):

$$\mathbf{G}\big|_{\theta=\mu_\theta} = \left\{ \left.\frac{\partial g(\boldsymbol{d},\boldsymbol{\theta})}{\partial \theta_1}\right|_{\theta=\mu_\theta}, \left.\frac{\partial g(\boldsymbol{d},\boldsymbol{\theta})}{\partial \theta_2}\right|_{\theta=\mu_\theta}, \ldots, \left.\frac{\partial g(\boldsymbol{d},\boldsymbol{\theta})}{\partial \theta_n}\right|_{\theta=\mu_\theta} \right\} \quad (4.$$

where n is the number of noise factors $\boldsymbol{\theta}$, and θ_i represents the ith noise factor.

It is well acknowledged that a design with a larger standard deviation of the sy tem performance would signal a lower degree of design robustness. However, in th absence of the covariance of noise factors $\boldsymbol{\theta}$, the estimate of the standard deviatio of the system performance cannot be realized. In reference to Eq. (4.1), the standar deviation of the system performance $\sigma_{g(d,\theta)}$ is mainly influenced by the noise fac tors' covariance $\boldsymbol{C}_{\theta}$ and the gradient vector $\mathbf{G}\big|_{\theta=\mu_\theta}$. In a given geotechnical design the covariance of the site-specific noise factors (or uncertain input parameters) $\boldsymbol{C}_{\theta}$ constant and would not be affected by the selection of design parameters $\boldsymbol{d}$. Thus, th variation of the system performance in the geotechnical design may be effectivel reduced by reducing the gradient $\mathbf{G}\big|_{\theta=\mu_\theta}$. In other words, the robustness of a desig could be raised by reducing this gradient without a complete characterization of th uncertainty of the hard-to-control and hard-to-characterize geotechnical parameter

The gradient vector $\mathbf{G}\big|_{\theta=\mu_\theta}$ formulated in Eq. (4.2) can be interpreted as the varia tion of the system performance caused by one unit change in the noise factors aroun the mean values μ_θ. The plots in Figure 4.1 illustrate this gradient-based robustnes concept: two designs (in terms of $\boldsymbol{d}_1$ and $\boldsymbol{d}_2$) with the same noise factors $\boldsymbol{\theta}$ exhib different patterns of system performance: one (i.e., $\boldsymbol{d}_1$ in Figure 4.1a) yields hig variation in the system performance while the other (i.e., $\boldsymbol{d}_2$ in Figure 4.1b) yield low variation in the system performance. The design that yields lower variation i the system performance, known as a more robust design by definition, is shown exhibit a lower gradient (**G**). Thus, the gradient is an effective measure of the desig robustness, and a robust design can be obtained by lowering the gradient of the sys tem performance to the noise factors without quantifying the uncertainties of nois

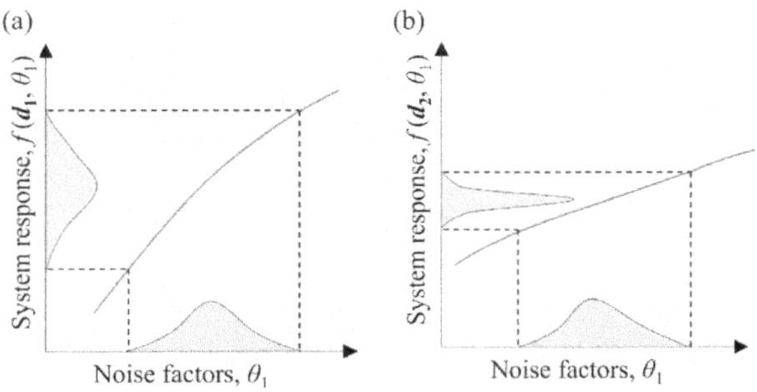

IGURE 4.1 Concept of gradient-based robustness measure: (a) System performance of a on-robust design d_1; (b) Reducing the variation in the system performance by adopting a obust design d_2 without reducing the variation in noise factors.

actors. For a geotechnical design where the mean values of the noise factors (μ_θ) annot be available, the gradient $\mathbf{G}|_{\theta=\mu_\theta}$ in Eq. (4.2) can be approximated by the term $\mathbf{G}|_{\theta=\mu_\theta}$ formulated below.

$$\mathbf{G}|_{\theta=\theta'} = \left\{ \left.\frac{\partial g(\boldsymbol{d},\boldsymbol{\theta})}{\partial \theta_1}\right|_{\theta=\theta'}, \left.\frac{\partial g(\boldsymbol{d},\boldsymbol{\theta})}{\partial \theta_2}\right|_{\theta=\theta'}, \ldots, \left.\frac{\partial g(\boldsymbol{d},\boldsymbol{\theta})}{\partial \theta_n}\right|_{\theta=\theta'} \right\} \tag{4.3}$$

vhere $\boldsymbol{\theta}'$ represents the characteristic or nominal values of the uncertain input vari- bles $\boldsymbol{\theta}$. This unique feature of this gradient-based robustness measure could be a erfect match with the conventional geotechnical practices where only the character- stic or nominal values of the site-specific geotechnical properties can be available, nd also an ideal match with the traditional FS-based design and LRFD.

Two illustrative examples are presented below to demonstrate the application and ffectiveness of this gradient-based robustness measure.

Example 1

A wall is to be supported on a strip foundation. The load from the upper structure (P) is 100 kN/m, while the bearing capacity (q_u) of the ground is characterized as a lognormal variable with a mean of 250 kPa and a coefficient of variation (COV) of 0.15, as shown in Figure 4.2. In the design of this strip foundation, the design performance (g) is formulated as follows.

$$g = q_u B - P$$

where B denotes the width of the strip foundation, which can be selected from the design space of {0.6 m, 0.7 m, 0.8 m, 0.9 m, 1.0 m, 1.1 m, 1.2 m, 1.3 m, 1.4 m, and 1.5 m}. The problem here is to analyze the relationship between the standard deviation of this design performance and the gradient of this design performance with respect to the bearing capacity.

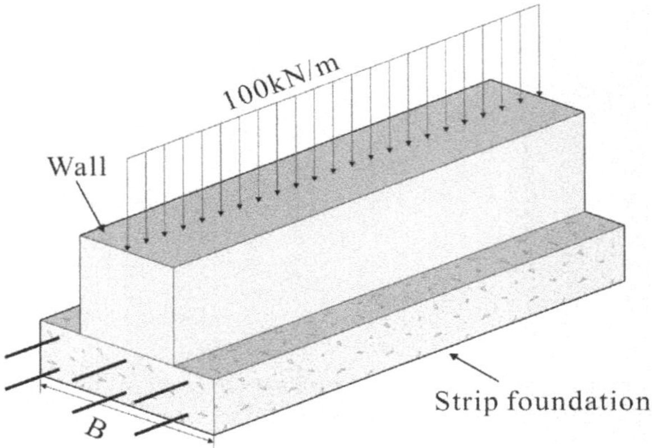

FIGURE 4.2 Schematic diagram of the strip foundation design problem in Example 1.

Solution:

Step 1: Calculate the standard deviation of the design performance for each design

The performance function (g) of the strip foundation is formulated below.

$$g = q_u B - P$$

Based on the formulated performance function (g), the standard deviation of the (σ_g) of performance function can be evaluated as follows.

$$\sigma_g = B \cdot \sigma_{q_u}$$

As an example, the standard deviation (σ_g) for a design of B = 0.6 m is estimated below. According to the statement "the bearing capacity (q_u) of the ground is characterized as a lognormal variable with a mean of 250 kPa and a COV of 0.15," the mean (μ_{q_u}) and standard deviation (σ_{q_u}) of the bearing capacity (q_u) are calculated as follows.

$$\mu_{q_u} = 250 \text{ (kPa)}, \ \mathrm{COV}_{q_u} = 0.15$$

$$\sigma_{q_u} = \mathrm{COV}_{q_u} \cdot \mu_{q_u} = 0.15 \times 250 \text{ (kPa)} = 37.5 \text{ (kPa)}$$

Thus, the standard deviation of design performance (g) can be computed as follows.

$$\sigma_g = B \cdot \sigma_{q_u} = 0.6 \text{ (m)} \times 37.5 \text{ (kPa)} = 22.5 \text{ (kN/m)}$$

For each candidate design in the design space, the corresponding standard deviation (σ_g) can readily be computed by applying the formulations above.

Step 2: Calculate the gradient of performance with respect to bearing capacity

The gradient of the design performance with respect to bearing capacity, denoted as G, represents the change in the performance function (g) when the ground bearing capacity q_u increases by one unit. Mathematically, the gradient term (G) is formulated as follows.

$$G = \frac{\partial g}{\partial q_u} = \frac{\partial (q_u B - P)}{\partial_{q_u}} = B$$

As an example, the gradient (G) for a design of $B = 0.6$ m is computed below.

$$G = \frac{\partial g}{\partial_{q_u}} = B = 0.6 \text{ (m)}$$

For each candidate design in the predefined design space, the corresponding gradient (G) can readily be computed by applying the formulations above.

Step 3: Analyze the relationship between the standard deviation and the gradient

The calculation results, in terms of the standard deviation and the gradient, of all candidate designs can lead to a relationship between the standard deviation (σ_g) and the gradient (G), as illustrated in Figure 4.3 (a). As can be seen, there exists a linear relationship between the standard deviation (σ_g) and the gradient (G), and the standard deviation increases linearly with the gradient, indicating that the gradient can effectively capture the standard deviation.

Further, the relationships between the standard deviation of design performance and the gradient for other COV values, such as 0.05, 0.10, 0.15, 0.20, 0.25, and 0.30, are analyzed, and the results are depicted in Figure 4.3(b). The results confirm that the linear relationship between the standard deviation and the gradient is not degraded by the selection of COV value of the uncertain variable, and the standard deviation always increases linearly with the gradient.

Example 2

This example also considers the strip foundation problem, and the load from the upper structure (P) is 100 kN/m. The bearing capacity of the ground (q_u) in this example can only be estimated from the soil properties with the equations below (Cherubini 2000).

$$q_u = 0.5\gamma_1 B N_\gamma + cN_c + \gamma_2 D N_q$$

where B is the width of the foundation; D is the depth of the foundation base with respect to the ground surface; γ_1 and γ_2 are the unit weight of the soil under the foundation base and that over the foundation base; c is the effective cohesion of

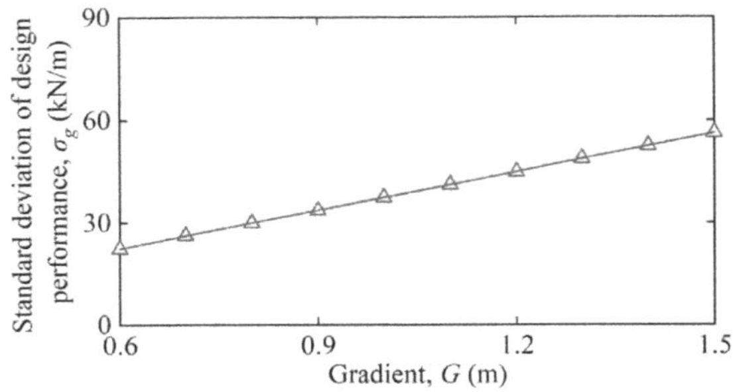

(a) Relationship between the standard deviation and the gradient for COV = 0.15

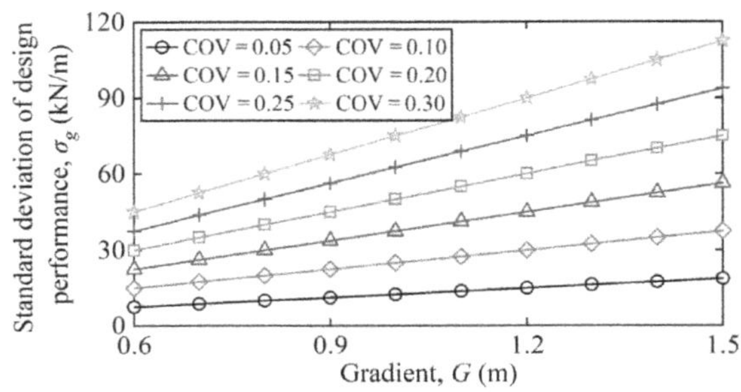

(b) Relationship between the standard deviation and the gradient for different COV values

FIGURE 4.3 Relationship between the standard deviation of design performance and the gradient in Example 1.

the soil under the foundation base; and, N_c, N_q, and N_γ are the bearing capacity factors, which are estimated as follows.

$$N_\gamma = 1.8(N_q - 1)\tan\varphi$$

$$N_q = \tan^2\left(\frac{\pi}{4} + \varphi\right)e^{\pi \tan\varphi}$$

$$N_c = (N_q - 1)\cot\varphi$$

where Φ is the effective friction angle of the soil under the foundation base. With the estimated ground bearing capacity (q_u), the performance function (g) in this strip foundation design can be established as follows.

$$g = q_u B - P$$

The unit weights of the soils in this example are characterized as $\gamma_1 = 18.0\ kN/m^3$ and $\gamma_2 = 18.0\ kN/m^3$, while the effective cohesion and the effective friction angle are characterized as two correlated random variables. The mean and COV of the cohesion (c) are 14.5 kPa and 0.30, respectively; the mean and

COV of the friction angle (ϕ) are 20.0° and 0.10, respectively; and the correlation coefficient between these two uncertain soil parameters is −0.5. The problem here is to analyze the relationship between the variation of the design performance and the gradient of the performance with respect to the uncertain soil properties.

Solution:

Step 1: Select the potential candidate designs

The objective in this example is to analyze the relationship between the variation of the design performance and the gradient of the design performance with respect to the uncertain soil properties. For illustration purposes, the design space in this example is taken as $B \in \{0.6$ m, 0.7 m, 0.8 m, . . . , 2.0 m$\}$ and $D \in \{2.0$ m, 2.2 m, 2.4 m$\}$.

Step 2: Estimate the standard deviation of design performance for each candidate design using Monte Carlo simulation

The standard deviation of the design performance in this example is calculated with the Monte Carlo simulation (MCS) method. Within the MCS approach, samples of the uncertain variables are first generated, and these samples are then input into the performance function to obtain the performance values, from which the standard deviation of the performance can be estimated. In this example, 100,000 samples are adopted to estimate this standard deviation.

The complete code of the MCS implementation is provided below.

```
%% MCS
function MCS
nsamples = 1e5;
xm = [14.5 20]';
xsd = [4.35 2]';
xr = [1 -0.5;
     – 0.5 1];
Cx = (xsd * transpose(xsd)).* xr;
X = mvnrnd(xm', Cx, nsamples);
for i = 1: nsamples
        G_samples(i, 1) = g(x(i,:), B, D);
end
stdG = std(G_samples);
end
```

The performance of the strip foundation can be evaluated as follows. For a design $\boldsymbol{d} = \{B, D\}$ with uncertain input parameters $\boldsymbol{\theta} = \{c, \phi\}$, the performance function can be expressed as a function of the variables of $\boldsymbol{d}$ and $\boldsymbol{\theta}$, in terms of $g(\boldsymbol{d}, \boldsymbol{\theta})$. With the estimated ground bearing capacity (q_u), the performance function (g) in this example can be established as follows.

$$g(\boldsymbol{d},\ \boldsymbol{\theta}) = q_u B - P$$

As an example, the performance function (g) for a design of B = 1.0 m and D = 2.0 m, with the mean values of uncertain variables (i.e., c = 14.5 kPa, Φ = 20°) as input, is calculated below. First, the bearing capacity factors (i.e., N_c, N_q and δ_D) are estimated as follows.

$$N_q = \tan^2\left(\frac{\pi}{4}+\frac{\varphi}{2}\right)e^{\pi\tan\varphi} = \tan^2\left(\frac{\pi}{4}+\frac{20^\circ}{2}\right)e^{\pi\tan 20^\circ} = 6.40$$

$$N_\gamma = 1.8(N_q-1)\tan\varphi = 1.8\times(6.40-1)\times\tan(20^\circ) = 3.54$$

$$N_c = (N_q-1)\cot\varphi = (6.40-1)\cot 20^\circ = 14.84$$

Based on the estimated ground bearing capacity factors (i.e., N_c, N_q and N_γ), the ground bearing capacity (q_u) can be calculated as follows.

$$\begin{aligned} q_u &= 0.5\gamma_1 B N_\gamma + cN_c + \gamma_2 D N_q \\ &= 0.5\times 18\ (\text{kN/m}^3)\times 1\ (\text{m})\times 3.54 + 14.5\ (\text{kPa})\times 14.84 + 18\ (\text{kN/m}^3)\times 2\ (\text{m})\times 6.40 \\ &= 477.44\ (\text{kPa}) \end{aligned}$$

where $\gamma_1 = 18.0\ kN/m^3$ and $\gamma_2 = 18.0\ kN/m^3$ are the unit weight of the soil under the foundation base and that over the foundation base. Finally, the performance (g) can readily be calculated:

$$g(\boldsymbol{d},\ \theta) = q_u B - P = 477.44\,(\text{kPa})\times 1\,(\text{m}) - 100\,(\text{kN/m}) = 377.44(\text{kN/m})$$

The performance function in this example can be computed with the following code.

```
function G =g(x, B, D)
% x: 1-c, cohesion;2-fi, friction angle
% design parameters: B, width; D, length
r1 = 18;
r2 = 18;
P = 100;
c = x(1);
fi =x(2) * pi/180;
Nq = tan(pi/4+fi/2)^2 * exp(pi * tan(fi));
Nc = (Nq – 1) * cot (fi);
Nr = 1.8 * (Nq – 1) * tan(fi);
qu = 0.5 * r1  * B * Nr + c * Nc + r2 * D * Nq;
G = qu * B- P;
end
```

For each candidate design, defined as a pair of parameters B and D, in the design space, the standard deviation of the design performance (σ_g) can readily be computed by applying the MCS method. As an example, the standard deviation for a candidate design of B = 1.0 m and D = 2.0 m is calculated as σ_g= 77.72 kN/m.

Step 3: Calculate the gradients of the performance function with respect to uncertain soil parameters

The gradient of the performance function (g) in this example is computed at the mean values of uncertain parameters, and a central difference scheme is adopted to approximate the partial derivatives. For each uncertain parameter θ_i (i.e., cohesion c or friction angle ϕ), the performance function (g) is evaluated with a small positive and negative perturbation, and the difference is then divided by the step size $\Delta\theta_i$ to obtain the partial gradient $\left.\frac{\partial g(\mathbf{d},\theta)}{\partial \theta_i}\right|_{\theta=u_\theta}$, where $\mu_\theta=[\mu_c,\mu_\varphi]$ represents the mean values of uncertain soil parameters $\boldsymbol{\theta}$.

$$G_c=\left.\frac{\partial g(\mathbf{d},\theta)}{\partial \theta_c}\right|_{\theta=\mu_\theta}\approx\frac{g\left(\mathbf{d},\mu_c+\frac{\Delta\theta_c}{2}\right)-g\left(\mathbf{d},\mu_c-\frac{\Delta\theta_c}{2}\right)}{\Delta\theta_c}$$

$$G_\varphi=\left.\frac{\partial g(\mathbf{d},\theta)}{\partial \theta_\varphi}\right|_{\theta=\mu_\theta}\approx\frac{g\left(\mathbf{d},\mu_\varphi+\frac{\Delta\theta_\varphi}{2}\right)-g\left(\mathbf{d},\mu_\varphi-\frac{\Delta\theta_\varphi}{2}\right)}{\Delta\theta_\varphi}$$

The step size $\Delta\boldsymbol{\theta}=[\Delta\theta_c,\Delta\theta_\varphi]$ is generally taken as a small value compared to its mean value. For illustration purposes, the step size $\Delta\boldsymbol{\theta}=[\Delta\theta_c,\Delta\theta_\varphi]$ in this example is set up as $\Delta\boldsymbol{\theta}=[0.1m_c,0.1m_\varphi]$. With the selected step size, the partial gradients can be computed as follows.

$$G_c=\left.\frac{\partial g(\mathbf{d},\theta)}{\partial \theta_c}\right|_{\theta=\mu_\theta}\approx\frac{g(\mathbf{d},\mu_c+0.05\mu_c)-g(\mathbf{d},\mu_c-0.05\mu_c)}{0.1\mu_c}$$

$$G_\varphi=\left.\frac{\partial g(\mathbf{d},\theta)}{\partial \theta_\varphi}\right|_{\theta=\mu_\theta}\approx\frac{g(\mathbf{d},\mu_\varphi+0.05\mu_\varphi)-g(\mathbf{d},\mu_\varphi-0.05\mu_\varphi)}{0.1\mu_\varphi}$$

For each candidate design in the design space, the corresponding gradients of the design performance (i.e., G_c and G_φ) can readily be computed by applying the formulations above.

Step 4: Analyze the relationship between the standard deviation and the gradients

The calculation results, in terms of the standard deviation of the performance (σ_g) and the partial gradients to uncertain parameters (i.e., G_c and G_φ) of all candidate designs, can lead to relationships between the standard deviation and the gradients, as shown in Figure 4.4.

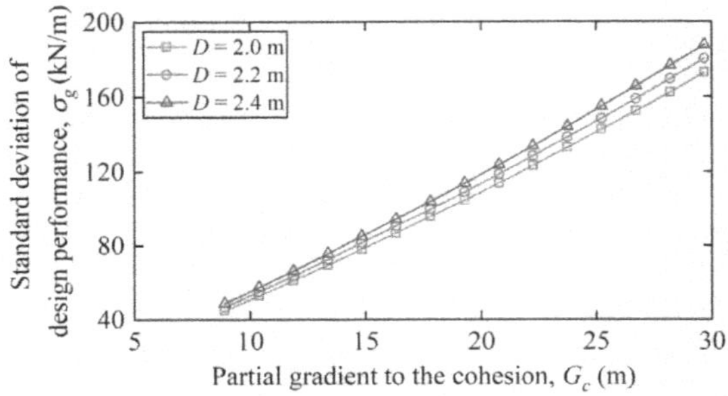

(a) Relationship between the standard deviation and the partial gradient to the cohesion

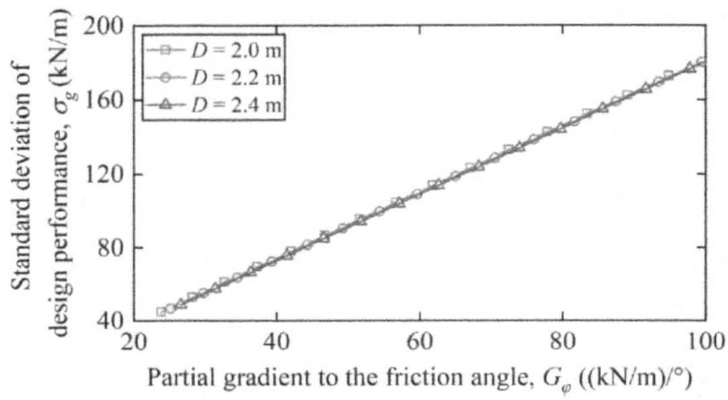

(b) Relationship between the standard deviation and the partial gradient to the friction angle

FIGURE 4.4 Relationships between the standard deviation of the performance and th partial gradients to uncertain soil parameters in Example 2.

4.2.2 Formulation of the SI-Based Design Robustness

Note that although the gradient $\mathbf{G}|_{\theta=\theta'}$ shown in Eq. (4.3) is an effective indicator o the design robustness, the following two issues need to be resolved before this gra dient-based robustness can be applied for the RGD optimization. First, this gradien is a dimensional vector. The units of the noise factors can be different, and thus, th optimization of this vector will be a problem. Second, this gradient is a vector rathe than a scalar, and thus, it might be inconvenient to screen the candidate design within an optimization framework. Hence, the gradient formulated in Eq. (4.3) mus be normalized with scaling factors and converted into a scalar for ease of applicatio in the RGD optimization.

To address the first problem, each partial derivative of this gradient vector $\mathbf{G}|_{\theta=\theta'}$ in terms of $\left.\frac{\partial g(\boldsymbol{d},\boldsymbol{\theta})}{\partial \theta_i}\right|_{\theta=\theta'}$, is multiplied by a scaling factor of $\frac{\theta_i'}{g(\boldsymbol{d},\boldsymbol{\theta}')}$ such that i becomes a dimensionless parameter. Similar normalization methods can be founc in Gonsamo (2011). The gradient vector $\mathbf{G}|_{\theta=\theta'}$ in Eq. (4.3) is then rewritten as a dimensionless vector of $\mathbf{J}|_{\theta=\theta'}$.

$$\mathbf{J}\big|_{\theta=\theta'} = \left\{ \frac{\theta_1'}{g(\boldsymbol{d},\boldsymbol{\theta}')} \frac{\partial g(\boldsymbol{d},\boldsymbol{\theta})}{\partial \theta_1}\bigg|_{\theta=\theta'}, \frac{\theta_2'}{g(\boldsymbol{d},\boldsymbol{\theta}')} \frac{\partial g(\boldsymbol{d},\boldsymbol{\theta})}{\partial \theta_2}\bigg|_{\theta=\theta'}, \ldots, \frac{\theta_n'}{g(\boldsymbol{d},\boldsymbol{\theta}')} \frac{\partial g(\boldsymbol{d},\boldsymbol{\theta})}{\partial \theta_n}\bigg|_{\theta=\theta'} \right\} \tag{4.4}$$

This scaled gradient vector $\mathbf{J}\big|_{\theta=\theta'}$ can be mathematically interpreted as a relative variation of the system performance due to the variation of the noise factors. To address the second problem, the Euclidean norm of the scaled dimensionless gradient vector $\mathbf{J}\big|_{\theta=\theta'}$, defined below can be calculated, and then the SI can be achieved.

$$SI = \left\| \mathbf{J}\big|_{\theta=\theta'} \right\| = \sqrt{\left(\mathbf{J}\big|_{\theta=\theta'}\right)\left(\mathbf{J}\big|_{\theta=\theta'}\right)^{\mathrm{T}}} \tag{4.5}$$

where $\left(\mathbf{J}\big|_{\theta=\theta'}\right)^{\mathrm{T}}$ represents the transpose vector of the gradient vector $\mathbf{J}\big|_{\theta=\theta'}$. It is known that in an n-dimensional space, the Euclidean norm indeed indicates the length of a vector. Thus, the formulated SI offers a single-value representation of the robustness. A lower SI indicates less performance sensitivity to noise factors and thus implies higher robustness.

It must be noted that the level of variation could vary with the geotechnical properties, and the noise factor that yields a larger variability could contribute more to the calculated robustness. For example, the undrained shear strength of the clay yields a larger variability in comparison to its unit weight. To explicitly consider this feature in geotechnical designs, a set of weighting factors is further included in the formulation of the scaled gradient vector $\mathbf{J}\big|_{\theta=\theta'}$ shown in Eq. (4.4), and the scaled gradient vector $\mathbf{J}\big|_{\theta=\theta'}$ is then updated as follows.

$$\mathbf{J}'\big|_{\theta=\theta'} = \left\{ w_1 \frac{\theta_1'}{g(\boldsymbol{d},\boldsymbol{\theta}')} \frac{\partial g(\boldsymbol{d},\boldsymbol{\theta})}{\partial \theta_1}\bigg|_{\theta=\theta'}, w_2 \frac{\theta_2'}{g(\boldsymbol{d},\boldsymbol{\theta}')} \frac{\partial g(\boldsymbol{d},\boldsymbol{\theta})}{\partial \theta_2}\bigg|_{\theta=\theta'}, \ldots, w_n \frac{\theta_n'}{g(\boldsymbol{d},\boldsymbol{\theta}')} \frac{\partial g(\boldsymbol{d},\boldsymbol{\theta})}{\partial \theta_n}\bigg|_{\theta=\theta'} \right\} \tag{4.6}$$

where $\mathbf{J}'\big|_{\theta=\theta'}$ represents the updated gradient vector, and w_i is the weighting factor adopted for the ith noise factor $\boldsymbol{\theta}_I$ (which indicates the contribution of the ith noise factor $\boldsymbol{\theta}_I$ to the design robustness). Here, the weighting factors may be estimated according to the levels of variation of the noise factors, which, in turn, could be estimated based on local experience, engineering judgment, and data reported in the literature.

With the aid of these weighting factors, a noise factor that exhibits a higher variability can now contribute more to the calculated design robustness. Similar to the estimate of the variation of the system performance shown in Eq. (4.1), the effect of the correlation among the noise factors can be further included in the derivation of the SI. Finally, this SI is updated as follows.

$$SI' = \sqrt{\left(\mathbf{J}'\big|_{\theta=\theta'}\right) \cdot \boldsymbol{R}_{,} \cdot \left(\mathbf{J}'\big|_{\theta=\theta'}\right)^{\mathrm{T}}} \tag{4.7}$$

where SI' represents the updated sensitivity index; $\left(\mathbf{J}'\big|_{\theta=\theta'}\right)^{\mathrm{T}}$ represents the transpose vector of the updated gradient vector $\mathbf{J}'\big|_{\theta=\theta'}$; and, $\boldsymbol{R}_{\theta}$ is a n by n matrix that depicts the correlation coefficients among the noise factors $\boldsymbol{\theta}$. Note that the Determination of $\boldsymbol{R}_{\theta}$ does not require a complete statistical characterization (i.e., the knowledge of covariance $\boldsymbol{C}_{\theta}$) of the noise factors $\boldsymbol{\theta}$. In the scenario where the knowledge of $\boldsymbol{C}_{\theta}$ is lacking, as in the geotechnical design with recognized but unquantified uncertainties, $\boldsymbol{R}_{\theta}$ might be estimated based on local experience, engineering judgment, and data reported in the literature, which is similar to the determination of the weighting factors shown in Eq. (4.6). Similarly, a lower SI' signals less sensitivity of the system performance to the noise factors and thus implies higher robustness. Two examples are presented to illustrate the application and effectiveness of this SI-based design robustness.

Example 3

The setting of this example is identical to that in Example 2; however, the problem here is to assess the relationship between the COV of the performance and the *SI* of the performance with respect to the uncertain soil parameters.

Solution:

Step 1: Select the potential candidate designs

The objective is to analyze the relationship between the COV of the design performance and the *SI* of the performance with respect to the uncertain soil parameters under different designs. For illustration purposes, the design space in this example is taken as $B \in \{0.6$ m, 0.7 m, 0.8 m, . . . , 2.0 m$\}$ and D $\boldsymbol{\theta}\{2.0$ m, 2.2 m, 2.4 m$\}$.

As mentioned in the problem, the effective cohesion (c) and the effective friction angle (φ) are characterized as two correlated random variables. The mean and COV of the cohesion (c) are characterized as 14.5 kPa and 0.30, respectively; the mean and COV of the friction angle (φ) are 20.0° and 0.10, respectively; and the correlation coefficient between these two uncertain soil parameters is −0.5.

Step 2: Estimate the coefficient of variation of the design performance for each candidate design using Monte Carlo simulation

The COV of the design performance in this example is estimated with the MCS method. Within the MCS approach, samples of the uncertain variables are first generated, and these samples are then input into the performance function to obtain the performance results, from which the COV of the performance can be estimated. In this example, 100,000 samples are drawn. The complete code of the MCS implementation is provided below.

```
%% MCS
function covG = cov_G(xmean, cov_matrix, B, D)
   % x: x(1), cohesion; x(2), friction angle
   % Design parameters: B, width; D, length
num_samples=1e5;
G_samples = zeros(num_samples, 1);
correlated_samples = mvnrnd(xmean, cov_matrix, num_samples);
   for i = 1:num_samples
       xi = correlated_samples(i,:);
       G_samples(i) = g(xi, B, D);
   end
stdG = std(G_samples);
meanG = mean(G_samples);
covG = stdG/meanG;
end
```

As presented in Example 2, the performance function for a candidate design $\boldsymbol{d} = \{B, D\}$ with uncertain input parameters $\boldsymbol{\theta} = \{c, \varphi\}$ can be established as follows.

$$g(\boldsymbol{d}, \ \theta) = q_u B - P$$

where (q_u) represents the ground bearing capacity. The performance function in this example can be computed with the following code.

```
function G =g(x, B, D)
% x: 1-c, cohesion;2-fi, friction angle
% design parameters: B, width; D, length
r1 = 18;
r2 = 18;
P = 100;
c = x(1);
fi =x(2) * pi/180;
Nq = tan(pi/4+fi/2)^2 * exp(pi * tan(fi));
Nc = (Nq – 1) * cot (fi);
Nr = 1.8 * (Nq – 1) * tan(fi);
qu = 0.5 * r1 * B * Nr + c * Nc + r2 * D * Nq;
G = qu * B- P;
end
```

For each candidate design, defined as a pair of parameters B and D, in the design space, the COV of the design performance (COV_g) can readily be computed by applying the MCS method. As an example, the COV of the design performance (COV_g) for a candidate design of B = 1.0 m and D = 2.0 m is calculated as $COV_g = 0.20$.

Step 3: Calculate the sensitivity index for each candidate design

For each candidate design, the *SI* of the design performance to the uncertain soil parameters can be calculated based on Eqs. (4.4–4.5).

$$\mathbf{J}\Big|_{\theta=\theta'} = \left\{ \frac{\theta_1'}{g(\mathbf{d},\boldsymbol{\theta}')} \frac{\partial g(\mathbf{d},\boldsymbol{\theta})}{\partial \theta_1}\bigg|_{\theta=\theta'}, \frac{\theta_2'}{g(\mathbf{d},\boldsymbol{\theta}')} \frac{\partial g(\mathbf{d},\boldsymbol{\theta})}{\partial \theta_2}\bigg|_{\theta=\theta'}, \ldots, \frac{\theta_n'}{g(\mathbf{d},\boldsymbol{\theta}')} \frac{\partial g(\mathbf{d},\boldsymbol{\theta})}{\partial \theta_n}\bigg|_{\theta=\theta'} \right\}$$

$$SI = \left\| \mathbf{J}\big|_{\theta=\theta'} \right\| = \sqrt{\left(\mathbf{J}\big|_{\theta=\theta'}\right)\left(\mathbf{J}\big|_{\theta=\theta'}\right)^{\mathrm{T}}}$$

The numerical gradient of the performance function (g) is computed at the mean values of the uncertain soil parameters, and a central difference scheme is adopted to approximate the partial derivatives. For each parameter θ_i (e.g., cohesion c or friction angle φ), the performance function (g) is evaluated with a small positive and negative perturbation, and the difference is then divided by the step size $\Delta\theta_i$ to obtain the partial derivatives $\frac{\partial g(\mathbf{d},\boldsymbol{\theta})}{\partial \theta_i}\Big|_{\theta=\mu_\theta}$, where $\mu_\theta = [\mu_c, \mu_\varphi]$ represents the mean values of uncertain soil parameters $\boldsymbol{\theta}$.

$$G_c = \frac{\partial g(\mathbf{d},\boldsymbol{\theta})}{\partial \theta_c}\bigg|_{\theta=\mu_\theta} \approx \frac{g\left(\mathbf{d}, \mu_c + \frac{\Delta\theta_c}{2}\right) - g\left(\mathbf{d}, \mu_c - \frac{\Delta\theta_c}{2}\right)}{\Delta\theta_c}$$

$$G_\varphi = \frac{\partial g(\mathbf{d},\boldsymbol{\theta})}{\partial \theta_\varphi}\bigg|_{\theta=\mu_\theta} \approx \frac{g\left(\mathbf{d}, \mu_\varphi + \frac{\Delta\theta_\varphi}{2}\right) - g\left(\mathbf{d}, \mu_\varphi - \frac{\Delta\theta_\varphi}{2}\right)}{\Delta\theta_\varphi}$$

The step size $\Delta\theta = [\Delta\theta_c, \Delta\theta_\varphi]$ is generally taken as a small value compared to its mean value. For illustration purposes, the step size $\Delta\theta = [\Delta\theta_c, \Delta\theta_\varphi]$ in this example is set up as $\Delta\theta = [0.1\mu_c, 0.1\mu_\varphi]$. With the selected step size, the partial derivatives can be computed as follows.

$$G_c = \frac{\partial g(\mathbf{d},\boldsymbol{\theta})}{\partial \theta_c}\bigg|_{\theta=\mu_\theta} \approx \frac{g(\mathbf{d}, \mu_c + 0.05\mu_c) - g(\mathbf{d}, \mu_c - 0.05\mu_c)}{0.1\mu_c}$$

$$G_\varphi = \frac{\partial g(\mathbf{d},\boldsymbol{\theta})}{\partial \theta_\varphi}\bigg|_{\theta=\mu_\theta} \approx \frac{g(\mathbf{d}, \mu_\varphi + 0.05\mu_\varphi) - g(\boldsymbol{d}, \mu_\varphi - 0.05\mu_\varphi)}{0.1\mu_\varphi}$$

Based on the computed partial derivatives, the scaled gradient vector can be obtained.

$$\mathbf{J}\Big|_{\theta=\mu_\theta} = \left\{ \frac{\mu_c}{g(\mathbf{d},\mu_{,})} \frac{\partial g(\mathbf{d},\boldsymbol{\theta})}{\partial \theta_c}\bigg|_{\theta=\mu_\theta}, \frac{\mu_\varphi}{g(\mathbf{d},\mu_{,})} \frac{\partial g(\mathbf{d},\boldsymbol{\theta})}{\partial \theta_\varphi}\bigg|_{\theta=\mu_\theta} \right\}$$

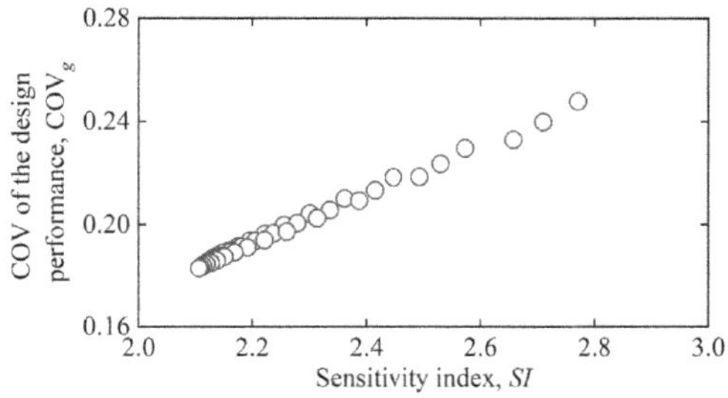

IGURE 4.5 Relationship between the coefficient of variation (COV) of the design erformance and the sensitivity index (*SI*) in Example 3.

Finally, the Euclidean norm of the scaled gradient vector can readily be computed, and the calculation of the *SI* is provided below.

$$SI = \left\| \mathbf{J}|_{\theta=\mu_\theta} \right\| = \sqrt{\left(\mathbf{J}|_{\theta=\mu_\theta}\right)\left(\mathbf{J}|_{\theta=\mu_\theta}\right)^{\mathrm{T}}}$$

For each candidate design in the design space, the corresponding *SI* of the design performance can readily be computed by applying the formulations above.

Step 4: Analyze the relationship between the coefficient of variation of the design performance and the sensitivity index

The calculation results, in terms of the COV of the performance and the *SI*, of all candidate designs are illustrated in Figure 4.5. As can be seen, the COV of the performance tends to increase linearly with the *SI*: a larger *SI* could always correspond to a larger COV of the design performance, indicating that the *SI* can effectively capture the variation of the design performance.

Further, the relationships between the COV of the performance and the *SI* are analyzed with other COV values of uncertain soil parameters: $COV_c = 0.2$, $COV_\varphi = 0.05$; $COV_c = 0.3$, $COV_\varphi = 0.10$; and $COV_c = 0.4$, $COV_\varphi = 0.15$. The results are depicted in Figure 4.6. The results confirm that the linear relationship between the COV of the performance and the *SI* is not degraded by the COV values of uncertain soil parameters: the COV of the performance tends to increase linearly with the *SI*.

Example 4

The setting of this example is identical to that in Example 2; however, the problem here is to assess the relationship between the COV of the performance and the updated sensitivity index (*SI'*) of the performance with respect to the uncertain soil properties.

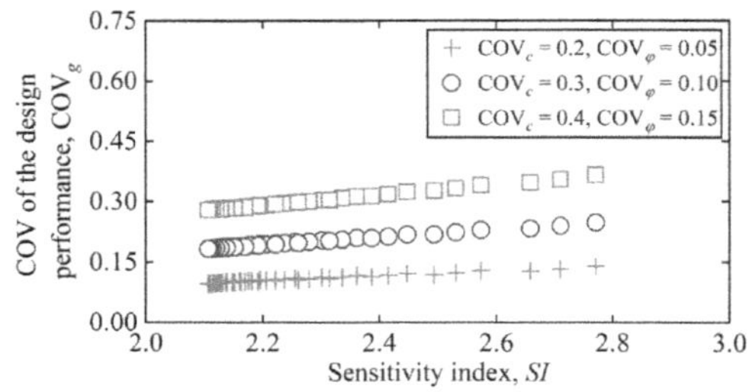

FIGURE 4.6 Relationship between the COV of the design performance and the sensitivity index (*SI*) obtained with different COV values of uncertain soil parameters in Example 3.

Solution:

Step 1: Select the potential candidate designs

The objective in this example is to analyze the relationship between the COV of the performance and the SI' of the performance with respect to the uncertain soil parameters. For illustration purposes, the design space in this example is taken as $B \in \{0.6 \text{ m}, 0.7 \text{ m}, 0.8 \text{ m}, \ldots, 2.0 \text{ m}\}$ and $D \in \{2.0 \text{ m}, 2.2 \text{ m}, 2.4 \text{ m}\}$.

As mentioned in the problem, the effective cohesion (c) and the effective friction angle (φ) are characterized as two correlated random variables. The mean and COV of the cohesion (c) are characterized as 14.5 kPa and 0.30, respectively; the mean and COV of the friction angle (φ) are 20.0° and 0.10, respectively; and the correlation coefficient between these two uncertain soil parameters ($\rho_{c\varphi}$) is -0.5.

Step 2: Estimate the coefficient of variation of the design performance for each candidate design using Monte Carlo simulation

The COV of the design performance in this example is estimated with the MCS method. Within the MCS approach, samples of the uncertain variables are first generated, and these samples are then input into the performance function to obtain the performance values, from which the COV of the performance can be estimated. In this example, 100,000 samples are drawn. The detailed procedures and code for calculating the COV of the design performance can be found in Example 3.

Step 3: Calculate the updated sensitivity index for each candidate design

For each candidate design, the SI' of the design performance to the uncertain soil parameters can be calculated as follows.

$$\mathbf{J}'\big|_{\theta=\theta'} = \left\{ w_1 \frac{\theta_1'}{g(\mathbf{d},\theta')} \frac{\partial g(\mathbf{d},\theta)}{\partial \theta_1}\bigg|_{\theta=\theta'}, w_2 \frac{\theta_2'}{g(\mathbf{d},\theta')} \frac{\partial g(\mathbf{d},\theta)}{\partial \theta_2}\bigg|_{\theta=\theta'}, \cdots, w_n \frac{\theta_n'}{g(\mathbf{d},\theta')} \frac{\partial g(\mathbf{d},\theta)}{\partial \theta_n}\bigg|_{\theta=\theta'} \right\}$$

$$SI' = \sqrt{\left(\mathbf{J}'|_{\theta=\theta'}\right) \cdot \boldsymbol{R}_{,} \cdot \left(\mathbf{J}'|_{\theta=\theta'}\right)^{\mathrm{T}}}$$

The gradient of the performance (g) is computed at the mean values of the uncertain soil parameters, and a central difference scheme is adopted to estimate the partial derivatives. For each uncertain parameter θ_i (i.e., cohesion c or friction angle φ), the performance function (g) is evaluated with a small positive and negative perturbation, and the difference is then divided by the step size $\Delta\theta_i$ to obtain the partial derivatives $\left.\frac{\partial g(\boldsymbol{d},\theta)}{\partial \theta_i}\right|_{\theta=\mu_\theta}$, where $\boldsymbol{\mu}_\theta = [\mu_c, \mu_\varphi]$ represents the mean values of uncertain soil parameters $\boldsymbol{\theta}$.

$$G_c = \left.\frac{\partial g(\boldsymbol{d},\theta)}{\partial \theta_c}\right|_{\theta=\mu_\theta} \approx \frac{g\left(\boldsymbol{d}, \mu_c + \frac{\Delta\theta_c}{2}\right) - g\left(\boldsymbol{d}, \mu_c - \frac{\Delta\theta_c}{2}\right)}{\Delta\theta_c}$$

$$G_\varphi = \left.\frac{\partial g(\boldsymbol{d},\theta)}{\partial \theta_\varphi}\right|_{\theta=\mu_\theta} \approx \frac{g\left(\boldsymbol{d}, \mu_\varphi + \frac{\Delta\theta_\varphi}{2}\right) - g\left(\boldsymbol{d}, \mu_\varphi - \frac{\Delta\theta_\varphi}{2}\right)}{\Delta\theta_\varphi}$$

The step size $\Delta\boldsymbol{\theta} = [\Delta\theta_c, \Delta\theta_\varphi]$ is generally taken as a small value compared to its mean value. For illustration purposes, the step size $\Delta\boldsymbol{\theta} = [\Delta\theta_c, \Delta\theta_\varphi]$ in this example is set up as $\Delta\boldsymbol{\theta} = [0.1\mu_c, 0.1\mu_\varphi]$. And, the updated gradient vector can be obtained with the following equation.

$$\mathbf{J}'|_{\theta=\mu_\theta} = \left\{ w_1 \frac{\mu_c}{g(\boldsymbol{d},...)} \left.\frac{\partial g(\boldsymbol{d},\theta)}{\partial \theta_c}\right|_{\theta=\mu_\theta}, w_2 \frac{\mu_\varphi}{g(\boldsymbol{d},...)} \left.\frac{\partial g(\boldsymbol{d},\theta)}{\partial \theta_\varphi}\right|_{\theta=\mu_\theta} \right\}$$

where w_1 is the weighting factor adopted for the cohesion, and w_2 is the weighting factor for the friction angle. These two weighting factors here are estimated according to the levels of variation of uncertain soil parameters. As the COV of the cohesion (c) is 0.30 and that of the friction angle (φ) is 0.10, the weighting factors are estimated as: $w_1 = 0.75$, and $w_2 = 0.25$.

Finally, the Euclidean norm of the scaled gradient vector ($\mathbf{J}'|_{\theta=...}$) can readily be computed, and the calculation of the *SI′* is provided below.

$$SI' = \sqrt{\left(\mathbf{J}|_{\theta=\mu_\theta}\right) \boldsymbol{R}_\theta \left(\mathbf{J}|_{\theta=\mu_\theta}\right)^{\mathrm{T}}}$$

where $\boldsymbol{R}_\theta$ takes the value of the correlation matrix between these two soil properties. Note that the correlation coefficient between these two uncertain soil parameters is -0.5.

For each candidate design in the design space, the *SI′* of the design performance can readily be computed by applying the formulations above.

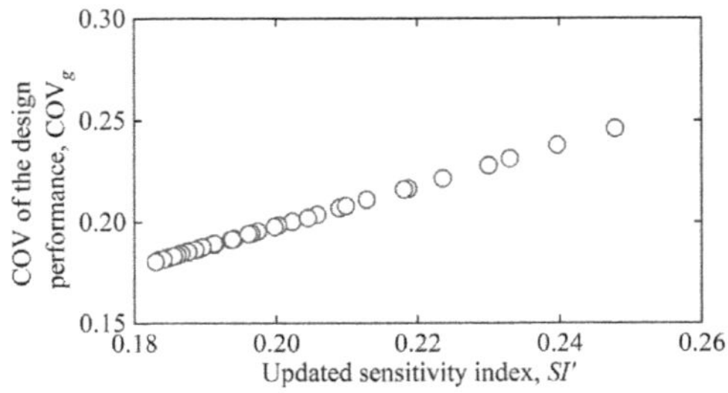

FIGURE 4.7 Relationship between the coefficient of variation (COV) of the design performance and the updated sensitivity index (SI') in Example 4.

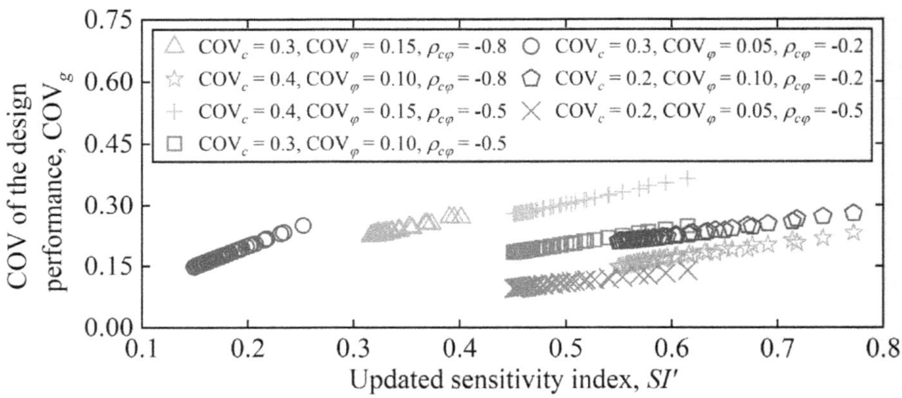

FIGURE 4.8 Relationship between the COV of the design performance and the updated sensitivity index (SI') obtained with different statistical values of uncertain soil parameters in Example 4.

Step 4: Analyze the relationship between the COV of the design performance and the updated sensitivity index

The calculation results, in terms of the COV of the design performance and the SI', of all candidate designs are illustrated in Figure 4.7. As can be seen, the COV of the performance tends to increase linearly with the SI': a larger SI' always corresponds to a larger COV of the performance. Thus, the SI' can effectively capture the variation of the performance.

Further, the relationships between the COV of the design performance and the SI' are analyzed with other statistical values of uncertain soil parameters: $COV_c = 0.3$, $COV_\varphi = 0.10$, $\rho_{c\varphi} = -0.5$; $COV_c = 0.3$, $COV_\varphi = 0.15$, $\rho_{c\varphi} = -0.8$; $COV_c = 0.3$, $COV_\varphi = 0.05$, $\rho_{c\varphi} = -0.2$; $COV_c = 0.4$, $COV_\varphi = 0.10$, $\rho_{c\varphi} = -0.8$; $COV_c = 0.2$, $COV_\varphi = 0.10$, $\rho_{c\varphi} = -0.2$; $COV_c = 0.4$, $COV_\varphi = 0.15$, $\rho_{c\varphi} = -0.5$; and $COV_c = 0.2$, $COV_\varphi = 0.05$, $\rho_{c\varphi} = -0.5$. The analysis results are illustrated in Figure 4.8. The plots confirm that the relationship bsetween the COV of the performance and the SI' is not degraded by the selection of the COV values of uncertain parameters: the COV of the design performance always increases linearly with the SI'.

.2.3 Coupling the Evaluation of Sensitivity Index with the Reliability Analysis

part from the FS-based design and LRFD, the robustness of the *SI* could also be ompatible with the existing reliability-based design (RBD). Within the context of e RBD, the statistical information of the site-specific geotechnical parameters first obtained. Then, the design safety can be assessed with the probability of ilure (or reliability index), and the variation of the system performance can be scertained. The variation of the system performance is an effective indicator of the esign robustness; however, the degree of "sensitivity" of the design performance to e noise factors cannot be captured by this variation-based design robustness. Thus, e SI could outperform the variation-based robustness in evaluating the degree of sensitivity" of the performance to noise factors.

The *SI′* may be improved in the following two aspects so that it can be more comatible with the conventional RBD:

1) The weighting factors shown in Eq. (4.6) are replaced by the coefficients of variation (COVs) of the noise factors, and in addition, the matrix of $\boldsymbol{R}_\theta$ shown in Eq. (4.7) is replaced by the correlation matrix of the noise factors;
2) The characteristic values of the noise factors shown in Eqs. (4.6–4.7) are replaced by the most probable point (MPP) of the noise factors that yields $g(\boldsymbol{d}, \boldsymbol{\theta}^*) = 0$, expressed as: $\boldsymbol{\theta}' = \boldsymbol{\theta}^*$. Note that this MPP is widely employed for deriving the reliability index utilizing the first-order reliability method (FORM) (Melchers 1987; Low and Tang 2007) and the advanced first-order reliability method (AFORM) (Ditlevsen and Madsen 1996; Melchers 1999). Additional details for the derivation of this MPP are available in Gong et al. (2014).

.3 ROBUST GEOTECHNICAL DESIGNS WITH SENSITIVITY INDEX AS ROBUSTNESS MEASURE

hree geotechnical design examples are presented in this section to demonstrate the pplication and effectiveness of the *SI*-based RGD.

.3.1 Design Example 1: Deep Excavation with Soldier Pile-Anchor Tieback Shoring System

he first geotechnical design example considers a real-world deep excavation project n Taiwan, which was studied by Gong et al. (2017). This deep excavation is located n Taipei, Taiwan, which was constructed for the student dormitory and underground arking of Wesley Girls High School. The plan layout of this excavation is approximately rectangular, with a length of 140 m and a width of 45 m. The excavation epths range from 6.9 m to 11.9 m, as illustrated in Figure 4.9. For illustration purposes, the robust design optimization focuses only on Section D, although other sections could be designed using the same RGD approach. With respect to Section D, ne excavation is carried out in layered soils with sand and rock. The soil profiles and

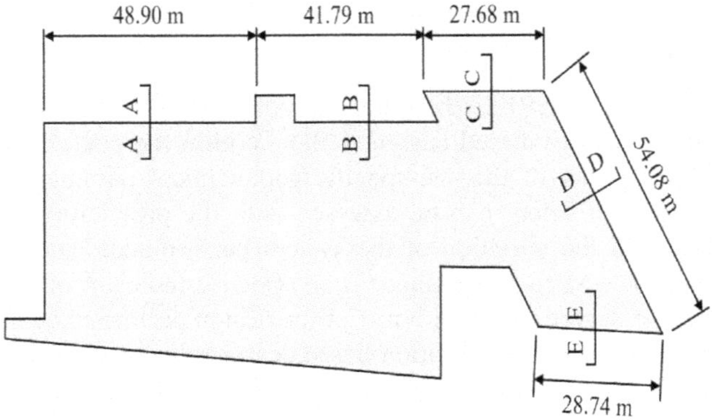

Section	Excavation dept
Section A	6.90 m
Section B	9.15 m
Section C	11.4 m
Section D	11.9 m
Section E	8.50 m

FIGURE 4.9 Plan layout of the excavation site in Design Example 1 (modified from Figure in Gong et al. 2017, with permission from Taylor & Francis Group).

TABLE 4.1
The Soil Layers and Soil Properties Characterized in Design Example 1

Layer No.	Soil type	Depth, (m)	Weight, γ_t (ton/m³)	Cohesion, c (ton/m²)	Friction angle, φ (°)	Modulus o horizonta subgrade reaction, k (ton/m³)
1	SM	0~3.35	1.80	0.0	28.0	500
2	GP	3.35~11.15	1.90	0.0	30.0	4,000
3	GP	11.15~15.50	2.30	0.0	36.0	5,000
4	Rock	15.50~20.0	2.40	2.0	34.0	5,000

properties for each soil layer are shown in Table 4.1. The effect of the groundwater i negligible in this example since the groundwater is far below the excavation deptl Based on the local experience of similar projects, a soldier pile wall that consist of steel-reinforced concrete piles and timber laggings is adopted as the retainin structure, and the 4-strand anchor (each with a diameter of 12.7 mm) is taken as th tieback of the retaining structure.

4.3.1.1 Deterministic Solution Model for Assessing the Performance of the Shoring System

To evaluate the design performance of the shoring system for this excavation prob lem, TORSA, a computer program developed by Trinity Foundation Engineerin Consultants (Sino-Geotechnics 2010), is adopted as the deterministic solutio model. TORSA has been validated as an effective tool in evaluating the stability (i.e push-in and basal heave) and serviceability (i.e., wall deflection) of the shoring sys tem for deep excavation through numerous projects in Taiwan. It must be mentione that other special-purpose computer programs such as DeepXcav (Deep Excavation 2014) or general finite element method (FEM)-based computer programs such as PLAXIS (Likitlersuang et al. 2013; Plaxis 2014) can also be employed.

TORSA is a FEM code based on the beam-on-elastic-foundation theory. The retaining structure, in terms of the soldier pile wall adopted in this deep excavation, is modeled as an elastic beam. The pressure acting on the back of the soldier pile wall is assumed to be the active earth pressure, and the resistance from soil inside the excavation is modeled as soil springs (Ou 2006; Sino-Geotechnics 2010). The water pressure, surcharge, tieback anchor (or internal strut) stiffness and preload, and the passive earth pressure caused by the plastic deformation of soil inside the excavation can all be considered. A limit equilibrium model that considers all forces and soil springs applied on the soldier pile wall can be formulated. This equilibrium model is solved via the FEM implemented in TORSA. For a given shoring system design, the factor of safety against push-in failure (FS_1), the factor of safety against basal heave failure (FS_2), and the maximum wall deflection (y) can be derived from the TORSA-based solution model.

4.3.1.2 Characterization of Noise Factors for the Robust Design Optimization

For this deep excavation problem, the main soil parameters that affect the stability and serviceability of the shoring system are the cohesion (c), friction angle (φ), and modulus of horizontal subgrade reaction (k_h). These soil parameters, along with the surcharge behind the wall (q_0), are the controlling input parameters. High variation in these parameters necessitates their treatment as noise factors in the robust design. The high variation of these soil parameters could be attributed to the inherent variability, measuring error, transformation uncertainty, and limited site investigation data availability (Phoon and Kulhawy 1999), while the variation of the surcharge behind the wall might come from the nearby transportation and heaping.

With limited data availability at a specified site, an accurate statistical characterization of the noise factors mentioned above cannot be possible; instead, these noise factors could only be characterized with their nominal values based on the available data and local experience (in a deterministic manner). Table 4.1 tabulates the characterized nominal values of these uncertain soil parameters in this example. Similarly, the surcharge behind the wall (q_0) is characterized by a nominal value of 1.0 ton/m^2. Based on the characterized nominal values of these noise factors, the deterministic analysis and design of the shoring system for this deep excavation can readily be conducted. In the context of the deterministic FS-based design, the nominal values shown in Table 4.1 are, indeed, the whole information available to the engineer.

4.3.1.3 Determination of Design Space for the Robust Design Optimization

Compared to the hard-to-control and hard-to-characterize noise factors discussed above, the diameter (D) and length (L) of the soldier pile, center-to-center spacing (I) between adjacent soldier piles, vertical spacing (V) and horizontal spacing (H) of the tieback anchors, and angle (α) of the tieback anchor with respect to the horizontal direction can be easily adjusted by the engineer, which are the easy-to-control input parameters in this design example. These easy-to-control geometry parameters are classified as the design parameters in the robust design of the shoring system for this deep excavation problem.

For illustration purposes, the preload of tieback anchors are all set up as 20 ton/tieback, and the free length and fixed length of tieback anchors are both designed as 8.0 m based on the local experience, even though the stability and serviceability of the shoring system can also be affected by these parameters. For ease of construction, a discrete design space, as tabulated in Table 4.2, is adopted in this robust design optimization of the shoring system. In the selected design space, a total number of 38,500 discrete candidate designs are possible, and the optimal design will be derived from these candidate designs. Given a specified vertical spacing of the anchor tiebacks, the excavation steps could be determined accordingly, as shown in Figure 4.10. In reference to Figure 4.10, the location of the tieback anchor is set at approximately 0.8 m above the excavation depth at each excavation stage, except for the last excavation stage.

4.3.1.4 Cost Estimate of the Shoring System for the Robust Design Optimization

According to the robust design optimization algorithm established in Chapter 3, cost efficiency and design robustness are the two design objectives to be optimized. For a given excavation project, the site dimension and excavation depths are fixed and constant. The major cost of concern in the design optimization is the cost of the shoring system (C), which is the summation of the cost of the soldier pile wall (C_w) and that of the tieback anchor system (C_t), expressed as:

$$C = C_w + C_t$$

Based on the geotechnical local practice, the cost of the soldier pile wall (C_w) could be approximated as follows (Hsii-Sheng Hsieh, personal communication 2013):

$$C_w = \left(\frac{SL}{I}\right)(L)\left(\frac{D}{0.6}\right)^2 (t_1)$$

TABLE 4.2
The Design Space Adopted in Design Example 1

Design parameter, *d*	**Design space, *DS* (by possible values of design parameters)**
Diameter of the solider pile, D (m)	{0.3 m, 0.4m, 0.5m, 0.6 m, 0.7 m}
Length of the solider pile, L (m)	{14 m, 15 m, 16 m, 17 m, 18 m, 19 m, 20 m}
Horizontal interval of the solider pile, I (m)	{D, D + 0.1 m, D + 0.2 m, . . . , D + 1.0 m}
Vertical spacing of tieback anchors, V (m)	{2.0 m, 2.5 m, 3.0 m, 3.5 m}
Horizontal spacing of tieback anchors, H (m)	{1.5 m, 2.0 m, 2.5 m, 3.0 m, 3.5 m}
Installed angle of the tieback anchor, α (°)	{10°, 15°, 20°, 25°, 30°}

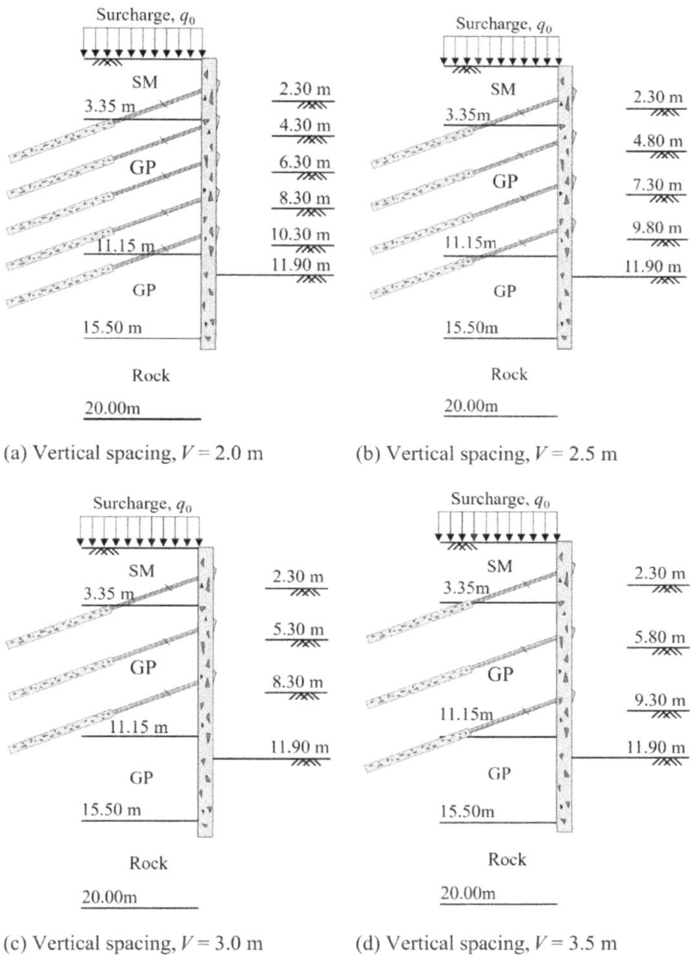

FIGURE 4.10 Possible tieback anchor layouts of the shoring system adopted in Design Example 1 (modified from Figure 5 in Gong et al. 2017, with permission from Taylor & Francis Group).

where SL is the perimeter of the excavation site (m). In this study, SL is taken as 54.08 m, as only Section D (shown in Figure 4.9) is considered. The parameter t_1 is the unit price (US dollar) for a 0.6 m diameter pile and in this study $t_1 = 66$ US dollars. Similarly, the cost of the tieback system (C_t) is estimated as follows (Hsii-Sheng Hsieh, personal communication 2013).

$$C_t = \left(\frac{SL}{H}\right)(VL)\ (t_2 \times l_2 + t_3)$$

where VL is the number of vertical levels of the tieback anchor system, which are determined based on the information shown in Figure 4.10. The parameter t_2 is the unit price (U.S. dollar) for the unit length of the anchor, l_2 is the length of the anchor, and t_3 is the unit price (US dollar) for the anchor head. In this study, these unit prices are: $t_2 = t_3 = 66$ US dollars (note: it is a coincidence that $t_1 = t_2 = t_3$ in this design example). It should be noted that the cost formulated above is the construction cost, and the material cost is already included in this cost estimate.

4.3.1.5 Robust Design Optimization Algorithm and Design Results

One significant feature of the RGD approach is that it treats the geotechnical design as a multi-objective optimization problem that explicitly and simultaneously considers the safety requirement, cost efficiency, and design robustness (Juang et al. 2013). Following the RGD framework developed, the safety requirement and design space are both considered compulsory design constraints that must be satisfied, while the design robustness and the cost efficiency are treated as design objectives to be optimized simultaneously. The robust design optimization algorithm for the design of this shoring system can be set up as follows.

Find: design parameters $\boldsymbol{d}$

Subject to: $\boldsymbol{d} \in$ design space $\boldsymbol{DS}$

$FS_1 > 1.5$ (factor of safety against push-in failure)

$FS_2 > 1.5$ (factor of safety against basal heave failure)

$y < 0.7\% H_f$ (maximum wall deflection)

Objectives: minimizing SI (sensitivity of maximum wall deflection to noise factors

minimizing C (cost)

where H_f is the excavation depth. The design constraint of $y < 0.7\% H_f$ is a limiting condition that prohibits the maximum wall deflection (y) from exceeding $0.7\% H_f$, as per Professional Standards Compilation Group (PSCG 2000). Here, the robust design is to seek an optimal design (represented by design parameters $\boldsymbol{d}$) in the design space ($\boldsymbol{DS}$; see Table 4.2) such that the design cost (C) is minimized and the design robustness is maximized (achieved by minimizing the SI), while the safety requirements with respect to both stability (in terms of FS_1 and FS_2) and serviceability (in terms of the maximum wall deflection y) are ensured. It should be mentioned that while other failure models in deep excavations may be present, such as the failure of the anchor systems and the retaining structure, they are not studied in this design example for simplicity.

Based on the design optimization algorithm formulated above, the robust design of the shoring system for this deep excavation problem can readily be conducted. For each of the candidate designs in the design space (see Table 4.2), the safety requirements (i.e., in terms of FS_1, FS_2, and the maximum wall deflection y), cost (C), and design robustness (in terms of the SI formulated in Eqs. 4.4–4.5) could be evaluated first. Then, a Pareto front that consists of 61 non-dominated designs can be established using the sorting algorithm of NSGA-II (Deb et al. 2002), as illustrated in Figure 4.11(a). Note that all the non-dominated designs on the Pareto ont are superior to all others in the feasible domain. Among these non-dominated esigns on the Pareto front, however, none of them is superior or inferior to others hen all design objectives are considered. That is to say, a tradeoff exists between esign robustness and cost efficiency. By applying the normal boundary intersection pproach (e.g., Deb and Gupta 2011; Juang et al. 2014), a knee point can be located, s shown in Figure 4.11(b). Indeed, the same knee point can be obtained utilizing the arginal utility function approach (Branke et al. 2004; Gong et al. 2014).

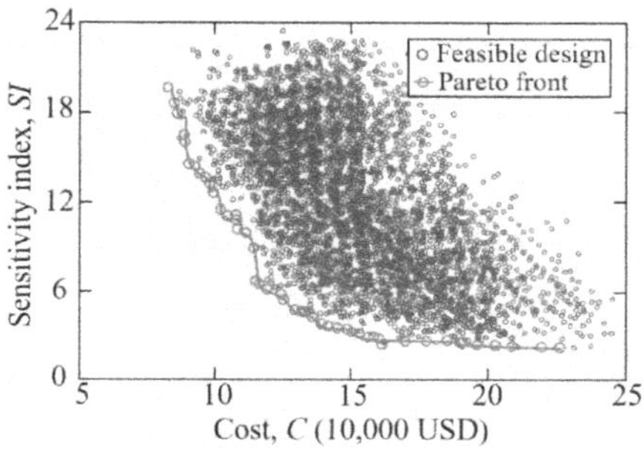

(a) Pareto front obtained from robust design optimization

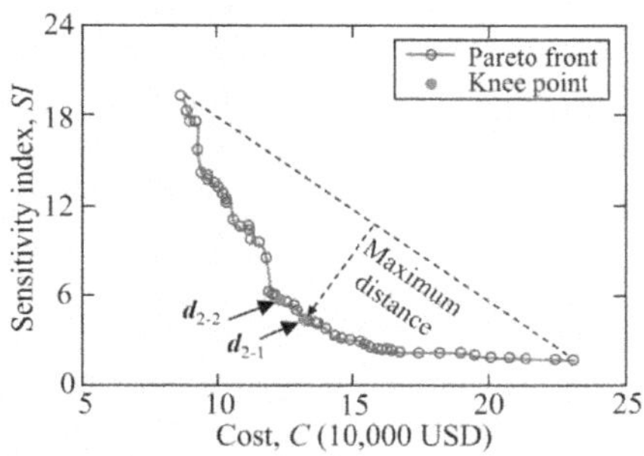

(b) Normal boundary intersection approach

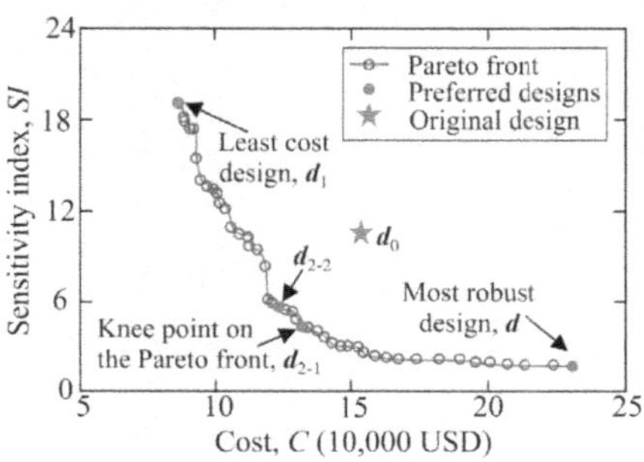

(c) Preferred designs on the Pareto front and the original design

IGURE 4.11 Results of the robust design optimization of the shoring system for deep xcavation in Design Example 1 (modified from Figure 6 in Gong et al. 2017, with permission om Taylor & Francis Group).

Using the simplified approach for locating the best compromise solution propose by the authors (Khoshnevisan et al. 2014; Gong et al. 2016, 2017), the most preferre design in the design pool can be obtained directly without developing and sortir the Pareto front, as also shown in Figure 4.11(b). The design parameters of this mo preferred design, denoted as $\boldsymbol{d}_{2-2}$, are listed in Table 4.3. As a comparison, the kne point design obtained by the original RGD method implemented with the norm. boundary intersection approach (or the marginal utility function approach), denote as $\boldsymbol{d}_{2-1}$, is also listed in Table 4.3. The difference between the design parameters $\boldsymbol{d}_{2-2}$ and that of $\boldsymbol{d}_{2-1}$ is negligible, indicating that the most preferred design obtaine with the simplified approach is practically the same as the knee point design obtaine with the RGD method that requires a two-step solution (developing a Pareto front l multi-objective optimization using algorithms such as NSGA-II, and then searc ing for knee point using, for example, the marginal utility function). This observ tion confirms the validity of the simplified method in locating the best compromi solution.

4.3.1.6 Superiority of the Design Obtained from the Robust Design Optimization

For comparison, two other designs on the Pareto front, one representing the least-co design and the other the most robust design (which is also the most costly desigr are depicted in Figure 4.11(c), along with the knee point design and the origin design. The original design is indeed one designed by an experienced engineerir firm (Hsii-Sheng Hsieh, personal communication 2013) without the knowledge the robust design optimization. Note that although the original design appears to be sound engineering practice, offering a compromise between the least cost design ar

TABLE 4.3
Comparison of Designs of the Shoring System Obtained in Design Example

Resulting design	Design parameters						Safety performance			Cost, C (USD 10,000)	Sensitivi index, S
	D (m)	L (m)	I (m)	V (m)	H (m)	α (°)	FS_1	FS_2	y (cm)		
Least cost design, $\boldsymbol{d}_1$	0.5	17	1.3	3.0	3.5	10	4.46	2.75	8.20	8.66	19.06
Most preferred design, $\boldsymbol{d}_{2-1}$	0.6	18	1.6	3.0	2.0	10	5.67	2.96	3.48	13.12	4.28
Proposed simplified approach, $\boldsymbol{d}_{2-2}$	0.5	18	1.4	3.0	2.0	10	5.67	2.96	4.94	12.31	5.58
Most robust design, $\boldsymbol{d}_3$	0.7	20	0.7	3.0	2.0	15	7.82	3.34	1.49	23.10	1.67
Original design, $\boldsymbol{d}_0$	0.5	17	0.6	3.0	2.5	20	4.46	2.75	4.13	14.42	11.22

the most robust design in this problem, it is inferior to the knee point design, as the latter is more robust and less costly. The design parameters, safety performance (i.e., FS_1, FS_2, and maximum wall deflection *y*), *SI*, and cost (*C*) of the least-cost design and those of the most robust design are tabulated in Table 4.3.

It can be noted that although all these five designs tabulated in Table 4.3 can meet the safety requirements, such as $FS_1 > 1.5$, $FS_2 > 1.5$, and $y < 0.7\% \; H_f = 8.33 \; cm$, different designs could yield different levels of design robustness and cost efficiency. In this deep excavation problem, the least-cost design among the pool of feasible designs is also the least robust design, in which a slight variation in noise factors can result in the highest variation in the system response (see Figure 4.11c), which is not desirable. On the other hand, the most robust design corresponds to the most costly, which is not desirable, either. The knee point design on the Pareto front (or the most compromise design obtained from the simplified approach) is shown as the best compromise between these two extremes on the Pareto front. Compared to the original design selected by the local engineering firm without robust design optimization, the knee point design (or the most compromise design) is cheaper and more robust.

In the design of the shoring system for this excavation problem, the serviceability of concern is the maximum wall deflection. An effective way to examine the sensitivity of this serviceability performance to each of the noise factors is to compare the normalized gradient vector (see Eq. 4.4), which could be interpreted as the variation of the maximum wall deflection caused by one unit change in each noise factor while maintaining constants for all other noise factors. This gradient could also be taken as a measure of the prediction error in the maximum wall deflection caused by one unit variation in the concerned noise factor. Figure 4.12 depicts a comparison of the variation of the maximum wall deflection caused by the relative variation in each noise factor in each of the designs tabulated in Table 4.3, designated as $\boldsymbol{d}_1$, $\boldsymbol{d}_{2-1}$, $\boldsymbol{d}_{2-2}$, $\boldsymbol{d}_3$, and $\boldsymbol{d}_0$ respectively. As can be seen, the maximum wall deflection is most affected by the friction angles of the top three soil layers and the surcharge behind the wall. This observation is reasonable since the excavation work is mainly carried out in the top three soil layers. The implication is that the prediction accuracy of the maximum wall deflection is dependent upon the accuracy of the characterization of the upper soil layers in this excavation problem. Also revealed in Figure 4.12 is that through the adoption of robust design optimization, the final design derived (i.e., $\boldsymbol{d}_{2-1}$ or $\boldsymbol{d}_{2-2}$) can yield a significant reduction of the sensitivity of the wall deflection to the noise factors, compared to the least-cost design ($\boldsymbol{d}_1$) and the original design ($\boldsymbol{d}_0$). Further reduction in the sensitivity could be achieved, for example, by adopting the most robust design (i.e., $\boldsymbol{d}_3$) in the design space; however, the cost efficiency will be sacrificed.

4.3.2 Design Example 2: A Drilled Shaft in Sand in Cohesionless Sand

The second design example considers the design of a drilled shaft in cohesionless sand (Wang et al. 2011; Juang et al. 2013), which was examined by Gong et al. (2016). As depicted in Figure 4.13, this drilled shaft is designed to bear a 50-year design load of 800 kN (i.e., $F_{50} = 800 \; kN$), and the allowable settlement of this drilled shaft is required to be less than 25 mm (i.e., $y_a = 25 \; mm$). Two fundamental requirements

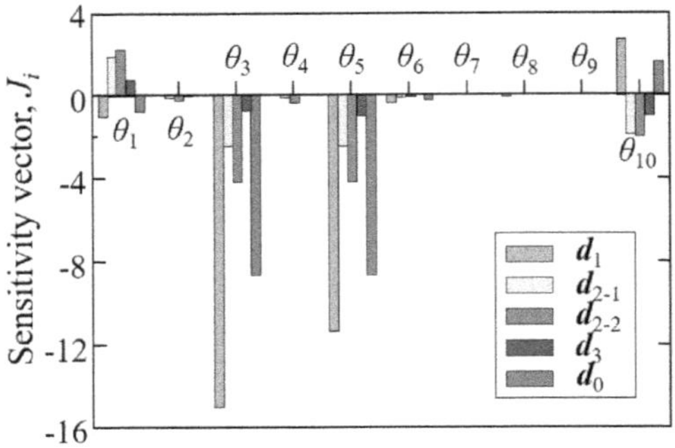

FIGURE 4.12 Sensitivity of the system response of concern (i.e., maximum wall deflection) with respect to noise factors (Note: $\theta_1 = \varphi$ of 1st soil layer; $\theta_2 = k_h$ of 1st soil layer; $\theta_3 = \varphi$ of 2nd soil layer; $\theta_4 = k_h$ of 2nd soil layer; $\theta_5 = \varphi$ of 3rd soil layer; $\theta_6 = k_h$ of 3rd soil layer; $\theta_7 = \varphi$ of 4th soil layer; $\theta_8 = k_h$ of 4th soil layer; $\theta_9 = s_u$ of 4th soil layer; $\theta_{10} = q_0$ behind the wall; the designs $\boldsymbol{d}_1$, $\boldsymbol{d}_{2-1}$, $\boldsymbol{d}_{2-2}$, $\boldsymbol{d}_3$, and $\boldsymbol{d}_0$ are listed in Table 4.3) (modified from Figure 8 in Gong et al. 2017, with permission from Taylor & Francis Group).

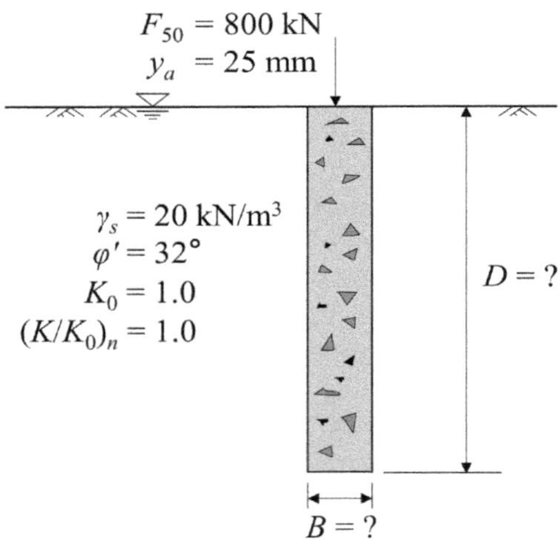

FIGURE 4.13 Schematic diagram of the design of a drilled shaft in sand in Design Example 2 (modified from Figure 3 in Gong et al. 2016, with permission from Computers and Geotechnics).

in this drilled shaft design problem are the bearing capacity and settlement. The former is referred to herein as the ultimate limit state (ULS) requirement, while the latter is referred to as the serviceability limit state (SLS) requirement. For this drilled shaft problem, the studies conducted by Wang et al. (2011) and Juang et al. (2013) indicate that the SLS requirement dominates this design; thus, only the SLS behavior of this drilled shaft is studied below, and the corresponding SLS performance function is formulated as follows.

$$g(\boldsymbol{d},\boldsymbol{\theta}) = Q_{\text{SLS}} - F_{load}$$

where Q_{SLS} is the SLS-based bearing capacity of the drilled shaft, the detailed formulation of which can be found in Phoon et al. (1995) and Wang et al. (2011); and, F_{load} is the un-factored load acting on the drilled shaft while F_{50} is the factored load.

The safety requirement in this design problem is assessed using the LRFD approach. In the context of the LRFD, the safety of the drilled shaft with respect to the SLS requirement can be assessed with the following equation (Phoon et al. 1995).

$$f(\boldsymbol{d},\boldsymbol{\theta}^{n}) = \psi_{C} Q_{\text{SLS}} - F_{50}$$

where ψ_c is the resistance factor, which is taken as 0.56 in this example (Phoon et al. 1995). If $f(\boldsymbol{d}, \boldsymbol{\theta}^{n}) > 0$, the design safety is satisfied; otherwise, the design safety is violated. In the standard LRFD approach, the resistance factor is mainly adopted to consider the effect of the recognized but unquantified uncertainty in the perceived uncertain parameters (i.e., effective friction angle φ' and coefficient of earth pressure at rest K_0 in this design example).

In the SI-based robust design of this drilled shaft, the uncertain soil parameters φ' and K_0 are referred to herein as the noise factors ($\boldsymbol{\theta}$), while the dimensions of the drilled shaft (i.e., shaft diameter B and shaft length D) are referred to as the design parameters ($\boldsymbol{d}$). Similar to that in Juang et al. (2013), a discrete design space ($\boldsymbol{DS}$) is adopted in this example: $\boldsymbol{DS} = \{(B, D) \mid B = \{0.9 \text{ m}, 1.2 \text{ m}, 1.5 \text{ m}\} \text{ and } D = \{2.0 \text{ m}, 2.2 \text{ m}, 2.4 \text{ m}, \ldots, 8.0 \text{ m}\}\}$. In the selected design space ($\boldsymbol{DS}$), a total number of 93 candidate designs are possible. Further, an additional constraint of $D/B \geq 4$ is adopted such that the drilled shaft design is consistent with the perception of deep foundation. Note that while a discrete design space is employed in this example, a continuous design space can also be handled, and similar results would be achieved. The cost of this drilled shaft (C), in US dollars (USD), can be expressed as a function of the design parameters ($\boldsymbol{d}$) as follows (R.S. Means Co. 2007; Wang et al. 2011).

$$C(\boldsymbol{d}) = \begin{cases} 77.5 \times \dfrac{D}{0.3} & \text{for } B = 0.9 \text{ m} \\ 116.0 \times \dfrac{D}{0.3} & \text{for } B = 1.2 \text{ m} \\ 157.0 \times \dfrac{D}{0.3} & \text{for } B = 1.5 \text{ m} \end{cases}$$

The effective friction angle φ' and the coefficient of earth pressure at rest K_0 are taken as the noise factors ($\boldsymbol{\theta}$) in this design example. In reference to Juang et al. (2013), the mean values of φ' and K_0 are taken as 32° and 1.0, respectively, the coefficients of variation (COVs) of φ' and K_0 are estimated as 5% to 10% and 20% to 90%, respectively, and the correlation coefficient between φ' and K_0 is estimated as -0.60 to -0.90. The nominal values of noise factors are characterized as their mean values. For instance, the nominal values of φ' and K_0 are taken as 32° and 1.0, respectively. As discussed above, the knowledge of COVs of φ' and

K_0 and the correlation coefficient between φ' and K_0 is not needed in the LRFD approach. However, to study the effects of the weighting factors and the correlation structure, which are included in the formulation of the *SI'* (see Eqs. 4.6–4.7), on the final design, the COVs of 7% and 50% for φ' and K_0, respectively, and the correlation coefficient of −0.75 are assumed.

4.3.2.1 Robust Design Optimization Results of This Drilled Shaft Problem

The safety, design robustness, and cost of all candidate designs in the design space (***DS***) are first evaluated. Then, the most preferred design in the design space (***DS***), in terms of the best compromise solution, for this drilled shaft could readily be identified using the simplified method based on the minimum distance approach described in Chapter 3 (e.g., Gong et al. 2016). The design parameters of the obtained most preferred design are $B = 0.9$ m and $D = 8.0$ m, and the corresponding cost and *SI* are $C =$ USD 2,067 and $SI = 0.340$, respectively. Table 4.4 contains the detailed analysis results of all candidate designs. As can be seen, the least cost design in the feasible domain yields $B = 0.9$ m and $D = 6.2$ m, which is associated with an *SI* of 0.482 (raw value) and a cost of $C =$ 1,602 USD (raw value). On the other hand, the most robust design in the feasible domain yields $B = 1.5$ m and $D = 8.0$ m, which is associated with a *SI* of 0.289 (raw value) and a cost of $C =$ USD 4,187. Compared to the least-cost design, the most preferred design could yield a significantly higher degree of robustness (normalized sensitivity index $SI^n = 0.262$ versus 1.0). In contrast, the increase in cost is only modest (normalized cost $C^n = 0.180$ versus 0.0). Meanwhile, compared to the most robust design, the most preferred design yields significant savings in cost (normalized cost $C^n = 0.180$ versus 1.0), while the design robustness is lowered only slightly ($SIn = 0.262$ versus 0.0). Thus, the design obtained from the robust design optimization (i.e., $B = 0.9$ m and $D = 8.0$ m) yields the best compromise between cost efficiency and design robustness in the design space selected.

The Pareto front is further obtained with the multi-objective optimization algorithm of NSGA-II (e.g., Deb et al. 2002). Figure 4.14 shows the derived Pareto front for this drilled shaft problem in both normalized design objectives space (Figure 4.14a) and unnormalized design objectives space (Figure 4.14b). It should be noted that the existence of three segments in this Pareto front, as opposed to a single smooth curve, is caused by the adoption of a discrete design space (i.e., with only three choices of the shaft diameter *B*). The knee point on the Pareto front is then identified using the marginal utility function approach (e.g., Branke et al. 2004). The design parameters of the located knee point are $B = 0.9$ m and $D = 8.0$ m, which are the same as those of the most preferred design obtained with the minimum distance approach.

4.3.2.2 Effects of Weighting Factors and Correlation Structure on the Final Design

In reference to the formulation of the updated sensitivity index (SI) formulated in Eqs. (4.6–4.7), the design robustness could be influenced by the *weighting factors* ($\boldsymbol{w}$) and the correlation structure ($\boldsymbol{R}_\theta$) of the noise factors ($\boldsymbol{\theta}$). The question is then whether the design results would be affected by the estimate of these parameters. To answer this question, let's recall the purpose of robust design, which is to locate the most preferred design (or best compromise design) in the selected design space.

ABLE 4.4
esults of the Robust Design Optimization for Design Example 2

	Design parameters						Normalized objectives		
o.	B (m)	D (m)	Safety, $g(d, \theta n)$	Cost, C (m³)	Sensitivity index, SI	Feasible or safe?	C^n	SI^n	Computed distance
	0.9	2.0	−561.64	516.67	0.309	No	–	–	–
	0.9	2.2	−536.54	568.33	0.383	No	–	–	–
	0.9	2.4	−511.15	620.00	0.482	No	–	–	–
	0.9	2.6	−485.50	671.67	0.618	No	–	–	–
	0.9	2.8	−459.58	723.33	0.821	No	–	–	–
	…	…	…	…	…	…	…	…	…
	0.9	5.8	−40.95	1,498.33	0.543	No	–	–	–
	0.9	6.0	−11.10	1,550.00	0.510	No	–	–	–
(1)	0.9	6.2	19.00	1,601.67	0.482	Yes	0.000	1.000	1.000
	0.9	6.4	49.34	1,653.33	0.458	Yes	0.020	0.874	0.875
	0.9	6.6	79.92	1,705.00	0.437	Yes	0.040	0.765	0.766
	…	…	…	…	…	…	…	…	…
	0.9	7.6	236.42	1,963.33	0.361	Yes	0.140	0.373	0.398
	0.9	7.8	268.45	2,015.00	0.350	Yes	0.160	0.315	0.353
(2)	0.9	8.0	300.71	2,066.67	0.340	Yes	0.180	0.262	0.318
	1.2	2.0	−426.13	773.33	1.419	No	–	–	–
	1.2	2.2	−387.68	850.67	3.217	No	–	–	–
	…	…	…	…	…	…	…	…	…
	1.5	7.2	1,259.29	3,768.00	0.306	Yes	0.838	0.084	0.842
	1.5	7.4	1,322.73	3,872.67	0.301	Yes	0.879	0.061	0.881
	1.5	7.6	1,386.49	3,977.33	0.297	Yes	0.919	0.040	0.920
	1.5	7.8	1,450.58	4,082.00	0.293	Yes	0.960	0.019	0.960
(3)	1.5	8.0	1,515.01	4,186.67	0.289	Yes	1.000	0.000	1.000

ote: (1) the least cost design in the feasible domain; (2) the most preferred design in the design space; and the most robust design in the feasible domain.

s the contribution of the gradient alone, as formulated in Eqs. (4.5–4.6) and shown Figure 4.1, sufficient to constrain the search for the most preferred design in the esign space? Or, is the quantitative knowledge of the statistical characterization f noise factors crucial in locating the most preferred design? A series of paramet-c analyses are performed below to investigate the effects of the assumption of the COVs (and thus weighting factors) and the correlation structure on the design out-ome (i.e., the most preferred design).

Table 4.5 lists the design outcomes obtained with various estimates of the weight-ng factors (expressed in terms of the COVs) and the correlation structure of noise

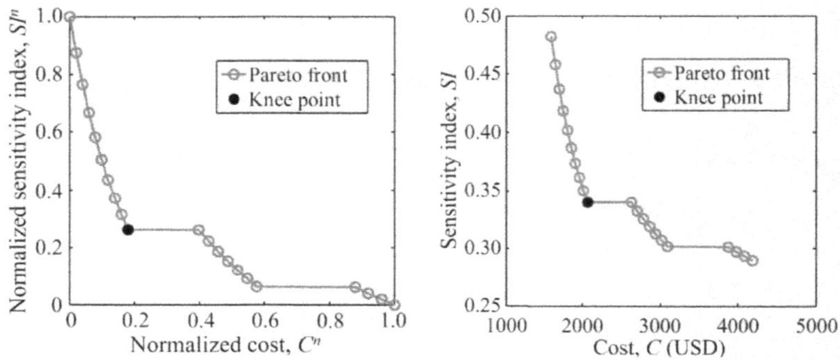

(a) Normalized design objectives (b) Unnormalized design objectives

FIGURE 4.14 The Pareto front and knee point obtained in Design Example 2 (modifi from Figure 4 in Gong et al. 2016, with permission from Computers and Geotechnics).

factors. As can be seen, the design outcome is insensitive to the weighting facto and correlation structure of noise factors over a wide range of possibilities. Thus, th effects of the weighting factors and correlation structure on the final design are n evident, and the robust design results are mainly dominated by the gradient tern In addition to the results shown in Table 4.5, 1,000 MCSs are conducted to genera the weighting factors and correlation structure. For these MCS runs, w_1 and w are assumed uniformly distributed over the possible ranges of [5%, 10%] and [20% 90%], respectively, and the correlation coefficient between φ' and K_0 is assumed be uniformly distributed in the range of [−0.6, −0.9]. Based on each of these MC samples, robust design optimization is applied, and an optimal design is obtaine It is noticed that the variation of the final designs among the 1,000 simulations relatively small and negligible. For example, the mean of the resulting shaft diamet (*B*) and shaft depth (*D*) are 0.9 m and 8.0 m, respectively, whereas the COVs of th resulting *B* and *D* values are only 1.05% and 0.75%, respectively. This observatic confirms that the effects of the weighting factors ($\boldsymbol{w}$) and the correlation structu ($\boldsymbol{R}_\theta$) of noise factors ($\boldsymbol{\theta}$) on the most preferred design (obtained from the robu design optimization) are small and negligible.

Given the statistical information of uncertain parameters shown in Table 4.5, th drilled shaft can readily be designed with the existing RBD approach, and the resul ing designs are also shown in Table 4.5. For illustration purposes, the target reliabi ity index (β_T) is set up as 2.6, and the reliability index (β) of each candidate desig is analyzed with the first-order reliability method (FORM) (Low and Tang 2007). A can be expected, the designs obtained with RBD are quite sensitive to the assumptio of the statistical information of uncertain parameters. In contrast, the robust desig outcome is insensitive to the assumption of the statistical information. The impl cation is that the outcomes obtained by RBD are meaningful only if the statistic information of noise factors (or uncertain parameters) is accurately quantified. O the other hand, robust design optimization could always yield a consistent desig in the face of recognized but unquantified uncertainty. The design results shown i Table 4.5 also depict that the robust design outcome tends to be more conservativ than the RBD outcome (indicated by a larger reliability index).

TABLE 4.5
Final Designs Obtained by Robust Design and RBD with Different Design Scenarios of the Statistical Characterization of Uncertain Input Parameters for Design Example 2

Design scenario	COV		Correlation structure R_θ	*R*-LRFD			RBD		
	φ'	K_0		B (m)	D (m)	β	B (m)	D (m)	β
(1)	7%	50%	$\begin{bmatrix}1.0 & -0.75\\ -0.75 & 1.0\end{bmatrix}$	0.9	8.0	6.29	0.9	5.6	2.65
(1)	5%	30%	$\begin{bmatrix}1.0 & -0.60\\ -0.60 & 1.0\end{bmatrix}$	0.9	8.0	7.48	0.9	5.2	2.65
(1)	5%	30%	$\begin{bmatrix}1.0 & -0.90\\ -0.90 & 1.0\end{bmatrix}$	0.9	8.0	13.60	0.9	5.0	2.85
(1)	5%	70%	$\begin{bmatrix}1.0 & -0.60\\ -0.60 & 1.0\end{bmatrix}$	0.9	8.0	6.77	0.9	5.2	2.63
(1)	5%	70%	$\begin{bmatrix}1.0 & -0.90\\ -0.90 & 1.0\end{bmatrix}$	0.9	7.8	12.30	0.9	4.8	3.10
(1)	10%	30%	$\begin{bmatrix}1.0 & -0.60\\ -0.60 & 1.0\end{bmatrix}$	0.9	8.0	3.60	0.9	7.0	2.70
(1)	10%	30%	$\begin{bmatrix}1.0 & -0.90\\ -0.90 & 1.0\end{bmatrix}$	0.9	8.0	5.36	0.9	6.6	2.87
(1)	10%	70%	$\begin{bmatrix}1.0 & -0.60\\ -0.60 & 1.0\end{bmatrix}$	0.9	8.0	3.50	0.9	7.0	2.70
(1)	10%	70%	$\begin{bmatrix}1.0 & -0.90\\ -0.90 & 1.0\end{bmatrix}$	0.9	8.0	6.27	0.9	6.0	2.79
(2)	–	–	$\begin{bmatrix}1.0 & 0\\ 0 & 1.0\end{bmatrix}$	0.9	8.0	–	–	–	–

Note: (1) the weighting factors of noise factors are taken as the COVs of noise factors for robust design; (2) the equal weights and an identity correlation matrix are adopted for robust design.

4.3.2.3 Effectiveness of the SI as a Design Robustness Measure

The *SI* is compared with the more traditional probabilistic analysis-based robustness measures (i.e., standard deviation of probability of failure σ_{P_f} or reliability-based feasibility index *FI*). For ease of calculating these rigorous probabilistic analysis-based robustness measures, a complete statistical characterization of the noise factors is demanded. Given the statistical characterization of noise factors shown in Table 4.6, the standard deviation of the probability of failure (σ_{P_f}) and the *FI* can readily be estimated for all these non-dominated designs shown in Figure 4.14. Comparisons of these two sophisticated robustness measures with the *SI* are depicted in Figure 4.15. The methodology and procedures for estimating the standard deviation of the probability of failure (σ_{P_f}), induced by the variation of COVs of noise factors, are outlined by Juang et al. (2013). The *FI* is defined as follows (Juang et al. 2013).

TABLE 4.6
A Complete Statistical Characterization of Noise Factors for Design Example 2

Soil parameter	Type of distribution	Mean	Coefficient of variation (COV): Approximate range	Mean of COV	Std. Dev. of COV
Effective friction angle, φ'	Lognormal	32°	5% to 10%	7%	1.25%
Coefficient of earth pressure at rest, K_0	Lognormal	1.0	20% to 90%	50%	17.50%

Note: the correlation coefficient between φ' and K_0 is assumed to be normally distributed with a mean of -0.75 and a standard deviation of 0.075.

$$FI = \Phi^{-1}\left[\Pr\left(P_f - P_{fT} \leq 0\right)\right]$$

where P_{fT} is the target probability of failure, which is taken as 0.0047 in this design example for illustration purpose; $\Pr(P_f - P_{fT} \leq 0)$ is the probability that the computed probability of failure (P_f) of the design performance is not greater than the target probability of failure (P_{fT}); and, $\Phi(\cdot)$ is the cumulative distribution function of the standard normal variable. The *FI* indeed measures the confidence level (or probability) that the safety constraint, in the form of the target failure probability, can be satisfied even with the uncertainty in the characterized statistics of noise factors (e.g. uncertain geotechnical parameters).

As depicted in Figure 4.15(a), the standard deviation of the probability of failure (σ_{P_f}) increases with the *SI*. That is to say, a design with a lower *SI* (indicating a higher degree of robustness) generally exhibits less variation in the estimated probability of failure. This is consistent with the concept of a robustly designed system. Similarly Figure 4.15(b) shows that the *FI* decreases with *SI*. Thus, a design with a higher *SI* (indicating a lower degree of robustness) yields a lower *FI*, suggesting a less robust design. This is, again, consistent with the concept of robust design. Note that both the standard deviation of probability of failure (σ_{P_f}) and the *FI* have been proven effective as a robustness measure (Juang et al. 2013), the plots in Figure 4.15 support the statement that *SI* is an effective robustness measure. Compared to the standard deviation of probability of failure (σ_{P_f}) and the *FI*, the evaluation of *SI* requires only deterministic analyses and is much more computationally efficient. This deterministic nature makes the *SI* an ideal robustness measure to be used in the RGD.

4.3.3 Design Example 3: A Shallow Foundation in Cohesionless Sand

The third design example considers the design of a shallow foundation in cohesionless sand, as shown in Figure 4.16. This design example was previously reported by Gong et al. (2014). In this example, the statistical information of the site-specific geotechnical parameters can be ascertained; thus, the design safety can be assessed with the reliability indexes.

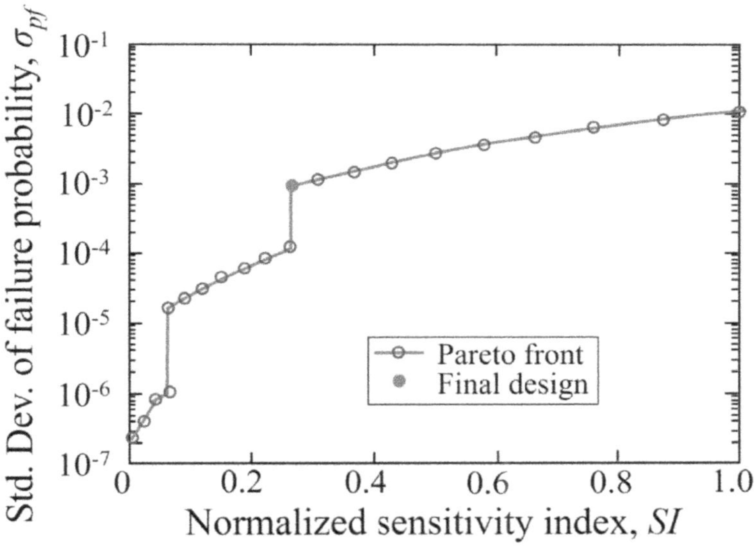

(a) σ_{P_f} versus *SI*

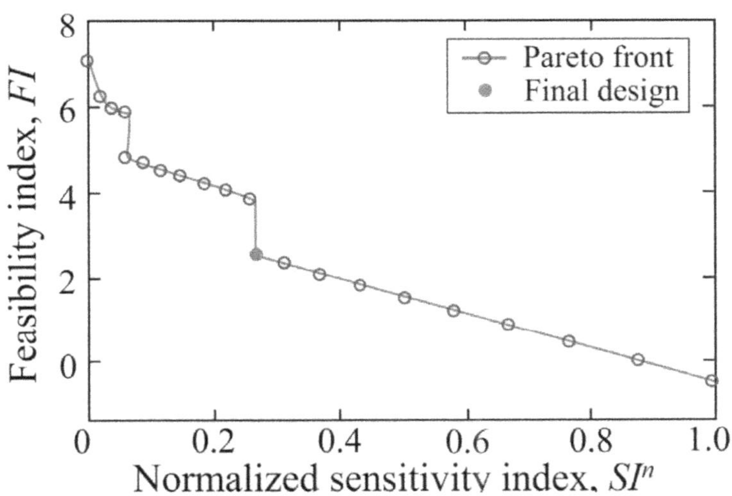

(b) *FI* versus *SI*

FIGURE 4.15 Effectiveness of the *SI* as a design robustness measure in Design Example 2 (modified from Figure 5 in Gong et al. 2016, with permission from Computers and Geotechnics).

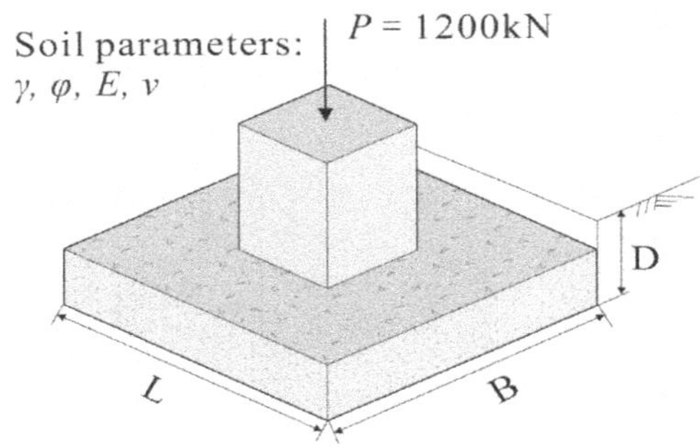

FIGURE 4.16 Schematic diagram of a shallow foundation design in Design Example 3 (modified from Figure 2 in Gong et al. 2014).

Two requirements of this shallow foundation design are the ultimate bearing capacity (safety requirement) and the settlement (serviceability requirement). These requirements can be expressed in the form of the ULS requirement and SLS requirement, respectively. The performance function with respect to ULS, denoted as $g_1(\boldsymbol{d},\boldsymbol{\theta})$, and that with regard to SLS, denoted as $g_2(\boldsymbol{d},\boldsymbol{\theta})$, are expressed as follows.

$$g_1(\boldsymbol{d},\boldsymbol{\theta}) = q_u BL - F_d - F_l$$

$$g_2(\boldsymbol{d},\boldsymbol{\theta}) = s_a - s$$

where q_u is the ultimate bearing pressure, a detailed formulation of which can be found in Gong et al. (2014); B and L are the width and length of the shallow foundation, respectively; F_d and F_l are the dead load and live load on the foundation, respectively; s_a is the maximum tolerable settlement, which is taken as 25 mm in this design example; and, s is the estimated settlement, a detailed formulation of which can be found in Gong et al. (2014). Both performance functions do not yield fixed values due to the uncertainties in geotechnical parameters (i.e., unit weight γ, friction angle φ, Young's modulus E, and Poisson ratio ν), loading parameters (i.e., dead load F_d and live load F_l), construction variation (through construction factors δ_B, δ_L, and δ_D for the width B, length L, depth D of the foundation, respectively), and model errors (i.e., model bias factor δ_1 and δ_2 for ULS and SLS models, respectively). In the framework of the RGD approach, all these uncertain parameters are treated as noise factors ($\boldsymbol{\theta}$). In contrast, the geometry parameters of the shallow foundation (i.e., B, L, and D) are treated as design parameters ($\boldsymbol{d}$). Figure 4.17 depicts all the elements considered in the robust design of this shallow foundation.

In reference to Figure 4.17, the construction variation is considered a noise factor. To account for this noise factor, the on-site geometry, denoted as B_T, L_T, and SI_1, of a candidate shallow foundation design, in the form of $\boldsymbol{d} = [B, L, D]$, is determined as follows.

$$B_T = B \cdot \delta_B,\ L_T = L \cdot \delta_L,\ D_T = D \cdot \delta_D$$

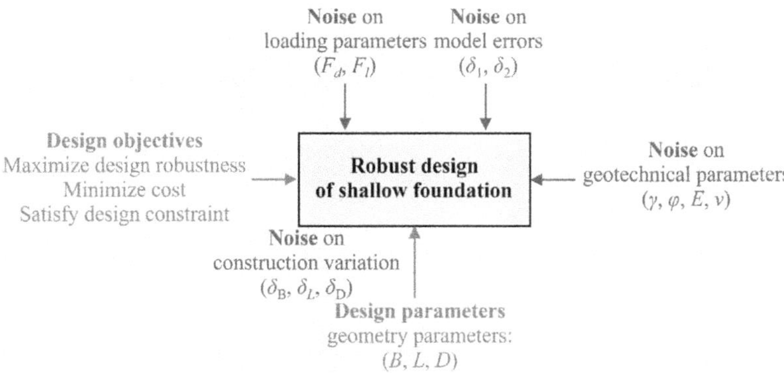

GURE 4.17 Illustration of the robust design of a shallow foundation in Design Example 3 nodified from Figure 3 in Gong et al. 2014).

he construction factors (in terms of δ_B, δ_L, and δ_D) are modeled as normally distrib-ed variables with a mean of 1.0 and a COV of 0.05, as listed in Table 4.7. This sta-stical information is selected only as an example to demonstrate the RGD approach. ertainly, other distribution types and statistical parameters of the construction fac-rs based on local data can be employed. The terms of on-site geometry parameters .e.,B_T, L_T, and D_T) are adopted for analyses of both the safety requirement and the esign robustness. Furthermore, to consider model errors, the performance functions rmulated above should be modified.

$$g_1(\boldsymbol{d},\boldsymbol{\theta}) = \frac{q_u}{\delta_1} B_T L_T - F_d - F_l$$

$$g_2(\boldsymbol{d},\boldsymbol{\theta}) = s_a - \frac{s}{\delta_2}$$

/ith the aid of the performance functions established above, the safety of candidate esigns can readily be assessed with probabilistic methods such as the advanced rst-order reliability method (AFORM) (Ditlevsen and Madsen 1996; Melchers 999). For the design of this shallow foundation problem, the target reliability indexes ith respect to ULS and SLS requirements, denoted as β_{T1} and β_{T2}, respectively, are et as 3.5 and 1.5, respectively (Zhang et al. 2011), which correspond to the target ailure probabilities (i.e., P_{T1} and P_{T2}) of 2.33×10^{-4} and 6.68×10^{-2}, respectively. The esign robustness, in terms of the *SI*, can also be derived with the methodology and rocedure described in Section 4.2, and the cost of the foundation design (C) can be stimated with the formulation in Wang and Kulhawy (2008).

The parameters setting in this design example can be summarized as follows. The oise factors ($\boldsymbol{\theta}$), including parameters γ, φ, E, ν, F_d, F_l, δ_B, δ_L, δ_D, δ_1, and δ_2, are eated as random variables and characterized in Table 4.7. The correlation coeffi-ients among these noise factors are assumed as: the construction factors (i.e., δ_B, δ_L, nd δ_D) are assumed to be correlated with a correlation coefficient of 0.5, and no

TABLE 4.7
Assumed Statistical Characterization of the Noise Factors in Design Example

Category	Parameters	Distribution	Mean (μ)	Coefficient o variation (COV
Noise on geotechnical parameters	Unit weight, γ (kN/m^3)	Normal	18.5	0.05
	Friction angle, φ (°)	Normal	38.0	0.11
	Young's modulus, E (Mpa)	Normal	40.0	0.20
	Poisson's ratio, ν	Normal	0.3	0.17
Noise on loading parameters	Dead load, F_d (kN)	Normal	1,200	0.10
	Live load, F_l (kN)	Normal	400	0.18
Noise on construction tolerances	Width, δ_B	Normal	1.0	0.05
	Length, δ_L	Normal	1.0	0.05
	Depth, δ_D	Normal	1.0	0.05
Noise on model errors	ULS solution, δ_L	Normal	1.0	0.05
	SLS solution, δ_2	Normal	1.0	0.05

correlation exists for all other pairs of noise factors. The statistical parameters show in Table 4.7 and the correlations assumed above are adopted only for illustration (the RGD approach.

The design space (***DS***) is set up as: $B \in \{0.5$ m, 0.6 m, 0.7 m, . . . , 5.0 m$\}$, $L \in \{B$, + 0.1 m, B + 0.2 m, . . . , 5.0 m$\}$, and $D \in \{1.0$ m$\}$, where the width (B) and the lengt (L) are both selected from a discrete domain for ease of construction, and the dept (D) is fixed as 1.0 m. Then, the optimization setting for this problem is formulate as follows:

Find: design parameters $\boldsymbol{d}$

Subject to: $\boldsymbol{d} \in$ design space ***DS***

$\beta_{\text{ULS}} \geq 3.5$ (ULS requirement)

$\beta_{\text{SLS}} \geq 1.5$ (SLS requirement)

Objectives: minimizing SI_1 (sensitivity of ULS performance to noise factors $\boldsymbol{\theta}$)

minimizing SI_2 (sensitivity of SLS performance to noise factors $\boldsymbol{\theta}$)

minimizing C (cost)

4.3.3.1 Robust Design Optimization Results of This Shallow Foundation Problem

On the basis of the evaluated design robustness (i.e., SI_1 for ULS performance func tion and SI_2 for SLS performance function), cost (C), design safety (i.e., β_1 for UL requirement and β_2 for SLS requirement) for all candidate designs in the design spac

DS), the Pareto front can be identified using the sorting algorithm of NSGA-II (Deb t al. 2002). Figure 4.18(a) depicts the derived Pareto front in a 3D plot, formed by a et of non-dominated optimal designs, with three design objectives: sensitivity index or the ULS performance (SI_1), sensitivity index for the SLS performance (SI_2), nd cost (C). Illustrated in Figures 4.18(b–c) are the 2D Pareto fronts that are the rojections of the 3D Pareto front shown in Figure 4.18(a), in the form of SI_1 versus C and SI_2 versus C, respectively. As illustrated in Figure 4.18, the sensitivity indexes f the design responses (SI_1 and SI_2) generally decrease with the cost (C). In other vords, the design robustness increases with the cost. Thus, a tradeoff relationship etween design robustness and cost efficiency can be revealed. It should be noted

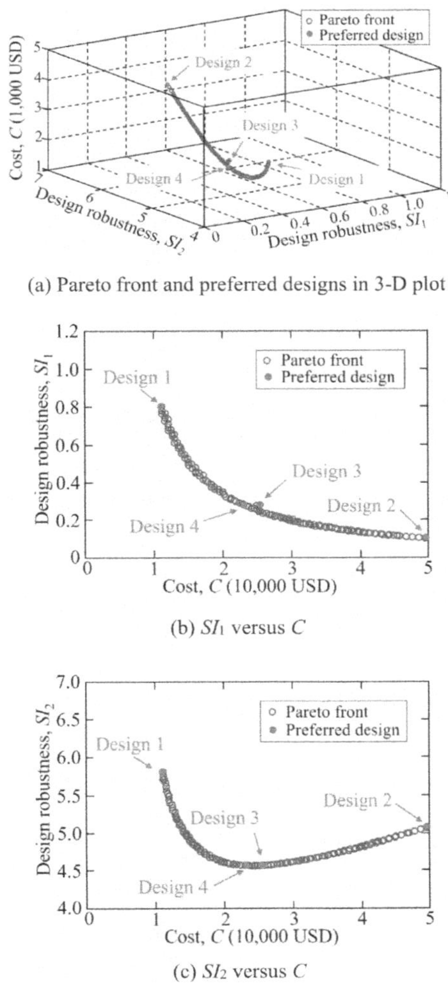

FIGURE 4.18 Results of the robust design optimization in Design Example 3 (modified rom Figure 4 in Gong et al. 2014).

that although a discrete design space (***DS***) is adopted in this design example, this is not a limitation: a continuous design space can also be employed, and similar design results (i.e., Pareto front) can be expected. To demonstrate this expectation, a continuous design space defined by $B \in [0.5 \text{ m}, 5.0 \text{ m}]$, $L \in [B, 5.0 \text{ m}]$, and $D \in \{1.0 \text{ m}\}$ is adopted. The same procedure for robust design is followed. Figure 4.19 shows a comparison of the Pareto front obtained using the discrete design space versus that obtained using the continuous design space.

The Pareto fronts shown in Figure 4.18 reveal a tradeoff relationship between design robustness and cost. It is interesting to note a few designs, keeping in mind that all designs on the Pareto front meet the design constraint (in terms of the ULS and SLS requirements and design space): Design 1 is the least cost design, which is also the least robust design against the variation in the noise factors; Design 2 is the most robust design with respect to the ULS performance function; and, Design 3 is the most robust design with respect to the SLS performance function. The most robust designs, naturally, are the most costly. Design 4 is the knee point on the Pareto front that is identified using the marginal utility function approach (Branke et al. 2004). As can be seen, on the upper side of the knee point (Design 4), a slight improvement in the robustness requires a large increase in cost (e.g., comparing Design 2 or 3 versus Design 4 in Table 4.8), which is not desirable. On the other side of this knee point, a slight reduction in the cost yields a large reduction of the design robustness (e.g., comparing Design 1 versus Design 4 in Table 4.8), which is also not desirable. In this example, Design 4 (knee point) is the most preferred design in the design space selected.

4.3.3.2 Effect of the Optimization Setting on the Derived Pareto Front

The effects of the optimization setting regarding the safety constraint and design space on the derived Pareto front are investigated below. In the previous robust design optimization, the target reliability index with respect to the SLS performance β_{T2} is set at 1.5, as per Zhang et al. (2011). Note that some engineers may wish to raise this

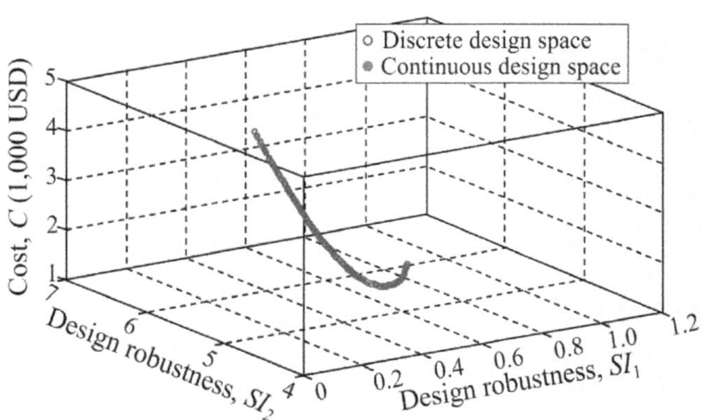

FIGURE 4.19 Discrete design space versus continuous design space in Design Example 3 (modified from Figure 5 in Gong et al. 2014).

TABLE 4.8
Examples of the Preferred Designs on the Derived Pareto Front in Design Example 3

Preferred design	Description	Design parameters: Width, B (m)	Length, L (m)	Cost, C (USD)	Sensitivity index, SI_1	Sensitivity index, SI_2
Design 1	Least cost design on the Pareto front	2.0	2.4	1,129	0.804	5.803
Design 2	Most robust design with respect to ULS performance	5.0	5.0	4,953	0.102	5.086
Design 3	Most robust design with respect to SLS performance	2.4	5.0	2,563	0.280	4.571
Design 4	Knee point (best compromise design) on the Pareto front	3.3	3.3	2,318	0.275	4.574

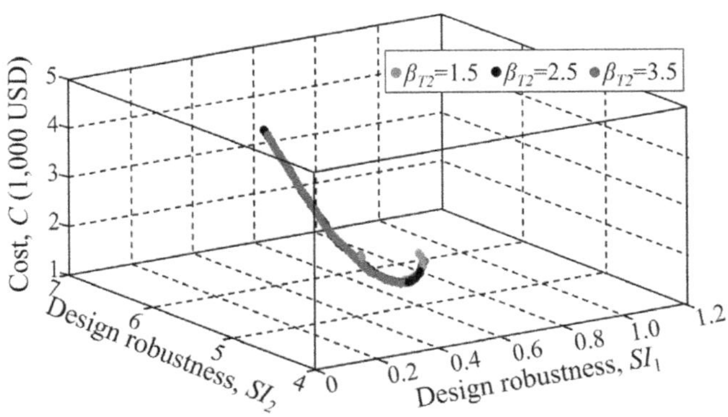

FIGURE 4.20 Pareto fronts derived with three levels of design safety in Design Example 3 (modified from Figure 8 in Gong et al. 2014).

performance requirement; thus, a question arises as to whether such a change will affect the derived Pareto front. Figure 4.20 shows the results of this investigation. Although some designs on the Pareto front obtained above with $\beta_{T2} = 1.5$no longer satisfy the design constraint when the target reliability index is increased to 2.5 and 3.5, the tradeoff relationship revealed by all three Pareto fronts remains unchanged. Similarly, three Pareto fronts can be derived with three design spaces (see Table 4.9). The results in Figure 4.21 indicate that the tradeoff relationship, revealed by the

TABLE 4.9
Different Design Spaces Adopted in the Robust Design in Design Example 3

Design space	Design space (defined by ranges of design parameters)
No. 1	$B \in \{0.5$ m, 0.6 m, 0.7 m, . . . , 5.0 m$\}$, $L \in \{B, B + 0.1$ m, $B + 0.2$ m, . . . , 5.0 m$\}$, and $D \in \{1.0$ m$\}$
No. 2	$B \in \{0.5$ m, 0.6 m, 0.7 m, . . . , 6.5 m$\}$, $L \in \{B, B + 0.1$ m, $B + 0.2$ m, . . . , 6.5 m$\}$, and $D \in \{1.0$ m$\}$
No. 3	$B \in \{0.5$ m, 0.6 m, 0.7 m, . . . , 8.0 m$\}$, $L \in \{B, B + 0.1$ m, $B + 0.2$ m, . . . , 8.0 m$\}$, and $D \in \{1.0$ m$\}$

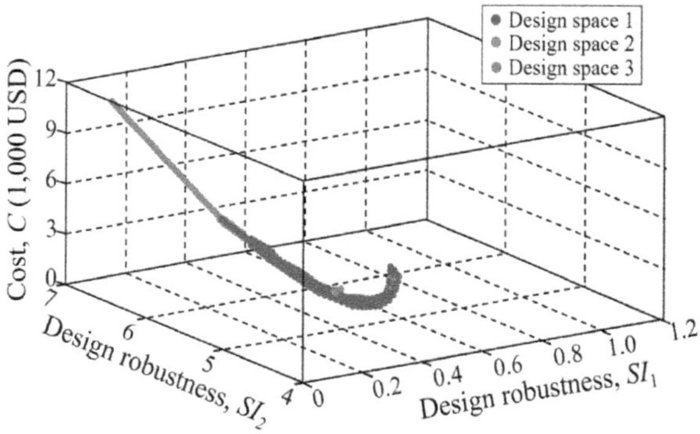

FIGURE 4.21 Pareto fronts derived with three design spaces in Design Example 3 (modified from Figure 9 in Gong et al. 2014).

Pareto front, is not affected by the setting of the design space adopted in the design optimization.

4.3.3.3 Effectiveness of the Sensitivity Index as a Design Robustness Measure

Similar to Design Example 2, the SI is compared with other design robustness measures, such as SNR and *FI*. Using the design of a shallow foundation with respect to ULS requirement as an example, the SNR of this system response, denoted as SNR_1, as a measure of the design robustness can be formulated as follows (Taguchi 1986; Phadke 1989).

$$SNR_1 = -10 \log\left\{\frac{1}{N}\sum_{i=1}^{N}\left(1\Big/\left(\frac{q_u}{\delta_1}B_T L_T\right)\right)^2\right\}$$

where N is the number of experiments. This SNR-based design robustness is indeed based on the concept that the performance is "better" with a larger bearing capacity

$\frac{q_u}{\sigma_1} B_T L_T$), which is adapted from one of the robust design concepts, “the larger, the etter,” as per Taguchi (1986) and Phadke (1989). This definition is appropriate for e shallow foundation design against the bearing capacity failure, although other NR definitions might be adopted.

For each design on the Pareto front shown in Figure 4.18(a), 1 million MCSs are rried out to simulate the joint distribution of noise factors so that the SNR of the oncerned design can be estimated. Here, the noise factors are assumed to be inde-endent from each other. As we are only interested in the general trend, this assump-on is deemed appropriate. For each design on the Pareto front, the derived SNR_1 is otted against the SI_1, and this process is repeated for all designs on the Pareto front. he results of this exercise, depicted in Figure 4.22(a), indicate that a higher *SNR* alue corresponds to a lower *SI* value (signaling a higher level of design robustness). s such, this *SI* is shown to be effective as a measure of design robustness.

Further, the *FI* is adopted for this shallow foundation design. According to arkinson et al. (1993), a system is considered to have “feasibility robustness” if it an remain feasible relative to the nominal constraints, even when the system under-oes variation. The *FI* of a shallow foundation design against the bearing capacity ilure, denoted as FI_1, is formulated as follows (Juang et al. 2013):

$$FI_1 = \Phi^{-1}\left\{\Pr\left[g_1(\boldsymbol{d},\theta) > 0\right]\right\}$$

here $Pr[g_1(\boldsymbol{d},\boldsymbol{q}) > 0]$ represents the probability that the design performance satis-es the ULS performance requirement. For each design on the Pareto front shown Figure 4.18(a), the FI_1 can readily be evaluated with the advanced first-order reli-ility method (AFORM) (Ditlevsen and Madsen 1996; Melchers 1999). The FI_1 of is design is then plotted against its $SI_1 - cu\imath$ value. This process is repeated for all esigns on the Pareto front, and the results, as shown in Figure 4.22(b), indicate that higher FI_1 corresponds to a lower SI_1 value, both signaling a higher level of design obustness. From there, this *SI* as an effective measure of design robustness is con-stent with the *FI*.

.4 SUMMARY AND REMARKS

o cope with the recognized but unquantified uncertainties in the current geotech-ical practice, we first illustrated the gradient-based SI to be an effective design obustness measure. This robustness measure is inspired by the gradient concept. owever, the multi-dimensional gradient vector cannot be directly applied to the obust design optimization. As outlined in this chapter, the multi-dimensional gradi-nt vector is transformed into the SI through scaling and Euclidean norm analyses. ince the statistical information of the noise factors is not required in the evaluation f this gradient-based SI, this robustness measure can be consistent with the existing esign approaches such as the FS and LRFD. The robustness of the design can be nproved by reducing the SI of the design performance to noise factors. For ease of lustrating the application of this SI in the RGD, three geotechnical design examples re presented.

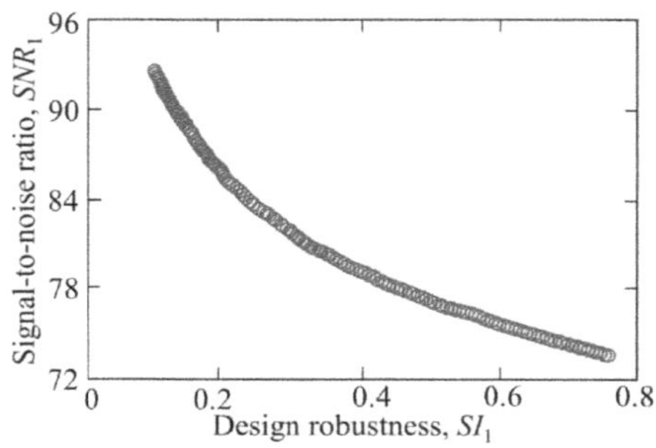

(a) SNR_1 versus SI_1 (ULS performance)

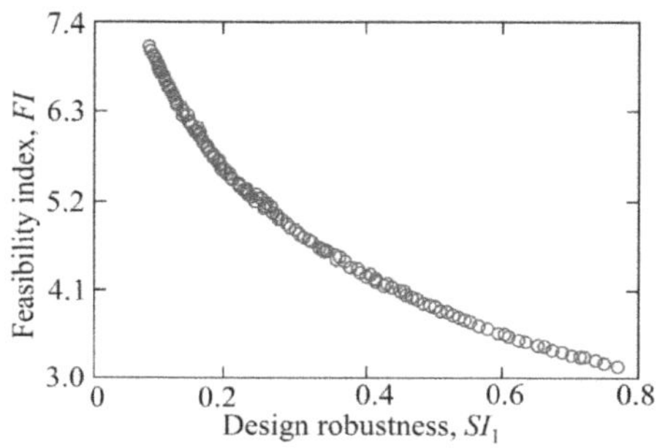

(b) FI versus SI_1 (ULS performance)

FIGURE 4.22 Sensitivity index compared with existing robustness measures in Desig Example 3 (modified from Figure 7 in Gong et al. 2014).

The design results of these geotechnical design examples illustrate that the gra dient-based SI is consistent with the more sophisticated robustness measures, suc as the reliability-based FI, standard deviation of probability of failure, and SNI However, the formulation of the gradient-based SI is more intuitive and mathemat cally sound, and the contribution of each noise factor to the design robustness coul be explicitly captured. Note that the level of variation can vary with the geotechn cal parameters, and the noise factor that yields a larger variability would contribu more to the evaluation of the design robustness. These features could be effectivel reflected in the gradient-based SI.

Finally, it should be noted that the tradeoff relationship between design robustnes and cost efficiency, revealed by the Pareto front, is not affected by the settings of th design space and the target safety requirement adopted in the design optimization.

REFERENCES

Ang, A.H.S., and Tang, W.H. (2007) *Probability concepts in engineering: Emphasis on appl cations to civil and environmental engineering* (2nd edition). Wiley, New York.

Beyer, H.G., and Sendhoff, B. (2007) Robust optimization – a comprehensive survey. *Com puter Methods in Applied Mechanics and Engineering*, 196(33), 3190–3218.

ranke, J., Deb, K., Dierolf, H., and Osswald, M. (2004) Finding knees in multi-objective optimization. *Parallel Problem Solving from Nature-PPSN*, VIII, 722–731.

hen, W., Allen, J.K., Tsui, K.L., and Mistree, F. (1996) A procedure for robust design: Minimizing variations caused by noise factors and control factors. *Journal of Mechanical Design, Transactions of the ASME*, 118(4), 478–485.

herubini, C. (2000) Reliability evaluation of shallow foundation bearing capacity on c', φ'soils. *Canadian Geotechnical Journal*, 37(1), 264–269.

eb, K., and Gupta, S. (2011) Understanding knee points in bicriteria problems and their implications as preferred solution principles. *Engineering Optimization*, 43(11), 1175–1204.

eb, K., Pratap, A., Agarwal, S., and Meyarivan, T.A.M.T. (2002) A fast and elitist multiobjective genetic algorithm: NSGA-II. *IEEE Transactions on Evolutionary Computation*, 6(2), 182–197.

eep Excavation LLC. (2014) *DeepXcav user's manual*. Astoria, New York.

itlevsen, O., and Madsen, H.O. (1996) *Structural reliability methods* (Vol. 178). Wiley, New York.

u, X., and Chen, W. (2000) Towards a better understanding of modeling feasibility robustness in engineering design. *Journal of Mechanical Design*, 122, 385–394.

ong, W., Huang, H., Juang, C.H., and Wang, L. (2017) Simplified-robust geotechnical design of soldier pile–anchor tieback shoring system for deep excavation. *Marine Georesources and Geotechnology*, 35(2), 157–169.

ong, W., Juang, C.H., Khoshnevisan, S., and Phoon, K.K. (2016) R-LRFD: Load and resistance factor design considering robustness. *Computers and Geotechnics*, 74, 74–87.

ong, W., Khoshnevisan, S., and Juang, C.H. (2014) Gradient-based design robustness measure for robust geotechnical design. *Canadian Geotechnical Journal*, 51(11), 1331–1342.

onsamo, A. (2011) Normalized sensitivity measures for leaf area index estimation using three-band spectral vegetation indices. *International Journal of Remote Sensing*, 32(7), 2069–2080.

uang, C.H., and Wang, L. (2013) Reliability-based robust geotechnical design of spread foundations using multi-objective genetic algorithm. *Computers and Geotechnics*, 48, 96–106.

uang, C.H., Wang, L., Hsieh, H.S., and Atamturktur, S. (2014) Robust geotechnical design of braced excavations in clays. *Structural Safety*, 49, 37–44.

uang, C.H., Wang, L., Liu, Z., Ravichandran, N., Huang, H., and Zhang, J. (2013) Robust Geotechnical Design of drilled shafts in sand – new design perspective. *Journal of Geotechnical and Geoenvironmental Engineering*, 139(12), 2007–2019.

hoshnevisan, S., Gong, W., Wang, L., and Juang, C.H. (2014) Robust design in geotechnical engineering–an update. *Georisk: Assessment and Management of Risk for Engineered Systems and Geohazards*, 8(4), 217–234.

ee, M.C., Kelly, D.W., Degenhardt, R., and Thomson, R.S. (2010) A study on the robustness of two stiffened composite fuselage panels. *Composite Structures*, 92(2), 223–232.

ikitlersuang, S., Surarak, C., Wanatowski, D., Oh, E., and Balasubramaniam, A. (2013) Finite element analysis of a deep excavation: A case study from the Bangkok MRT. *Soils and Foundations*, 53(5), 756–773.

ow, B.K., and Tang, W.H. (2007) Efficient spreadsheet algorithm for first-order reliability method. *Journal of Engineering Mechanics*, 133(12), 1378–1387.

Marano, G.C., Sgobba, S., Greco, R., and Mezzina, M. (2008) Robust optimum design of tuned mass dampers devices in random vibrations mitigation. *Journal of Sound and Vibration*, 313(3), 472–492.

Melchers, R.E. (1987) *Structural reliability: Analysis and prediction*. Ellis Horwood Ltd., Chichester, West Sussex, UK.

Melchers, R.E. (1999) *Structural reliability analysis and prediction* (2nd edition). John Wiley, New York.

Nagy, Z.K., and Braatz, R.D. (2004) Open-loop and closed-loop robust optimal control c batch processes using distributional and worst-case analysis. *Journal of Process Contro* 14(4), 411–422.

Ou, C.Y. (2006) *Deep excavation-theory and practice*. Taylor & Francis Group, England.

Park, G.J., Lee, T.H., Lee, K.H., and Hwang, K.H. (2006) Robust design: An overview. *AIA Journal*, 44(1), 181–191.

Parkinson, A., Sorensen, C., and Pourhassan, N. (1993) A general approach for robust optima design. *Journal of Mechanical Design*, 115(1), 74–80.

Phadke, M.S. (1989) *Quality engineering using robust design*. Prentice Hall, New Jersey.

Phoon, K.K., and Kulhawy, F.H. (1999) Characterization of geotechnical variability. *Canadia Geotechnical Journal*, 36(4), 612–24.

Phoon, K.K., Kulhawy, F.H., and Grigoriu, M.D. (1995) *Reliability-based design of foundations for transmission line structures*. Rep. TR-105000. Electric Power Research Institute, Palo Alto, CA.

Plaxis bv. PLAXIS-2D (2014) *User's manual*. Delft, the Netherlands.

Professional Standards Compilation Group (PSCG). (2000) *Specification for excavation i shanghai metro construction*. Professional Standards Compilation Group, Shanghai China.

R.S. Means Co. (2007) *R.S. Means building construction cost data*. Kingston, MA.

Sino-Geotechnics. (2010) *User manual of Taiwan originated retaining structure analysi for deep excavation*. Sino-geotechnics Research and Development Foundation, Taipei Taiwan.

Taguchi, G. (1986) *Introduction to quality engineering: Designing quality into products an processes*. Quality Resources, White Plains, New York.

Tsui, K.L. (1992) An overview of Taguchi method and newly developed statistical methods fo robust design. *IIE Transactions*, 24(5), 44–57.

Wang, Y., Au, S.K., and Kulhawy, F.H. (2011) Expanded reliability-based design approac for drilled shafts. *Journal of Geotechnical and Geoenvironmental Engineering*, 137(2) 140–149.

Wang, Y., and Kulhawy, F.H. (2008) Economic design optimization of foundations. *Journal o Geotechnical and Geoenvironmental Engineering*, 134(8), 1097–1105.

Zhang, J., Zhang, L.M., and Tang, W.H. (2011) Reliability-based optimization of geotechnical systems. *Journal of Geotechnical and Geoenvironmental Engineering*, 137(12) 1211–1221.

Zhang, W., Chen, B.S., and Tseng, C.S. (2005) Robust H∞ filtering for nonlinear stochasti systems. *Signal Processing, IEEE Transactions on*, 53(2), 589–598.

Zhu, S., and Fukushima, M. (2009) Worst-case conditional value-at-risk with application t robust portfolio management. *Operations Research*, 57(5), 1155–1168.

5 Signal-to-Noise Ratio-Based Robust Geotechnical Design

5.1 INTRODUCTION

The inherent variability, testing errors, transformation uncertainty, and the limited site investigation effort make it challenging to characterize the geotechnical parameters at a site. When there is minimal data at a site, the engineer may have to characterize the geotechnical parameters by the characteristic or nominal values. To consider the perceived uncertainty in the face of limited data availability, the engineer frequently relies on information available from the literature, local experiences, and engineering judgment. In many situations, the distribution intervals of the concerned geotechnical parameter could be determined through an estimate of its highest conceivable value (HCV) and lowest conceivable value (LCV). Compared to the nominal values, these distribution intervals provide information on the level of uncertainty of the concerned geotechnical properties. For example, with the adoption of three-sigma bounds (or its variant), the variation or standard deviation of the concerned geotechnical parameter may be derived from its HCV and LCV (Duncan 2000; Christian 2004).

As a matter of fact, the design situation where only HCV and LCV are available is quite common in geotechnical practices. Note that while these distribution intervals cannot allow for a complete or full statistical characterization, they could provide an alternative approach to the uncertainty characterization of the geotechnical parameters. For example, the geotechnical parameters may be modeled using the fuzzy set theory (Zadeh 1965). In the context of the fuzzy set theory, an uncertain variable can be modeled with only the knowledge of its HCS and LCV, which are generally easier to determine even with limited data availability. Indeed, the application of fuzzy set theory has a track record in geotechnical practices, particularly when the site-specific data is limited (e.g., Juang et al. 1992, 1998; Giasi et al. 2003; Sonmez et al. 2003; Luo et al. 2011; Park et al. 2012). For a geotechnical design with fuzzy data as inputs, the design performance will be a fuzzy set. For ease of processing fuzzy set data, the vertex method proposed by Dong and Wong (1987) is usually adopted. The fuzzy design performance provides more information about the state of design safety and thus allows for informed decision-making; and, if desired, the probability of failure might also be interpreted from this fuzzy performance (e.g., Shrestha and Duckstein 1998; Park et al. 2012).

In reference to the robust geotechnical design (RGD) methodology detailed in Chapter 3, design robustness (of the performance against the variation in noise

DOI: 10.1201/9781003564454-5

factors) is an additional design objective that should be explicitly considered in the robust design. The selection of the robustness measure is a vital issue in the RGD, which is strongly affected by the level of the characterized uncertainty of noise factors (e.g., hard-to-control geotechnical parameters). For example, the gradient-based sensitivity index (SI) is suitable for the design situation where only characteristic or nominal values of noise factors can be estimated, whereas the variation of design performance or reliability-based feasibility index (FI) (Chen et al. 1996; Du and Chen 2000; Park et al. 2006; Juang and Wang 2013) is more applicable to the situation where the statistical information of noise factors could be calibrated. The worst-case-analysis-based robustness measure (e.g., Nagy and Braatz 2004; Zhu and Fukushima 2009) is more applicable to the situation where the worst-case characteristic values of noise factors can be estimated. Finally, the Taguchi signal-to-noise ratio (SNR) (Taguchi 1986; Tsui 1992) may be adopted for evaluating the design robustness in a situation where the noise factors can only be characterized by their potential distribution intervals (or fuzzy numbers).

This chapter focuses on the development of the robustness measure of fuzzy set-based SNR and its application in RGD. This RGD methodology can be applicable to situations where the hard-to-control and hard-to-characterize geotechnical properties can only be characterized by their distribution intervals. Within the context of this RGD methodology, the site-specific geotechnical parameters are characterized by the fuzzy set theory, and the uncertainty propagation analysis in the geotechnical design is conducted utilizing the vertex method (Dong and Wong 1987; Juang et al. 1998). Based on the fuzzy design performance, the probability of failure and the statistics of the design performance could both be assessed. Then, the design safety can be evaluated with the estimated probability of failure. At the same time, the design robustness can be measured with the SNR derived from the statistics of the design performance.

This chapter consists of three parts. First, the modeling of the geotechnical parameters with fuzzy sets and the establishment of the fuzzy set-based SNR are presented. Next, three geotechnical design examples are analyzed to illustrate the application and effectiveness of the SNR-based RGD. Finally, the concluding remarks are drawn.

5.2 ROBUSTNESS MEASURE OF FUZZY SET-BASED SIGNAL-TO-NOISE RATIO

The fuzzy set-based uncertainty characterization of the geotechnical parameters is first introduced in this section, and the robustness measure of fuzzy set-based SNR is then formulated. Finally, several illustrative examples are presented to demonstrate the concept and effectiveness of this robustness measure.

5.2.1 Characterization of Noise Factors with Fuzzy Sets and the Uncertainty Propagation

A fuzzy set is a set of ordered pairs, denoted as $[x, \mu(x)]$, where a member x belongs to this set with a certain level of confidence, called membership grade $\mu(x)$. This set

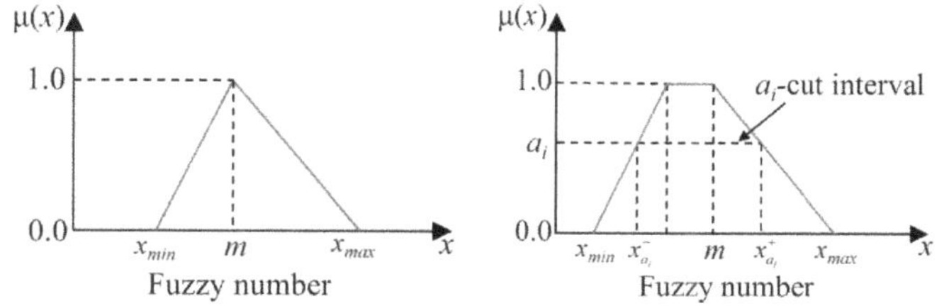

(a) Triangular membership function (b) Trapezoidal membership function

GURE 5.1 Illustrative examples of the membership function in fuzzy set theory.

' ordered pairs collectively defines a membership function that specifies a mem-ership grade for each member (Zadeh 1965). Although the membership function is ot a probability density function (PDF), it does provide a degree of confidence or kelihood that a member x belongs to the set. A fuzzy set with a membership func-on that is convex in shape and with its highest membership grade equal to 1.0 is a ecial fuzzy set called a fuzzy number. Illustrated in Figure 5.1 are the triangular ıd trapezoidal membership functions, which are the most widely used membership ınctions for characterizing the uncertain geotechnical parameters in the literature .g., Juang et al. 1998; Dodagoudar and Venkatachalam 2000; Park et al. 2012; Gong al. 2015a).

As an example, the friction angle (φ) of sand characterized as "about 32°" based pon limited site-specific data indicates the uncertainty about the statement of = 32°. Although this friction angle may be modeled as a random variable with ı assumed distribution, the limited data availability cannot allow for such a pre-se statistical characterization. Alternatively, the statement of "about 32°" could be ıaracterized as a fuzzy number of $\overline{32°}$, where the highest membership grade is qual to 1.0 for $\varphi = 32°$. Often, in geotechnical practice, only the HCV and LCV of geotechnical parameter can be estimated based on the combination of the limited te-specific data, local experiences, and engineering judgment. In such a situation, ıis parameter could be conveniently modeled using a triangular fuzzy number. This easily done by setting the membership grade $\mu(x)$ at x = HCV and LCV to 0, hile the membership grade $\mu(x)$ at x = most likely value (or mode) to 1.0, where the ıode could be estimated separately or taken as the average of HCV and LCV. This iangular membership function is adopted in this chapter for illustration purposes, ıough other membership functions (e.g., trapezoidal membership function) may also e adopted.

In design situations where the uncertain geotechnical parameters are modeled as ızzy numbers, the performance analysis of geotechnical designs involves the prop-gation of fuzzy numbers through the solution model. The vertex method proposed y Dong and Wong (1987) can be adopted to deal with the propagation of fuzzy umbers. This method is based on the α-cut concept. In reference to Figure 5.2(a), n interval with a lower bound of $x^-_{a_i}$ and an upper bound of $x^+_{a_i}$ could be formed at ıe membership grade of α_i; mathematically, a fuzzy number could be represented y a set of α-cut intervals with α ranging from 0 to 1. Through the vertex method,

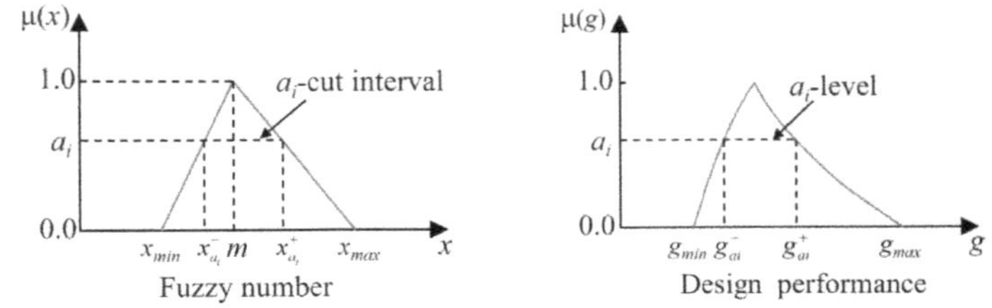

(a) $_i$-cut interval of an input fuzzy number (b) Fuzzy design performance at $_i$-cut level

FIGURE 5.2 α-cut intervals for uncertainty propagation analysis using the vertex method.

the uncertainty propagation in the design performance analysis can be realized wi the following steps (e.g., Juang et al. 1998; Luo et al. 2011; Gong et al. 2014b, 2015a

1) The input fuzzy numbers are discretized into a set of α-cut intervals. For example, taking $\Delta\alpha = 0.2$ in this chapter yields six different levels (i.e., $\alpha =$ 0, 0.2, 0.4, 0.6, 0.8, and 1.0). The step size of $\Delta\alpha = 0.2$ is adequate to achieve a converged design performance analysis for most geotechnical problems. For significant problems, a smaller step size may be demanded.
2) At each α_i-cut level, an interval of each input fuzzy number can be obtained, and the combinations of the vertices (i.e., lower bounds $x^-_{a_i}$ and upper bounds $x^+_{a_i}$ of these α_i-cut intervals of all fuzzy input parameters) can be established. The number of vertex combinations is $2n$, where n indicates the number of fuzzy input parameters to the solution model.
3) At each α_i-cut level, a vertex combination represents a set of inputs to the solution model, with which the design performance can be evaluated. This process is repeated for all $2n$ vertex combinations, yielding $2n$ values of the performance. Let's take only the minimum and maximum values of the performance; an interval of the design performance could be formed, representing the design performance at this specified α_i-cut level, as shown in Figure 5.2(b).
4) Based on the intervals of the design performance obtained for all α-cut levels, a fuzzy number that represents the design performance with fuzzy inputs can be established. This fuzzy number is the outcome of the propagation of fuzzy numbers through the solution model, which offers detailed information about the state of the design performance.

Two examples are presented below to illustrate the application of this fuzzy s theory and the propagation of fuzzy numbers in geotechnical engineering.

Example 1

The first example considers the characterization of the undrained shear strength of the clay at a site. Given the limited project budget, no test data can be available at the concerned site. The only information that can be available is from local experience, which suggests that the undrained shear strength is about 30 kPa, and the minimum value is no less than 15 kPa. The problem here is to characterize the undrained shear strength with the fuzzy set theory.

Solution:

Based on the limited information provided, a triangular membership function is selected in this example to characterize the fuzzy undrained shear strength of the clay at a site.

According to the statement "the undrained shear strength is about 30 kPa," we can set the membership at 30 kPa as 1.0; and, according to the statement "the minimum value is no less than 15 kPa," we know that the LCV of the undrained shear strength is 15 kPa. Let's further assume that 30 kPa is the average of the HCV and LCV, the HCV of the undrained shear strength can be computed as 45 kPa.

From there, a triangular membership function can readily be constructed, as shown in Figure 5.3, where the membership is 0 when the shear strength is 15 kPa or 45 kPa, and the membership is 1 when the shear strength is 30 kPa.

Example 2

A small structure is to be supported on a strip foundation, as shown in Figure 5.4. The load (P) from the structure and the bearing capacity (q) of the ground are both characterized as fuzzy numbers. The local experience indicates that the HCV and LCV of the bearing capacity (q) are estimated to be 180 kPa and 320 kPa, respectively; and, the HCV and LCV of the load (P) are characterized as 100 kN/m and 200 kN/m, respectively. For illustration purposes, the triangular membership function is assumed in this example, and the width of this strip foundation (B) is set to 1.2 m. The problem here is to evaluate the design safety of this strip foundation using the vertex method.

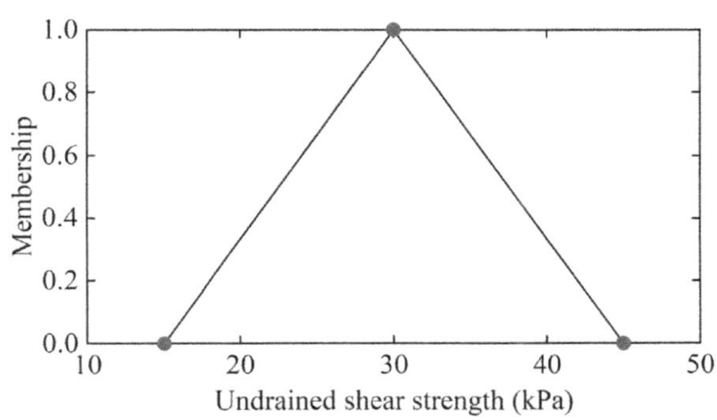

FIGURE 5.3 The characterized membership function in Example 1.

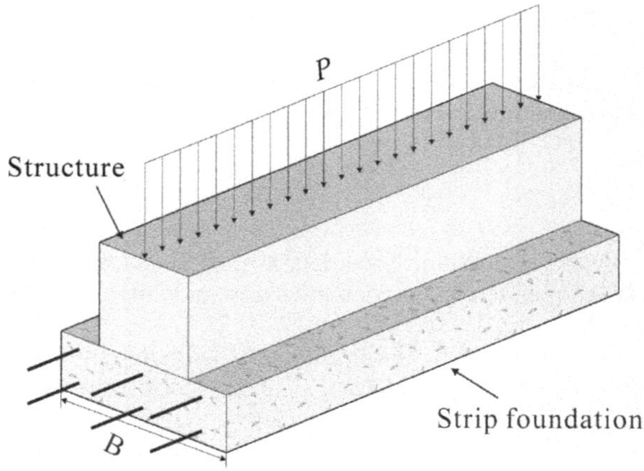

FIGURE 5.4 Schematic diagram of the strip foundation design in Example 2.

Solution:

Step 1: Determine the membership functions of the fuzzy numbers

The objective in this example is to evaluate the design safety of this strip foundation using the vertex method. As mentioned in the problem, a triangular membership function is adopted in this example. According to the statement "the highest conceivable value (HCV) and LCV of the bearing capacity (q) are estimated to be 180 kPa and 320 kPa," the mode value of the bearing capacity (q) can be calculated as follows.

$$q_{HCV} = 320 \text{ (kPa)}, \; q_{LCV} = 180 \text{ (kPa)}$$

$$q_{mode} = \frac{q_{HCV} + q_{LCV}}{2} = \frac{320 \text{ (kPa)} + 180 \text{ (kPa)}}{2} = 250 \text{ (kPa)}$$

where q_{HCV}, q_{LCV}, and q_{mode} represent the HCV, LCV, and mode of the ground bearing capacity (q), respectively. The triangular membership function built for the ground bearing capacity is shown in Figure 5.5(a). According to the statement "the HCV and LCV of the load (P) are characterized as 100 kN/m and 200 kN/m," the mode of the load (P) can be calculated.

$$P_{HCV} = 200 \text{ (kN/m)}, \; P_{LCV} = 100 \text{ (kN/m)}$$

$$P_{mode} = \frac{P_{HCV} + P_{LCV}}{2} = \frac{200 \text{ (kN/m)} + 100 \text{ (kN/m)}}{2} = 150 \text{ (kN/m)}$$

where P_{HCV}, P_{LCV}, and P_{mode} represent the HCV, LCV, and mode of the load (P), respectively. The triangular membership function built for the load is shown in Figure 5.5(b).

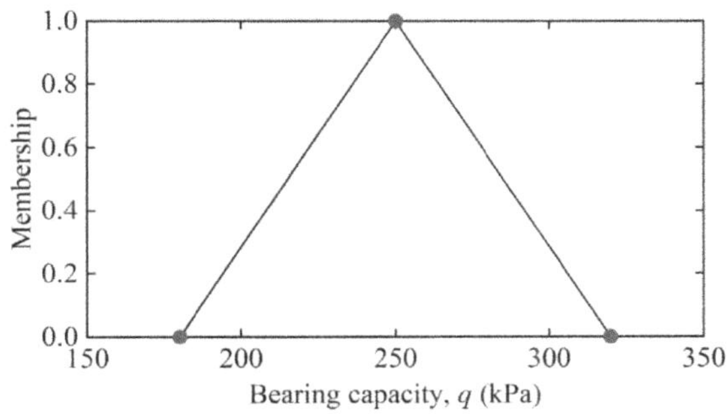

(a) The triangular membership function of the ground bearing capacity

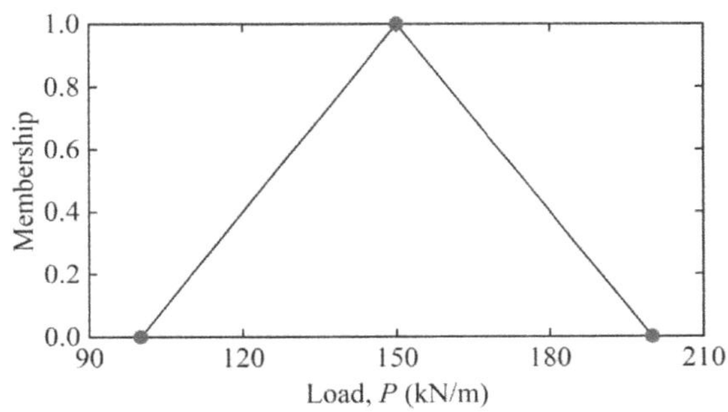

(b) The triangular membership function of the load

FIGURE 5.5 The triangular membership functions of the ground bearing capacity and the load.

Step 2: Discretize fuzzy numbers with α-cuts

For ease of applying the vertex method, the input fuzzy numbers should be discretized in a set of α-cut intervals. For illustration purposes, the step size of $\Delta\alpha = 0.2$ is adopted in this example for the discretization of the fuzzy input numbers (i.e., ground bearing capacity q and load P). The selected step size yields six different levels (i.e., $\alpha_i = 0, 0.2, 0.4, 0.6, 0.8$, or 1.0).

At each α_i-cut level, an interval of each input fuzzy number of the bearing capacity (q) and the load (P) can be obtained. The calculation methods are provided below.

$$[q_{imin}, q_{imax}] = [q_{LCV} + \alpha_i (q_{mode} - q_{LCV}), q_{HCV} - \alpha_i (q_{HCV} - q_{mode})]$$

$$[P_{imin}, P_{imax}] = [P_{LCV} + \alpha_i (P_{mode} - P_{LCV}), P_{HCV} - \alpha_i (P_{HCV} - P_{mode})]$$

where q_{imin} and q_{imax} represent the minimum and maximum value of the α_i-cut interval of the ground bearing capacity (q), respectively; and, P_{imin} and P_{imax}

represent the minimum and maximum value of the α_i-cut interval of the load (P), respectively.

As an example, the α_i-cut intervals of the ground bearing capacity (q) and the load (P) for the cut level of $\alpha_i = 0.2$ can be calculated as follows.

$$q_{min} = q_{LCV} + \alpha\left(q_{mode} - q_{LCV}\right) = 180\ (\text{kPa}) + 0.2 \times (250\ (\text{kPa}) - 180\ (\text{kPa})) = 194\ (\text{kPa})$$

$$q_{max} = q_{HCV} - \alpha\left(q_{HCV} - q_{mode}\right) = 320\ (\text{kPa}) - 0.2 \times (320(\text{kPa}) - 250\ (\text{kPa})) = 306\ (\text{kPa})$$

$$P_{min} = P_{LCV} + \alpha\left(P_{mode} - P_{LCV}\right) = 100\ (\text{kN/m}) + 0.2 \times (150\ (\text{kN/m}) - 100\ (\text{kN/m})) = 110\ (\text{kN/m})$$

$$P_{max} = P_{HCV} - \alpha\left(P_{HCV} - P_{mode}\right) = 200(\text{kN/m}) - 0.2 \times (200\ (\text{kN/m}) - 150\ (\text{kN/m})) = 190\ (\text{kN/m})$$

For each α_i-cut level, the corresponding intervals of the ground bearing capacity (q) and the load (P) are readily calculated by applying the formulations above, and the calculation results are tabulated in Table 5.1.

Step 3: Calculate the factor of safety for each α_i-cut level

The factor of safety (FS) of the strip foundation can be evaluated as follows:

$$FS = \frac{qB}{P}$$

The vertex method is employed here to construct the fuzzy FS of the strip foundation. The essence of the vertex method is to derive the minimum and maximum values of the FS considering all possible combinations of the α_i-cut vertices of the

TABLE 5.1
Intervals of the Ground Bearing Capacity and the Load at α-Cut Levels in Example 2

	Bearing capacity, q(kPa)		Load, P (kN/m)	
α_i-cut level	q_{imin}(kPa)	q_{imax}(kPa)	P_{imin}(kN/m)	P_{imax}(kN/m)
0.0	180	320	100	200
0.2	194	306	110	190
0.4	208	292	120	180
0.6	222	278	130	170
0.8	236	264	140	160
1.0	250	250	150	150

fuzzy input parameters. For each α_i-cut level, there are four vertex combinations, and the related *FS* of the strip foundation can be calculated as follows.

Vertex combination 1: $FS_{i1} = \frac{q_{imin}B}{P_{imin}}$, Vertex combination 2: $FS_{i2} = \frac{q_{imin}B}{P_{imax}}$

Vertex combination 3: $FS_{i3} = \frac{q_{imax}B}{P_{imin}}$, Vertex combination 4: $FS_{i4} = \frac{q_{imax}B}{P_{imax}}$

Based on the minimum and maximum values of the *FS*, denoted as FS_{imin} and FS_{imax}, respectively, an interval of the design performance can be established as $[FS^-_{a_i} = FS_{imin}, FS^+_{a_i} = FS_{imax}]$, representing the performance at this specified α_i-cut level. As an example, the *FS* for the cut level of $\alpha_i = 0.2$ can be calculated as follows.

Vertex combination 1: $FS_1 = \frac{q_{min}B}{P_{min}} = \frac{194\ (\text{kPa}) \times 1.2\ (\text{m})}{110\ (\text{kN/m})} = 2.12$

Vertex combination 2: $FS_2 = \frac{q_{min}B}{P_{max}} = \frac{194\ (\text{kPa}) \times 1.2\ (\text{m})}{190\ (\text{kN/m})} = 1.23$

Vertex combination 3: $FS_3 = \frac{q_{max}B}{P_{min}} = \frac{306\ (\text{kPa}) \times 1.2\ (\text{m})}{110\ (\text{kN/m})} = 3.34$

Vertex combination 4: $FS_4 = \frac{q_{max}B}{P_{max}} = \frac{306\ (\text{kPa}) \times 1.2\ (\text{m})}{190\ (\text{kN/m})} = 1.93$

Based on the calculation results listed above, the fuzzy design performance of this strip foundation at this specified α_i-cut level can be obtained as [1.23, 3.34].

For each α_i-cut level, the corresponding intervals of the *FS* can readily be computed by applying the formulations above, and the calculation results are listed in Table 5.2. Figure 5.6 depicts the derived fuzzy performance of this strip foundation.

.2.2 Formulation of Fuzzy Set-Based Signal-to-Noise Ratio

he fuzzy design performance, such as the one shown in Figure 5.2(b), can provide uch information about the state of design performance (and safety), including lower ound, upper bound, mode, and the degree of support of these values and any other alue in the range defined. It is more desirable to have a single value representation f the design performance such that it can be incorporated into the robust design amework. To this end, an index of the design safety, such as a reliability index or robability of failure adopted in the conventional reliability-based design (RBD), is esirable. Indeed, various approaches have been established to estimate the probaility of failure (P_f) of a geotechnical system with fuzzy performance (e.g., Shrestha nd Duckstein 1998; Dodagoudar and Venkatachalam 2000; Giasi et al. 2003; Park t al. 2012). However, most of such methods are based on the normalization of the iembership function of the fuzzy performance into a PDF, assuming that the PDF

TABLE 5.2
The Derived Intervals of Design Safety at each α_i-cut Levels in Example 2

α_i-cut level	Vertex combination				α_i-cut interval of desig performance	
	FS_{i1}	FS_{i2}	FS_{i3}	FS_{i4}	$FS^-_{a_i}$	FS
0.0	2.16	1.08	3.84	1.92	1.08	3.8
0.2	2.12	1.23	3.34	1.93	1.23	3.
0.4	2.08	1.39	2.92	1.95	1.39	2.9
0.6	2.05	1.57	2.57	1.96	1.57	2.
0.8	2.02	1.77	2.26	1.98	1.77	2.
1.0	2.00	2.00	2.00	2.00	2.00	2.0

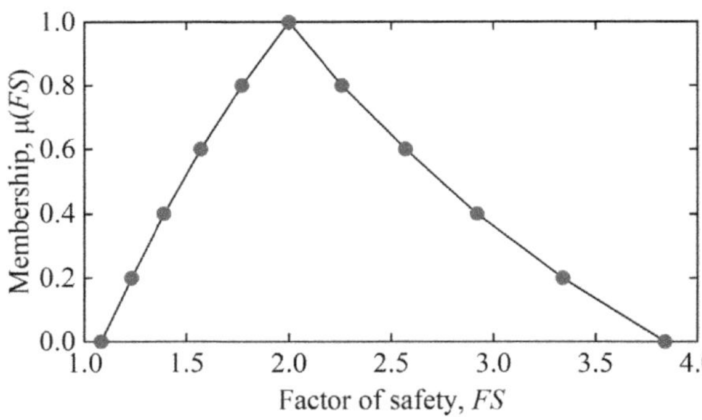

FIGURE 5.6 The derived fuzzy design performance of the strip foundation.

is proportional to the membership function. This procedure is not rigorous from probabilistic point of view. For example, for each point [g, $\mu(g)$] on the membershi function, as shown in Figure 5.2(b), the term $\mu(g)$ measures the membership grad or the degree of confidence for this performance value (g), which is not a probabi ity. From a probabilistic point of view, the likelihood of occurrence of the possib outcome or performance value, for example, $g^-_{a_i}$ (or $g^+_{a_i}$) relies upon the membershi grades of the n input fuzzy numbers. Because each of the n input fuzzy numbers ha the same membership grade of α_i, the chance of occurrence of $g^-_{a_i}$ (or $g^+_{a_i}$) would b $(\alpha_i)^n$. This is similar to the analogy of finding the joint probability of occurrence c a series of n independent events, each with a chance of α_i.

To satisfy the axiom of probability, the chance of occurrence for the design pe formance $g = g^-_{a_i}$ (or $g^+_{a_i}$), which is $(\alpha_i)^n$, must be transformed into a probabilit such that this fuzzy membership function can be converted into a probability mas

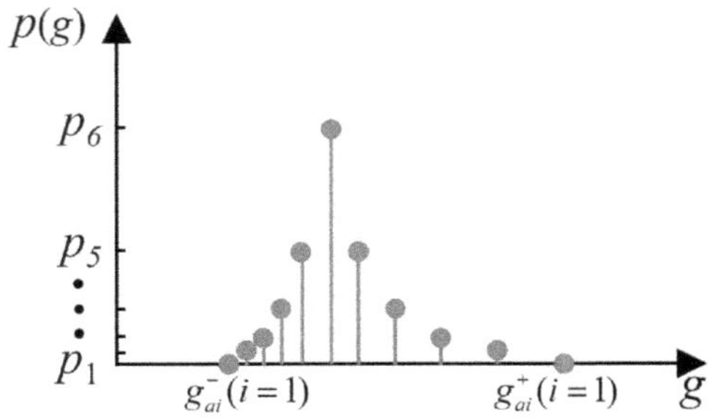

IGURE 5.7 Interpretation of the fuzzy design performance as a discrete random variable.

unction (PMF), as depicted in Figure 5.7. The equation below can be adopted for this ransformation (Gong et al. 2014c, 2014c).

$$p_i = \frac{(\alpha_i)^n}{2\sum_{i=1}^{i=m-1} (\alpha_i)^n + (\alpha_m)^n} \tag{5.1}$$

vhere p_i represents the probability or likelihood of occurrence of the design per-ormance $g = g_{\alpha_i}^{-}$ (or $g_{\alpha_i}^{+}$), in which α_i is a membership grade (i = 1, 2, 3, …, m) nd m is the number of the α-cut intervals discretized; and, n is the number of nput fuzzy parameters. With the aid of this transformation, the following axiom of robability can be satisfied: $2(p_1 + p_2 + p_3 + \ldots + p_{(m-1)}) + p_m = 1$. Then, the mean μ_g) and standard deviation (σ_g) of the design performance (g) can be estimated s follows.

$$\mu_g = \sum_{i=1}^{i=m-1} p_i\left(g_{\alpha_i}^{-} + g_{\alpha_i}^{+}\right) + p_m g_{\alpha_m} \tag{5.2}$$

$$\sigma_g^{\ 2} = \sum_{i=1}^{i=m-1} p_i\left[\left(g_{\alpha_i}^{-} - \mu_g\right)^2 + \left(g_{\alpha_i}^{+} - \mu_g\right)^2\right] + p_m\left(g_{\alpha_m} - \mu_g\right)^2 \tag{5.3}$$

Based on the estimated mean (μ_g) and standard deviation (σ_g) of the design erformance, the reliability index (β) of the geotechnical design could readily be valuated as follows, with an assumption that the design performance is normally listributed.

$$\beta = \frac{\mu_g}{\sigma_g} \tag{5.4}$$

When the FS is taken as the design performance, it can be assumed to follow a log-normal distribution because of the non-negative feature of the FS.

In the context of the proposed RGD, design robustness is considered an additional design objective that must be explicitly considered and optimized. As the hard-to-control and hard-to-characterize geotechnical parameters are characterized as fuzzy sets, the theoretically sound robustness measures such as the reliability-based FI (e.g., Juang and Wang 2013), which are based on strict probabilistic analyses, could not be applicable, while the simpler gradient-based SI cannot take the ranges of the variation of the noise factors into account. The Taguchi SNR (Taguchi 1986; Tsui 1992), formulated below, is adopted herein as a measure of design robustness.

$$SNR = 10\log_{10}\left(\frac{{\mu_g}^2}{{\sigma_g}^2}\right) \tag{5.5}$$

According to the definition shown in Eq. (5.5), a higher *SNR* means less variation in the design performance, and thus, higher design robustness is acquired. Although the *SNR* formulated in Eq. (5.5) can be considered approximate for most geotechnical problems (especially when *FS* is taken as the design performance), other *SNR* definitions may be adopted in special design situations. For example, in the design situation where the design performance is labeled "the larger, the better," the following *SNR* formulation could be adopted.

$$SNR = 10\log_{10}\left({\mu_g}^2\right) \tag{5.6}$$

In the situation where the design performance is labeled "the smaller, the better," the following *SNR* formulation can be adopted.

$$SNR = -10\log_{10}\left({\mu_g}^2\right) \tag{5.7}$$

Two examples are presented below to illustrate the uncertainty propagation in the geotechnical design with fuzzy sets as inputs and the SNR-based design robustness.

Example 3

The background and the parameters setting of this example are identical to those in Example 2; however, the problem in this example is to interpret the fuzzy design performance. The question is to determine the reliability index of this strip foundation.

Meanwhile, the load (*P*) and the ground bearing capacity (*q*) are both assumed to follow normal distributions, and their statistics can be estimated with the two-sigma bound. Please derive the reliability index of this strip foundation and then compare this reliability index with that interpreted from the fuzzy design performance.

Solution:

Step 1: Convert fuzzy membership function to probability mass function

As presented in Example 2, the fuzzy membership function for the *FS* is obtained, as illustrated in Figure 5.6. The corresponding fuzzy membership function can be transformed into a PMF by converting the chance of occurrence of the *FS* into a probability. The calculation process is presented as follows.

According to the formulation in Eq. (5.1), the probability or likelihood of occurrence of $FS = FS^-_{a_i}$ (or $FS^+_{a_i}$), denoted as $p_i\,(i = 1, 2, 3\ldots6)$, can be calculated as follows.

$$p_1 = \frac{(\alpha_1)^2}{2\sum_{i=1}^{i=5}(\alpha_i)^2 + (\alpha_6)^2} = \frac{(0)^2}{2\times(0^2+0.2^2+0.4^2+0.6^2+0.8^2)+(1.0)^2} = 0$$

$$p_2 = \frac{(\alpha_2)^2}{2\sum_{i=1}^{i=5}(\alpha_i)^2 + (\alpha_6)^2} = \frac{(0.2)^2}{2\times(0^2+0.2^2+0.4^2+0.6^2+0.8^2)+(1.0)^2} = 0.012$$

$$p_3 = \frac{(\alpha_3)^2}{2\sum_{i=1}^{i=5}(\alpha_i)^2 + (\alpha_6)^2} = \frac{(0.4)^2}{2\times(0^2+0.2^2+0.4^2+0.6^2+0.8^2)+(1.0)^2} = 0.047$$

$$p_4 = \frac{(\alpha_4)^2}{2\sum_{i=1}^{i=5}(\alpha_i)^2 + (\alpha_6)^2} = \frac{(0.6)^2}{2\times(0^2+0.2^2+0.4^2+0.6^2+0.8^2)+(1.0)^2} = 0.106$$

$$p_5 = \frac{(\alpha_5)^2}{2\sum_{i=1}^{i=5}(\alpha_i)^2 + (\alpha_6)^2} = \frac{(0.8)^2}{2\times(0^2+0.2^2+0.4^2+0.6^2+0.8^2)+(1.0)^2} = 0.188$$

$$p_6 = \frac{(\alpha_6)^2}{2\sum_{i=1}^{i=5}(\alpha_i)^2 + (\alpha_6)^2} = \frac{(1)^2}{2\times(0^2+0.2^2+0.4^2+0.6^2+0.8^2)+(1.0)^2} = 0.294$$

With the aid of this transformation, the following axiom of probability can be satisfied: $2(p_1 + p_2 + p_3 + \ldots + p_5) + p_6 = 1$.

Step 2: Estimate the mean and standard deviation of the factor of safety

Based on the occurrence probability of the *FS* for each α_i-cut level, the mean (μ_{FS}) and standard deviation (σ_{FS}) of the *FS* can be estimated as follows. The interval of the *FS* at each α_i-cut level is listed in Table 5.2.

$$\mu_{FS} = \sum_{i=1}^{i=5} p_i (FS_{\alpha_i}^- + FS_{\alpha_i}^+) + p_6 FS_{\alpha_6}$$
$$= 0.012 \times (1.23 + 3.34) + 0.047 \times (1.39 + 2.92) + 0.106 \times (1.57 + 2.57)$$
$$+ 0.188 \times (1.77 + 2.26) + 0.294 \times 2$$
$$= 2.042$$

$$\sigma_{FS}{}^2 = \sum_{i=1}^{i=5} p_i \left[\left(FS_{\alpha_i}^- - \mu_{FS} \right)^2 + \left(FS_{\alpha_i}^+ - \mu_{FS} \right)^2 \right] + p_6 \left(FS_{\alpha_6} - \mu_{FS} \right)^2$$
$$= 0.012 \times \left[(1.23 - 2.042)^2 + (3.34 - 2.042)^2 \right]$$
$$+ 0.047 \times \left[(1.39 - 2.042)^2 + (2.92 - 2.042)^2 \right]$$
$$+ 0.106 \times \left[(1.57 - 2.042)^2 + (2.57 - 2.042)^2 \right]$$
$$+ 0.188 \times \left[(1.77 - 2.042)^2 + (2.26 - 2.042)^2 \right]$$
$$+ 0.294 \times (2 - 2.042)^2$$
$$= 0.161$$

$$\sigma_{FS} = \sqrt{\sigma_{FS}{}^2} = \sqrt{0.161} = 0.401$$

Step 3: Estimate the reliability index based on the fuzzy factor of safety

The reliability index (β) of the strip foundation can be approximated based on the mean (μ_{FS}) and standard deviation (σ_{FS}) of the *FS*. The *FS* in this example is assumed to follow a lognormal distribution because of its non-negative feature. The reliability index (β) of the strip foundation can then be estimated as follows:

$$\beta = \frac{\ln\left(\dfrac{\mu_{FS}}{\sqrt{1 + (\sigma_{FS}/\mu_{FS})^2}} \right)}{\sqrt{\ln\left(1 + (\sigma_{FS}/\mu_{FS})^2\right)}} = \frac{\ln\left(\dfrac{2.042}{\sqrt{1 + (0.401/2.042)^2}} \right)}{\sqrt{\ln\left(1 + (0.401/2.042)^2\right)}} = 3.57$$

Meanwhile, the reliability index of the strip foundation can be estimated with the Monte Carlo simulation (MCS) method, given the statistical information of uncertain input parameters. Here, the load (P) and the ground bearing capacity (q) are both assumed to follow normal distributions, and their statistics can be estimated with the two-sigma bound as follows.

$$\mu_q = \frac{q_{HCV} + q_{LCV}}{2} = \frac{320\ (\text{kPa}) + 180\ (\text{kPa})}{2} = 250\ (\text{kPa})$$

$$\sigma_q = \frac{q_{HCV} - q_{LCV}}{4} = \frac{320\ (\text{kPa}) - 180\ (\text{kPa})}{4} = 35\ (\text{kPa})$$

$$\mu_P = \frac{P_{HCV} + P_{LCV}}{2} = \frac{200\ (\text{kN/m}) + 100\ (\text{kN/m})}{2} = 150\ (\text{kN/m})$$

$$\sigma_P = \frac{P_{HCV} - P_{LCV}}{4} = \frac{200\ (\text{kN/m}) - 100\ (\text{kN/m})}{4} = 25\ (\text{kN/m})$$

where μ_q and σ_q represent the mean and standard deviation of the ground bearing capacity (q), respectively; and, μ_q and σ_q are the mean and standard deviation of the load (P), respectively.

With the aid of the failure probability (P_f) estimated by the MCS method, the reliability index (β) can readily be calculated as follows.

$$\beta = \Phi^{-1}(1 - P_f)$$

where $\Phi^{-1}()$ is the inverse cumulative distribution function (CDF) of the standard normal distribution. In this example, 1,000,000 samples of uncertain parameters are generated, and the reliability index (β) is estimated as $\beta = 3.07$. It should be noted that the reliability index of this strip foundation interpreted from the fuzzy *FS* is comparable to that estimated with the MCS method (based on the statistics of uncertain input parameters characterized with the two-sigma bound). The code of the adopted MCS method is provided here.

```
% Monte Carlo simulation of FS = q * B/P based on normal distribution sampling

clear; clc;

%% Parameter settings
n_samples = 1e6;                    % Number of samples
q_mean = 250;
q_std = 35;                         % Mean and standard deviation of q
P_mean = 150;
P_std = 25;                         % Mean and standard deviation of P
B = 1.2;                            % Coefficient (modifiable)

%% Generate samples based on normal distribution
q_samples = q_mean + q_std .* randn(n_samples,1);    % q ~ N(250, 23.33)
P_samples = P_mean + P_std .* randn(n_samples,1);    % P ~ N(150, 16.67)

%% Calculate FS (Factor of Safety) samples
FS_samples = B * q_samples ./ P_samples;

%% Calculate failure probability
% Failure is defined as FS < 1
failure_count = sum(FS_samples < 1);
failure_probability = failure_count/n_samples;

%% Calculate reliability index beta
beta_index = -norminv(failure_probability);

%% Display results
fprintf('Monte Carlo Simulation Results (Samples: %d):\n', n_samples);
fprintf('Failure probability: %.6e\n', failure_probability);
fprintf('Reliability index β: %.4f\n', beta_index);
```

Example 4

The background and the parameters setting of this example are identical to those in Example 2; however, this problem is to estimate the reliability index and robustness of this strip foundation design from the fuzzy design performance. Furthermore, the following widths of this strip foundation design should be analyzed: $B = 0.7, 0.8, 0.9, \ldots, 2.5$ (m). Please plot the relationship between the reliability index and the foundation width, that between the robustness and the foundation width, and that between the reliability index and the design robustness.

Solution:

Step 1: Estimate the reliability indexes of candidate designs with the vertex method

The reliability index (β) of the strip foundation for each candidate design is estimated using the vertex method. According to the information in Example 2, the HCV, LCV, and mode of the bearing capacity (q) and those of the load (P) are presented below.

$$q_{HCV} = 320 \text{ (kPa)}, \; q_{LCV} = 180 \text{ (kPa)}$$

$$q_{mode} = \frac{q_{HCV} + q_{LCV}}{2} = \frac{320 \text{ (kPa)} + 180 \text{ (kPa)}}{2} = 250 \text{ (kPa)}$$

$$P_{HCV} = 200 \text{ (kN/m)}, \; P_{LCV} = 100 \text{ (kN/m)}$$

$$P_{mode} = \frac{P_{HCV} + P_{LCV}}{2} = \frac{200 \text{ (kN/m)} + 100 \text{ (kN/m)}}{2} = 150 \text{ (kN/m)}$$

where q_{HCV}, q_{LCV}, and q_{mode} denote the HCV, LCV, and mode of the ground bearing capacity (q), respectively; and, P_{HCV}, P_{LCV}, and P_{mode} denote the HCV, LCV, and mode of the load (P), respectively.

For ease of applying the vertex method, the input fuzzy numbers should be discretized in a set of α-cut intervals. For illustration purposes, the step size of $\Delta\alpha = 0.2$ is adopted in this example. The selected step size yields six different levels (i.e., α_i = 0, 0.2, 0.4, 0.6, 0.8, or 1.0).

At each α_i-cut level, an interval of each input fuzzy number of the bearing capacity (q) and the load (P) can be obtained. The calculation methods are provided below.

$$[q_{imin}, q_{imax}] = [q_{LCV} + \alpha_i (q_{mode} - q_{LCV}), \; q_{HCV} - \alpha_i (q_{HCV} - q_{mode})]$$

$$[P_{imin}, P_{imax}] = [P_{LCV} + \alpha_i (P_{mode} - P_{LCV}), \; P_{HCV} - \alpha_i (P_{HCV} - P_{mode})]$$

where q_{imin} and q_{imax} represent the minimum and maximum value of the α_i-cut interval of the bearing capacity (q), respectively; and, P_{imin} and P_{imax} represent the minimum and maximum value of the α_i-cut interval of the load (P), respectively.

The vertex method is employed here to construct the fuzzy FS of the strip foundation. For each α_i-cut level, there are four vertex combinations, and the related FS of the strip foundation can be calculated as follows.

Vertex combination 1: $FS_{i1}=\dfrac{q_{imin}B}{P_{imin}}$, Vertex combination 2: $FS_{i2}=\dfrac{q_{imin}B}{P_{imax}}$

Vertex combination 3: $FS_{i3}=\dfrac{q_{imax}B}{P_{imin}}$, Vertex combination 4: $FS_{i4}=\dfrac{q_{imax}B}{P_{imax}}$

Based on the minimum and maximum values of the FS, denoted as FS_{imin} and FS_{imax}, respectively, an interval of the design safety can be established as [$FS_{a_i}^- = FS_{imin}$, $FS_{a_i}^+ = FS_{imax}$], representing the fuzzy performance at this specified α_i-cut level. The fuzzy membership function of the FS can be converted into a PMF. The probability or likelihood of occurrence of $FS = FS_{a_i}^-$ (or $FS_{a_i}^+$), denoted as $p_i\,(i = 1, 2, 3 \ldots 6)$, can be approximated as follows.

$$p_i = \frac{(\alpha_i)^n}{2\sum_{i=1}^{i=m-1}(\alpha_i)^n + (\alpha_m)^n}$$

Then, the mean (μ_{FS}) and standard deviation (σ_{FS}) of the FS can be estimated as follows.

$$\mu_{FS} = \sum_{i=1}^{i=5} p_i(FS_{\alpha_i}^- + FS_{\alpha_i}^+) + p_6 FS_{\alpha_6}$$

$$\sigma_{FS}^2 = \sum_{i=1}^{i=5} p_i\left[\left(FS_{\alpha_i}^- - \mu_{FS}\right)^2 + \left(FS_{\alpha_i}^+ - \mu_{FS}\right)^2\right] + p_6\left(FS_{\alpha_6} - \mu_{FS}\right)^2$$

The FS in this example is assumed to follow a lognormal distribution because of its non-negative feature. Based on the estimated mean (μ_{FS}) and standard deviation (σ_{FS}), the reliability index (β) of this strip foundation can be estimated as follows:

$$\beta = \frac{\ln\left(\dfrac{\mu_{FS}}{\sqrt{1+(\sigma_{FS}/\mu_{FS})^2}}\right)}{\sqrt{\ln\left(1+(\sigma_{FS}/\mu_{FS})^2\right)}}$$

For each candidate design, the reliability index (β) can be computed by applying the formulations above. As an example, the reliability index for a candidate design of B = 1.2 m is estimated as β = 3.57. Figure 5.8 shows the resulting relationship between the reliability index (β) and the width (B) of the strip foundation. As can be seen, the reliability index (β) increases with the width (B) of the strip foundation.

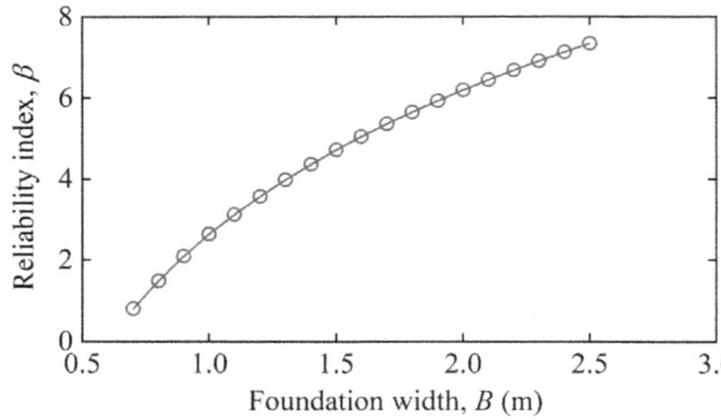

FIGURE 5.8 Relationship between reliability index and foundation width.

Step 2: Estimate the design robustness from the fuzzy design performance

The design robustness of this strip foundation design in this example is evaluated using the Taguchi SNR, a higher SNR means less variation in the design performance. The design performance is specified as $g = FS-1$ in this example. Based on the estimated mean (μ_{FS}) and standard deviation (σ_{FS}) of the *FS* of the foundation, the mean and standard deviation of the performance (g), denoted as μ_g and σ_g, respectively, can be estimated as follows.

$$\mu_g = \mu_{FS} - 1, \ \sigma_g = \sigma_{FS}$$

Based on the estimated mean (μ_g) and standard deviation (σ_g) of the design performance (g), the design robustness (*SNR*) of the strip foundation could readily be evaluated as follows.

$$SNR = 10 \log_{10}\left(\frac{\mu_g^{\ 2}}{\sigma_g^{\ 2}}\right)$$

As an example, the *SNR* for a candidate design of $B = 1.2$ m is calculated below.

$$SNR = 10 \log_{10}\left(\frac{\mu_g^{\ 2}}{\sigma_g^{\ 2}}\right) = 10 \log_{10}\left(\frac{(\mu_{FS} - 1)^2}{\sigma_{FS}^{\ 2}}\right) = 10 \log_{10}\left(\frac{(2.042\text{-}1)^2}{0.401^2}\right) = 6.76$$

For each candidate design in the selected design space, the design robustness (*SNR*) can readily be computed by applying the formulations above. Shown in Figure 5.9 is the resulting relationship between the design robustness (*SNR*) and the width (*B*) of the strip foundation. As can be seen, the design robustness (*SNR*) increases with the width (*B*) of the strip foundation.

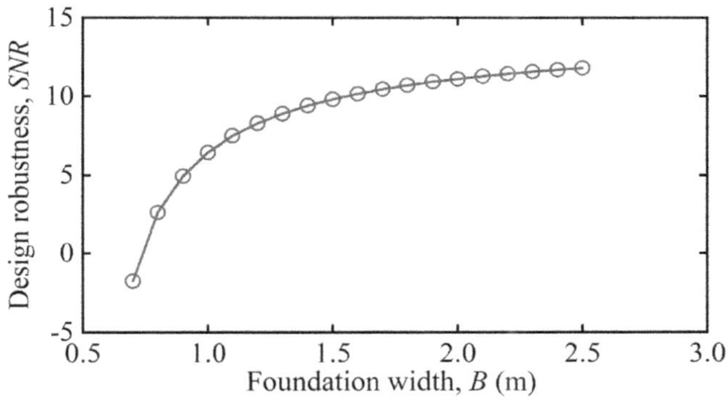

FIGURE 5.9 Relationship between design robustness and foundation width.

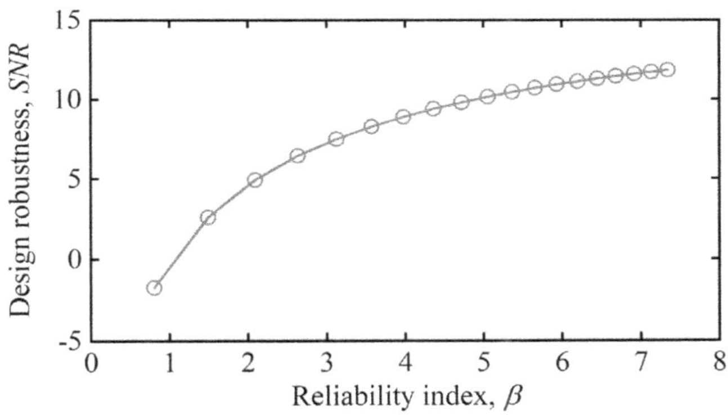

FIGURE 5.10 Relationship between design robustness and reliability index in Example 4.

Step 3: Analyze the relationship between design robustness and reliability index

Illustrated in Figure 5.10 is the relationship between the design robustness (*SNR*) and the reliability index (β) obtained from this example. As can be expected, there exists a positive relationship between the design robustness (*SNR*) and the reliability index (β), and a design with a higher reliability index could always correspond to higher design robustness.

5.2.3 Coupling the Evaluation of Signal-to-Noise Ratio with the Reliability Analysis

Note that the robustness measure of SNR, shown in Eqs. (5.5–5.7), are established as functions of the statistics of the design performance. This robustness measure can also be applied in design situations where the statistics of noise factors can be characterized. Given the statistical information of the hard-to-control noise factors (θ), the statistics of the design performance (g) can be evaluated with various probabilistic methods such as the MCS (e.g., Binder and Heermann 2010), point estimate method (PEM) (e.g., Zhao and Ono 2000), and mean-value first-order second-moment (MVFOSM) method (e.g., Ang and Tang 2004; Zhang et al. 2011). Here, the MVFOSM is introduced. The mean (μ_g) of the design performance can be approximated as follows.

$$\mu_g \approx g(\boldsymbol{d}, \mu_\theta) \tag{5.8}$$

where $g(\boldsymbol{d}, \boldsymbol{\theta})$ represents the performance function for a candidate design $\boldsymbol{d}$ with noise factors of $\boldsymbol{\theta}$, which is formulated as follows: the design $\boldsymbol{d}$ is deemed safe or satisfactory if the value of this performance function $g(\boldsymbol{d}, \boldsymbol{\theta})$ is greater than 0; otherwise, the design $\boldsymbol{d}$ is unsatisfactory. The term μ_θ represents the mean of noise factors $\boldsymbol{\theta}$. And, based on the Taylor series expansion, the standard deviation of the design performance (σ_g) can be approximated as:

$$\sigma_g \approx \sqrt{\left(\mathbf{G}\big|_{\theta=\mu_\theta}\right) \cdot \mathbf{C}_\theta \cdot \left(\mathbf{G}\big|_{\theta=\mu_\theta}\right)^{\mathrm{T}}} \tag{5.9}$$

where C_θ is the covariance matrix of the noise factors $\boldsymbol{\theta}$; $\mathbf{G}\big|_{\theta=\mu_\theta}$ is the gradient of the design performance g($\boldsymbol{d}$, $\boldsymbol{\theta}$) to the noise factors $\boldsymbol{\theta}$ evaluated at the mean of noise factors μ_θ, which is formulated as follows (Gong et al. 2014a).

$$\mathbf{G}\big|_{\theta=\mu_\theta} = \left\{ \frac{\partial g(\boldsymbol{d},\boldsymbol{\theta})}{\partial \theta_1}\bigg|_{\theta=\mu_\theta}, \frac{\partial g(\boldsymbol{d},\boldsymbol{\theta})}{\partial \theta_2}\bigg|_{\theta=\mu_\theta}, \ldots, \frac{\partial g(\boldsymbol{d},\boldsymbol{\theta})}{\partial \theta_n}\bigg|_{\theta=\mu_\theta} \right\} \tag{5.10}$$

where θ_i represents the ith noise factor (i = 1, 2, 3, …, n) and n is the number of noise factors. Based on the estimated statistics of design performance, the robustness measure of SNR, shown in Eqs. (5.5–5.7), can readily be computed. Thus, the evaluation of SNR can be coupled with the conventional reliability analysis

5.3 ROBUST GEOTECHNICAL DESIGNS WITH SIGNAL-TO-NOISE RATIO AS DESIGN ROBUSTNESS

Three geotechnical designs are undertaken in this section to demonstrate the application and effectiveness of the SNR-based RGD.

5.3.1 Design Example 1: Road Cut Design in a Two-Layer Earth Slope

The first design example considers a road cut problem in a two-layer earth slope, which was examined by Gong et al. (2014c). In reference to Figure 5.11, ABCD is the prior ground surface before the road construction, while AFED is the posterior ground surface after the cut. Assuming that the height of the slope is known, the design of this road cut problem reduces to the problem of determining slope angles of α_1 and α_2. The unit weights of two soil layers (i.e., γ_1 and γ_2) are characterized

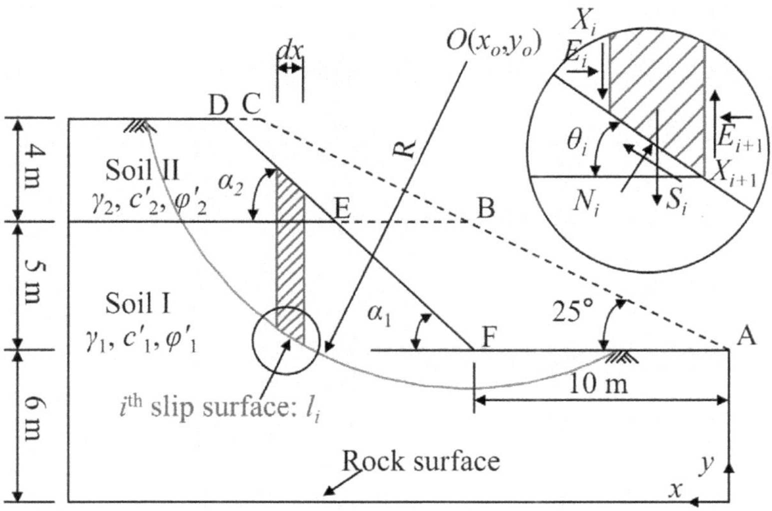

GURE 5.11 Schematic diagram of the road cut in a two-layer earth slope in Design xample 1 (modified from Figure 1 in Gong et al. 2014c, with permission from American ociety of Civil Engineers).

deterministic values, taken as 20.00 kN/m^3 and 18.20 kN/m^3, respectively. In con-ast, the uncertain soil parameters, in terms of the effective cohesions and effective iction angles of two soil layers (i.e., c'_1, φ'_1, c'_2, φ'_2) and the average ratio of ore water pressure to the total earth pressure (r_u), are treated as noise factors. The wer bounds, upper bounds, and most likely values (or modes) of these noise factors e estimated based on limited test data with engineering judgment, as tabulated in able 5.3. These values can be used to construct the triangular fuzzy number for each ncertain soil parameter.

.3.1.1 Deterministic Solution Model for Assessing the Performance of the Road Cut Slope

he stability of an earth slope can be assessed using either finite element method .g., Kim and Lee 1997; Griffiths and Lane 1999) or limit equilibrium method (e.g., ishop 1955; Spencer 1967). The critical slip surface could be circular or noncircu-r (e.g., Bishop 1955; Greco 1996; Kim and Lee 1997). In this geotechnical design xample, the simplified Bishop's method with a circular slip surface is adopted for arth slope stability analysis. Of course, other solution models may also be employed. or simplicity, the forces from two adjacent slices are assumed to be balanced for ach slice of the slope shown in Figure 5.11. Note that these assumptions would not gnificantly affect the overall stability of the earth slope. Symbolically,

$$E_{i+1} - E_i = 0$$

$$X_{i+1} - X_i = 0$$

TABLE 5.3
Assumed Membership Functions of Fuzzy Soil Parameters in Design Example

Soil layer	Soil parameter	Lower bound (a)	Mode (m)	Upper bound
I	c'_1 (kN/m^2)	10.0	15.0	20.0
	φ'_1 (°)	28.0	34.0	40.0
II	C_2 (kN/m^2)	13.0	20.0	27.0
	φ'_2 (°)	20.0	26.0	32.0
I and II	r_u	0.1	0.3	0.5

where E_{i+1} and E_i are the total horizontal forces from slice (i+1) and slice (i– acting on slice i, respectively; and, X_{i+1} and X_i are the total shear forces fro slice (i+1) and slice (i–1) on slice i, respectively. With the assumptions made, tl simplified Bishop's method (Bishop 1955) is reduced to the ordinary method slices (e.g., Fellenius 1936; Turnbull and Hvorslev 1967). The stability of slice is only influenced by the self-weight (W_i), shearing resistance (S_i) from the sl surface, and normal load (N_i) from the slip surface. Since the normal load (N will not yield a rotating moment around the circular center (O) of the slip surfac the rotating moment (M_{Si}) of the slope generated by the slice i can be compute as follows (Fellenius 1936).

$$M_{Si} = (R)(W_i)\sin\theta_i$$

where R is the radius of the slip surface, and θ_i is the angle of the slip surface slice i measured from the horizontal plane. The resistance moment (M_{Ri}) generate by slice i is computed as:

$$M_{Ri} = \left\{c'_i l_i + \left[W_i \cos\theta_i - r_u (W_i/dx) l_i\right]\tan\varphi'_i\right\} R$$

where c'_i is the effective cohesion along the slip surface of slice i; l_i is the arc lengt along the slip surface of slice i, approximated as $l_i = dx/\cos\theta_i$; dx = width of slic i; φ' is the effective friction angle along the slip surface of slice i; and r_u is the por water pressure ratio, defined as the average ratio of pore water pressure to the tot earth pressure along the slip surface (e.g., Bishop 1955; El-Ramly et al. 2005). The the overall FS for this earth slope can be integrated as follows (e.g., Fellenius 193 Ching et al. 2009).

$$FS = \frac{\sum_{i=1}^{i=n}\left\{c'_i(dx/\cos\theta_i) + \left[\cos\theta_i - (r_u/\cos\theta_i)\right]W_i \tan\varphi'_i\right\}}{\sum_{i=1}^{i=n}\left(W_i \sin\theta_i\right)}$$

vhere n is the total number of slices, which is taken as 50 in this design example. Jaturally, the computed FS varies with the choice of the slip surface, and only the lip surface that yields the minimum FS could reflect the stability of earth slopes. Iere, the search for such a critical slip surface can be conducted based on the desire o minimize FS, which is defined as follows.

$$FS = f\left(\boldsymbol{SS}, \boldsymbol{G}, \boldsymbol{S}\right)$$

vhere $\boldsymbol{SS}$ indicates the parameters of the slip surface (x_o, y_o, and R, see Figure 5.11); $\boldsymbol{G}$ indicates the geometry parameters of the cut slope or design parameters in this xample (α_1 and α_2, see Figure 5.11); and, $\boldsymbol{S}$ represents soil parameters (γ_1, c'_1, φ'_1, γ_2, c'_2, φ_2, and r_u, see Figure 5.11). The minimization of FS could be performed vith ***fmincon***, a built-in function in MATLAB version R2010a (MathWorks 2010) hat returns the minimum value of a constrained nonlinear multivariable function, nd the derived minimum value of FS captures the stability of the road cut design.

.3.1.2 Parameters and Optimization Setting of the Robust Geotechnical Design

Vithin the context of the RGD of this road cut problem, the geometry parameters i.e., α_1 and α_2) are taken as the design parameters. In contrast, the uncertain soil parameters (i.e., c'_1, φ'_1, c'_2, φ'_2, and r_u) are treated as noise factors. RGD aims to elect the optimal design parameters (i.e., α_1 and α_2) such that the variation of the lesign performance (i.e., slope stability in this example) induced by the variation of noise factors (i.e., c'_1, φ'_1, c'_1, φ'_2, and r_u) can be minimized, or equivalently the lesign robustness in terms of SNR is maximized, while safety and cost efficiency equirements are satisfied. In this design example, the robustness of the design (SNR) s formulated as follows:

$$SNR = 10\log_{10}\left(\frac{\mu_{FS}^{2}}{\sigma_{FS}^{2}}\right)$$

vhere μ_{FS} and σ_{FS} are the mean and standard deviation of the FS, respectively.

For simplicity, the construction cost (C) of the candidate designs (represented by lesign parameters α_1 and α_2) is captured herein with the excavation volume per longitudinal length (m^3/m), which is the area of the polygon ABCDEFA (see Figure 5.11), lthough other sophisticated models for estimating the cost of the candidate road cut lesigns may be adopted. The cost of a candidate slope design with design parameters of α_1 and $\alpha_1 = 46.0°, \alpha_2 = 32.5°$ is computed as follows.

$$C = \left(90 + \frac{32.5}{\tan\alpha_1} + \frac{8}{\tan\alpha_2} - \frac{40.5}{\tan 25°}\right)\mathrm{C}_0$$

vhere C_0 represents the cost per unit of slope excavation.

To ensure the feasibility of the construction, the design parameters (i.e., α_1 and α_2) are taken from the following discrete design space: $\boldsymbol{DS} = \{(\alpha_1, \alpha_2) \mid \alpha_1 \in \{25.0°, 26.5°, 28.0°, \ldots, 46.0°$ and $\alpha_2 \in \{25.0°, 26.5°,$

28.0°, … , 46.0°}}. This discrete design space is mainly obtained from a preliminary assessment with engineering judgment; of course, other design spaces may be adopted, and the results could be similar. The robust design optimization of this earth slope is limited to this discrete domain to ensure the feasibility of the construction. In this design example, the design space consists of 225 candidate designs. The RGD can then be undertaken with either a tri-objective optimization setting that considers design safety, robustness, and cost as objectives or a bi-objective optimization setting that considers design robustness and cost as objectives plus safety as constraint. These two optimization settings are formulated below.

1) Tri-objective optimization algorithm

Find:	design parameters $\boldsymbol{d}=\{\alpha_1,\alpha_2\}$
Subject to:	$\boldsymbol{d} \in$ design space $\boldsymbol{DS}$
Objectives:	minimizing probability of failure P_f
	maximizing design robustness SNR
	minimizing cost C

2) Bi-objective optimization algorithm

Find:	design parameters $\boldsymbol{d}=\{\alpha_1,\alpha_2\}$
Subject to:	$\boldsymbol{d} \in$ design space $\boldsymbol{DS}$
	$P_f <$ target failure probability P_T
Objectives:	maximizing design robustness SNR
	minimizing cost C

where P_f is the probability of failure of the geotechnical design, and P_T is the targeted failure probability that is generally adopted in the conventional RBD.

5.3.1.3 Relationships Among Safety, Robustness, and Cost in the Selected Design Space

As a discrete design space is adopted in this design example, the safety, robustness and cost for each candidate design can be analyzed, from which the relationships among the safety, robustness, and cost can be studied.

Figure 5.12(a) depicts the relationship between the safety (in terms of the probability of failure P_f) and cost (C) by plotting those values for all 225 candidate designs. Note that this plot is prepared only to provide a baseline reference, in which the design robustness is not checked, and the target failure probability is not adopted to screen the candidate designs. Nevertheless, the results in Figure 5.12(a) demonstrate that there exists a tradeoff relationship between safety and cost: the probability of failure tends to decrease with the cost. As can be seen, the boundary of all discrete design points shown in Figure 5.12(a) visually establishes a Pareto front, which is indeed a set of non-dominated optimal designs with respect to these

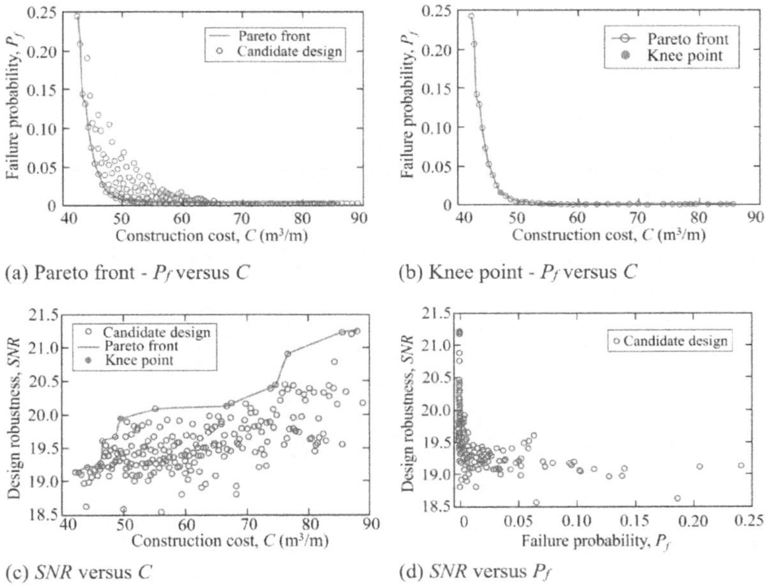

(a) Pareto front - P_f versus C

(b) Knee point - P_f versus C

(c) SNR versus C

(d) SNR versus P_f

FIGURE 5.12 Relationships among the safety, design robustness, and cost of candidate designs in Design Example 1 (modified from Figure 3 in Gong et al. 2014c, with permission from American Society of Civil Engineers).

two design objectives (i.e., to maximize safety and to minimize cost). Among these 225 discrete designs, 33 designs are identified as non-dominated optimal designs that belong to the Pareto front. To establish the Pareto front automatically, we may adopt the sorting algorithm NSGA-II (Deb et al. 2002). Further, the knee point concept (e.g., Branke et al. 2004; Deb and Gupta 2011) is applied to find the most preferred design (or best compromise design) from the sorted Pareto front. In Figure 5.12(b), a knee point is identified with the marginal utility function approach suggested by Branke et al. (2004). On the left side of this knee point, a slight reduction in cost can drastically increase the probability of failure, which is not desirable. On the right side of this knee point, a slight decrease in the probability of failure can yield a considerable increase in cost, which is also not desirable. As such, this knee point (i.e., k_v) represents the best compromise between these two objectives.

Similarly, Figure 5.12(c) depicts the tradeoff relationship between design robustness and cost efficiency. This tradeoff relationship can be observed with the Pareto front, which is defined by 15 non-dominated optimal designs without reference to design safety. Following the knee point concept, a knee point (i.e., $\alpha_1 = 46.0°$, $\alpha_2 = 28.0°$) can be located, which yields the best compromise design. Finally, the relationship between design robustness and safety in the design space is obtained and plotted in Figure 5.12(d). The design robustness generally decreases the probability of failure; thus, a higher level of design safety is always accompanied by a more RGD.

5.3.1.4 Robust Design Optimization Results of This Road Cut Problem

Figure 5.13(a) illustrates the design safety (in terms of the probability of failure P_f), design robustness (in terms of the *SNR*), and cost (*C*) of all 225 discrete candidate designs in a three-dimensional (3D) plot. Applying the NSGA-II sorting algorithm to maximize the safety (or to minimize the probability of failure), robustness (or to maximize *SNR*), and cost efficiency (or to minimize the cost) using the tri-objective optimization setting established, 47 non-dominated optimal designs can be obtained, as shown in Figure 5.13(b). Though presented in a 3D graph, the tradeoff relationship between safety and cost efficiency, and that between design robustness and cost efficiency, can be easily observed. Also plotted in Figure 5.13(b) is the knee point located among these non-dominated optimal designs, in which the design parameters are $\alpha_1 = 46.0°$ and $\alpha_2 = 28.0°$. This knee point yields the best compromise among the design safety, robustness, and cost efficiency. At this knee point, the probability of failure, robustness, and cost are $P_f = 4.38\times10^{-3}$, *SNR* = 19.94, and *C* = 49.58 m^3/m, respectively.

Next, the RGD of the road cut problem that adopts the bi-objective optimization setting is conducted. As an example, at a target failure probability of $P_T = 1.0\times10^{-2}$, the optimization with the bi-objective optimization setting yields a Pareto front

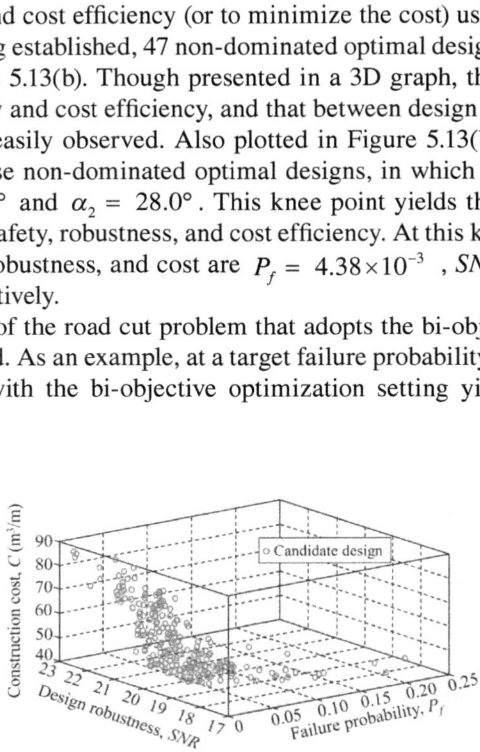

(a) All 225 candidate designs in 3-D plot

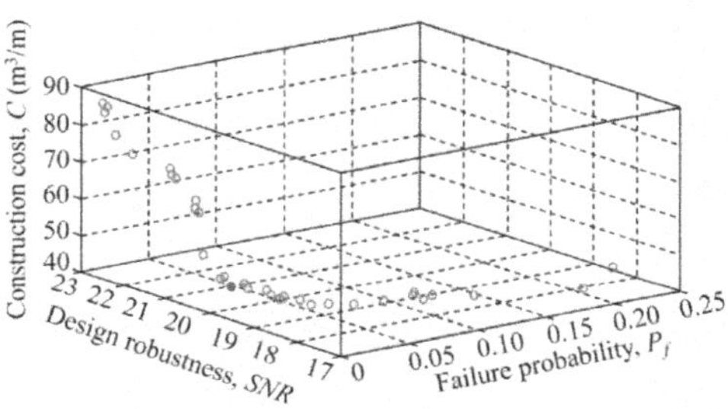

(b) Pareto front and knee point in a 3-D plot

FIGURE 5.13 Tri-objective robust optimization of the earth slope design (modified from Figure 4 in Gong et al. 2014c, with permission from American Society of Civil Engineers).

ith 10 non-dominated optimal designs, as shown in Figure 5.14(a). Among all the signs that are on this Pareto front, design A is the most economical (with the west cost), and design C is the most robust (with the highest design robustness). ne rest of the designs can be considered a compromise between these two designs. ne final design could be selected based on the desired level of cost or robustness. lternatively, the final design can be chosen based on the best compromise of cost

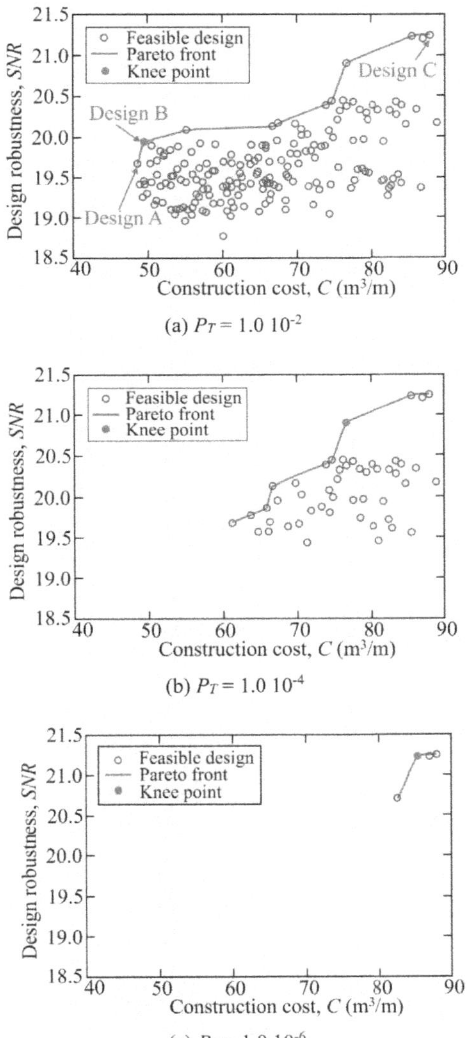

IGURE 5.14 Bi-objective robust optimization of the earth slope design in Design Example (modified from Figure 5 in Gong et al. 2014c, with permission from American Society of ivil Engineers).

TABLE 5.4
Results of RGD with Various Target Failure Probabilities in Design Example

Target failure probability (P_T)	No. of feasible designs	No. of non-dominated optimal designs	Knee point on the Pareto front		Three objectives at the knee point		
			α_1 (°)	α_2 (°)	Actual P_f	*SNR*	C (m³/
1.0×10^{-2}	163	10	46.0	28.0	4.38×10^{-3}	19.94	49.58
1.0×10^{-4}	48	9	29.5	26.5	1.21×10^{-6}	20.90	76.64
1.0×10^{-6}	5	3	26.5	25.0	4.78×10^{-8}	21.23	85.49

efficiency and robustness, recognizing that the designs on the Pareto front are tho non-dominated optimal designs that meet the safety requirement. To this end, th knee point (design B) can be selected. It is interesting to note that the knee point (i.e $\alpha_1 = 46.0°$, $\alpha_2 = 28.0°$) in this bi-objective optimization with the constraint $P_f < P_T = 1.0\times10^{-2}$ coincides with the knee point obtained from the tri-objecti optimization setting without the constraint on the safety.

It can be expected that the resulting feasible (or safe) designs, Pareto front, ar knee point are all affected by the choice of the target failure probability. To exami this effect, the same road cut slope is designed using the same bi-objective optim zation setting with different failure probabilities (i.e., $P_T = 1.0\times10^{-2}$, 1.0×10^{-4}, ar 1.0×10^{-6}), and the final designs are listed in Table 5.4. The results confirm that the target failure probability is reduced (or a higher level of safety is demanded fewer feasible designs and non-dominated optimal designs are secured, and the mo preferred design (or knee point on the resulting Pareto front) will have a higher co and higher design robustness. Illustrated in Figures 5.14(b–c) are the Pareto fron and knee points obtained for the scenarios of $P_T = 1.0\times10^{-4}$ and $P_T = 1.0\times10^{-}$ respectively.

5.3.2 Design Example 2: Tunnel Cross-Section Design with Fuzzy Sets as Inputs

The second design example considers the design of a tunnel cross-section, whic was studied by Gong et al. (2014b). In reference to Figure 5.15, a shield tunnel is con structed in soils with an embedded depth of $H = 15.0$ m, and the tunnel's inner radiu is set up as $R_{in} = 2.75$ m. The basic design performances considered in the RGD c this shield tunnel cross-section are the segment strength and convergence deforma tion, known as the ultimate limit state (ULS) and serviceability limit state (SLS requirements, respectively. For illustration purposes, the target reliability indexe with respect to ULS and SLS requirements, denoted as β_{T1} and β_{T2}, respectivel are set as 4.2 and 2.7, respectively. The design parameters that are optimized are th segment thickness (t), steel reinforcement ratio (ρ), and diameter of joint bolts (D_j

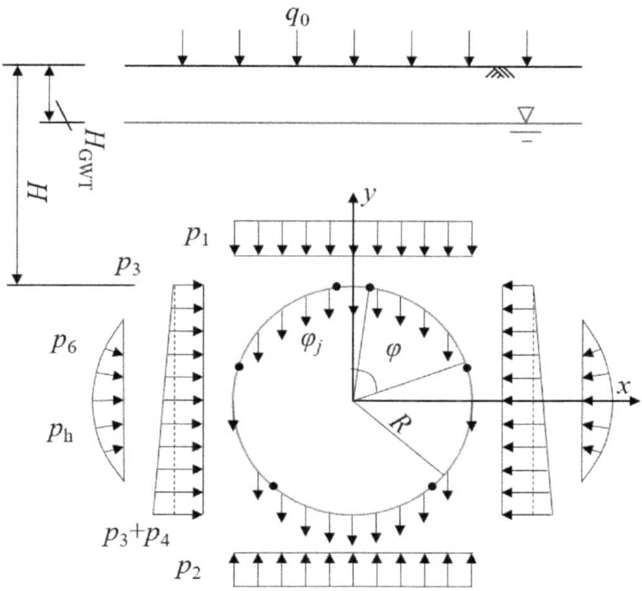

IGURE 5.15 Schematic diagram of loads on the shield-driven tunnel cross-section in Design Example 2 (modified from Figure 1 in Gong et al. 2014b, with permission from Elsevier).

5.3.2.1 Deterministic Solution Model for Assessing the Performance of Tunnel Cross-Sections

As the emphasis of this example is on the design of tunnel cross-sections, the effect of the tunnel longitudinal behavior on the performance of the tunnel cross-section is not included for simplicity; however, it is considered in the subsequent example (i.e., Design Example 3). The analysis of the performance of a tunnel cross-section is simplified as a plane strain problem in this design example. Among the various available approaches to analyze the internal forces and the convergence deformation of jointed shield tunnels (e.g., Wood 1975; Lee et al. 2001; Koyama 2003), the analytical solution model proposed by Lee et al. (2001) is adopted in this design example for its simplicity and wide acceptance. Figure 5.15 shows the loads acting on the shield tunnel cross-section, including the earth pressure, water pressure, dead load, ground surface surcharge, and subgrade reaction. The formulations of these loads can be available in Gong et al. (2014b). According to the formulations in Lee et al. (2001), the internal forces and convergence deformation of jointed shield tunnels are mainly affected by the design parameters of tunnel segment thickness (t), segment steel reinforcement ratio (ρ), and diameter of joint bolts (D_j), through the factors of tunnel segment stiffness and joint stiffness.

The outputs of the analytical model by Lee et al. (2001) include the internal forces and convergence deformation of the concerned tunnel cross-section (with inputs of load conditions, segment stiffness, and joint stiffness), which are readily

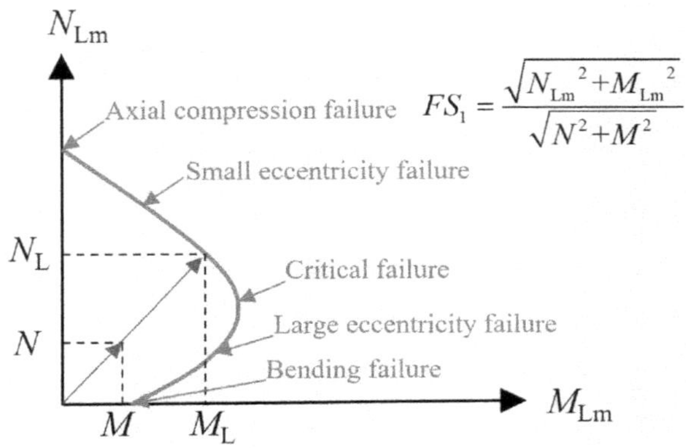

FIGURE 5.16 Structure safety assessment of concrete tunnel segment using plasticity theory (modified from Figure 2 in Gong et al. 2014b, with permission from Elsevier).

adopted to assess the segment structure safety (ULS requirement) and serviceability (SLS requirement). Plasticity theory is applied to evaluate the structure safety of the segment. In reference to Figure 5.16, the structure failure of the segment is said to occur only when the combination of internal forces, in terms of bending moment (M) and axial force (N), exceeds the related limit state (M_L, N_L) on the ultimate bearing envelope of the tunnel segment, derived from the plasticity theory. For simplicity, the FS for the structure safety, denoted as FS_1, in a deterministic approach is computed as:

$$FS_1 = \frac{\sqrt{N_{Lm}{}^2 + M_{Lm}{}^2}}{\sqrt{N^2 + M^2}}$$

It is noted that the value of FS_1 varies with the circumferential position within the tunnel ring of concern. Thus, the ULS is governed by the position with the minimum value of FS_1.

The maximum tunnel convergence deformation derived can readily be adopted to assess the tunnel serviceability (SLS requirement). As specified in the Chinese metro code (MCPRC 2003), the maximum convergence of a shield tunnel cross-section should be controlled under 0.4%D to 0.6%D to prevent operational distress, where D represents the outer diameter of the tunnel cross-section. As such, in a deterministic approach, the FS against the tunnel serviceability failure, denoted as FS_2, can be conservatively evaluated as follows.

$$FS_2 = \frac{0.4\% D}{\max(\Delta_v, 2\Delta_h)}$$

where Δ_v and $2\Delta_h$ are the calculated tunnel convergence deformation in the vertical direction and horizontal direction, respectively.

5.3.2.2 Noise Factors Characterization and Design Space Selection for the Robust Design

In this design example, the unit weight of soil (γ_s) and water (γ_w) are both treated as deterministic or fixed parameters due to their negligible variation compared with the other geotechnical parameters, such as soil resistance coefficient (*K*s), soil cohesion strength (*c*), soil friction angle (φ), and groundwater table (H_{GWT}). In addition to the uncertain geotechnical parameters (i.e., *K*s, *c*, φ, and H_{GWT}), the surcharge (q_0) on the ground also involves significant uncertainty, and its effect on the tunnel performance cannot be ignored. Collectively, these five parameters are taken as noise factors in this design example. These noise factors are represented with fuzzy numbers, and detailed parameters to characterize the membership functions of these fuzzy numbers are tabulated in Table 5.5. The upper and lower bounds (HCV and LCV) of these noise factors, as shown in Table 5.5, are mainly determined based on local experience, literature reports, and engineering judgment. The design parameters in the robust design of this tunnel cross-section, including segment thickness (*t*), steel reinforcement ratio (ρ), and diameter of joint bolt (D_j), are optimized in a pre-assigned continuous design space. Based on local practices in Shanghai, China, the design space can be determined, as shown in Table 5.6.

Apart from the noise factors in Table 5.5 and the design parameters shown in Table 5.6, the design performance of a shield tunnel cross-section can also be influenced by some other parameters. However, the levels of variation of these parameters are small and could be treated as fixed-value parameters. The fixed parameters involved are listed in Table 5.7.

TABLE 5.6
Design Space for the Robust Design of the Tunnel Cross-Section in Design Example 2

Design parameter	Assigned ranges	Value
Segment thickness (*t*)	Lower limit (t_l: m)	0.200
	Upper limit (t_u: m)	0.500
Steel reinforcement ratio (ρ)	Lower limit (ρ_l: %)	0.50
	Upper limit (ρ_u: %)	4.00
Diameter of joint bolt (D_j)	Lower limit (D_{jl}: mm)	10.0
	Upper limit (D_{ju}: mm)	50.0

TABLE 5.5
Characterization of the Membership Functions of Noise Factors in Design Example 2

Noise factors	Lower bound (a)	Mode [$m = (a + b)/2$]	Upper bound (b)
Soil resistance coefficient (Ks: kN/m^3)	3,500	9,250	15,000
Soil cohesion strength (c: kN/m^2)	0	7.5	15
Soil friction angle (ρ: °)	30	32.65	35.3
Ground water table (H_{GWT}: m)	0.5	1.25	2
Ground surcharge (q_0: kN/m^2)	0	10	20

TABLE 5.7
Deterministic Parameters for Assessing Tunnel Performance in Design Example 2

Category	Parameter	Value
Soil	Unit weight (γ_s: kN/m^3)	18.0
Tunnel	Embedded depth (H: m)	15.0
	Tunnel inner radius (R_{in}: m)	2.75
	With of tunnel ring (b: m)	1.0
	Joint position of half structure (φ_i: °)	8, 73, 138
Concrete segment	Unit weight of concrete (γ_c: kN/m^3)	25.0
	Elastic modulus of concrete (E_c: kN/m^2)	35×10^6
	Compression strength of concrete (f_c: kN/m^2)	39×10^3
	Ultimate plastic strain of concrete (ε_p)	0.0033
Steel reinforcement	Elastic modulus of steel (E_s: kN/m^2)	210×10^6
	Yielding strength of steel bar (f_y: kN/m^2)	345×10^3
	Thickness of protective cover (a: m)	0.05
Joint bolt	Bolt length (l_b: m)	0.4
	Number of bolts at each joint	2
	Distance from bolts center to tunnel inside surface (h)	$t/3$
Water	Unit weight (γ_w: kN/m^3)	9.8

3.2.3 Optimization Setting of the Robust Design of This Tunnel Cross-Section

this design example, the optimization setting for the RGD of the concerned tunnel oss-section is set up as follows.

$$\begin{aligned}
&\text{Find:} && \text{design parameters } \mathbf{d}=\{t,\rho,D_j\} \\
&\text{Subject to:} && t_l \le t \le t_u;\ \rho_l \le \rho \le \rho_u;\ D_{jl} \le D_j \le D_{ju}; \\
& && \beta_1 \ge \beta_{T1};\ \beta_2 \ge \beta_{T2} \\
&\text{Objectives:} && \text{maximizing robustness } SNR_1 \\
& && \text{maximizing robustness } SNR_2 \\
& && \text{minimizing cost } C(t,\rho,D_j)
\end{aligned}$$

where β_{T1} and β_{T2} are the target reliability indexes with respect to ULS and SLS quirements, respectively; $C(t,\rho,D_j)$ is the cost, which is a function of the design rameters (i.e., segment thickness t, steel reinforcement ratio ρ, and diameter of int bolt D_j). As can be seen, the design space and safety requirement (i.e., reliabily index β_1 with respect to ULS requirement and reliability index β_2 with respect SLS requirement) are set as design constraints, while the design robustness (i.e., NR_1 for ULS performance and SNR_2 for SLS performance) and the cost $C(t,\rho,D_j)$ e both set as design objectives. For ease of comparison, the optimization setting at is adopted in the conventional RBD is also given below.

$$\begin{aligned}
&\text{Find:} && \text{design parameters } \boldsymbol{d}=\{t,\rho,D_j\} \\
&\text{Subject to:} && t_l \le t \le t_u;\ \rho_l \le \rho \le \rho_u;\ D_{jl} \le D_j \le D_{ju}; \\
& && \beta_1 \ge \beta_{T1};\ \beta_2 \ge \beta_{T2} \\
&\text{Objective:} && \text{minimizing cost } C(t,\rho,D_j)
\end{aligned}$$

It is noted that only the cost $C(t,\rho,D_j)$ is taken as a design objective in the traitional RBD. Thus, the main difference between RGD and RBD is that the design bustness is treated as an additional design objective. As in the RBD, the safety quirement of a design in the RGD is satisfied with the following settings: $\beta_1 \ge \beta_{T1}$ nd $\beta_2 \ge \beta_{T2}$. This safety constraint assures that the resulting designs are compulsoly brought to a specified target level of design safety while both design robustness nd cost efficiency are optimized.

For illustration purposes, only the cost of one tunnel ring is investigated in this esign example, which consists of tunnel segment concrete cost, steel reinforcement ost, and joint bolt cost. According to the market survey in Shanghai, the unit prices f segment concrete (c_c), reinforcement steel (c_s), and joint bolt (c_b) are 600 RMB/^{3} (97.77 USD/m^3), 4,000 RMB/10^{3}kg (645.16 USD/10^{3}kg), and 10 RMB/kg (1.61 SD/kg), respectively. In this design example, the cost function $C(t,\rho,D_j)$ in the GD of the tunnel cross-section is computed as follows.

$$C(t,\rho,D_j) = c_c Q_c + c_s Q_s + c_b Q_b$$

where Q_c, Q_s, and Q_b represent the quantity of segment concrete (m^3), reinforc ment steel bar (10^{3}kg), and joint bolt (kg) of the shield-driven tunnel per ring, respe tively. In addition to the cost, the other objective in the RGD is the robustness of t design. In this design example, the robustness of the design is measured by the SN formulated below.

$$SNR_1 = 10\log_{10}\left(\frac{E^2[Fs_1]}{\sigma^2[Fs_1]}\right)$$

$$SNR_2 = 10\log_{10}\left(\frac{E^2[Fs_2]}{\sigma^2[Fs_2]}\right)$$

where $E[Fs_1]$ and $\sigma[Fs_1]$ represent the mean and standard deviation of the fact of safety for the structure safety (Fs_1), respectively; $E[Fs_1]$ and $\sigma[Fs_1]$ represe the mean and standard deviation of the FS against the serviceability failure (Fs_1 and, SNR_1 and SNR_2 represent the design robustness for the ULS and SLS perfo mance, respectively.

5.3.2.4 Robust Design Optimization Results of This Tunnel Cross-Section

With the aid of the NSGA-II optimization algorithm (Deb et al. 2002), the RGD this tunnel cross-section could readily be carried out using the robust design optim zation setting established above. The population size adopted in the NSGA-II is s at 50, and the generation number is set at 100. Figure 5.17(a) shows the Pareto fro (a set of non-dominated optimal designs) obtained with the objectives of robustne SNR_1, robustness SNR_2, and cost (*C*). The design parameters, in terms of tunn segment thickness (*t*), steel reinforcement ratio (ρ), and diameter of joint bolt (D_j), these non-dominated optimal designs are plotted in Figure 5.17(b).

Depicted in Figure 5.17(c) are the two-dimensional (2D) Pareto fronts, which a the projections of the three-dimensional (3D) Pareto front shown in Figure 5.17(a indicating the trade-off relationships between design robustness (both SNR_1 an SNR_2) and cost (*C*) in the 2D graphs. Both robustness indexes of SNR_1 and SNR tend to increase the cost, showing that the robustness of the tunnel design can b enhanced through more investment. The trade-off relationship (trend line) betwee SNR_1 and *C* appears more pronounced than that between SNR_2 and *C*. Here, the va ues of SNR_1 and SNR_2 are in the range of 13–30 and 22–27, respectively. Such rang values represent the possible values of SNR_1 and SNR_2 within the design space.

The observed more obvious trade-off effect between SNR_1 and cost is most likel due to the fact that there is a larger variation of SNR_1 in the design space selecte The less pronounced trade-off relationship between SNR_2 and cost suggests th SNR_2 involves less variability in the design space. However, this does not nece sarily imply that the SNR_2 does not affect the optimal design. Indeed, the Paret front should be a surface that is a function of SNR_1, SNR_2, and cost simultan ously, as illustrated in Figure 5.17(a). It should be mentioned that the phenomeno in Figure 5.17 might be problem-specific; in other problems, the robustness index SNR_2 may play a more critical role in robust design optimization.

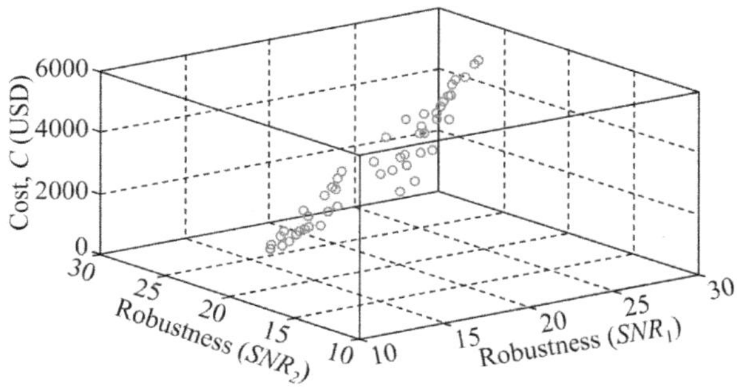

(a) All non-dominated solutions (Pareto front) shown in 3-D graph of objectives

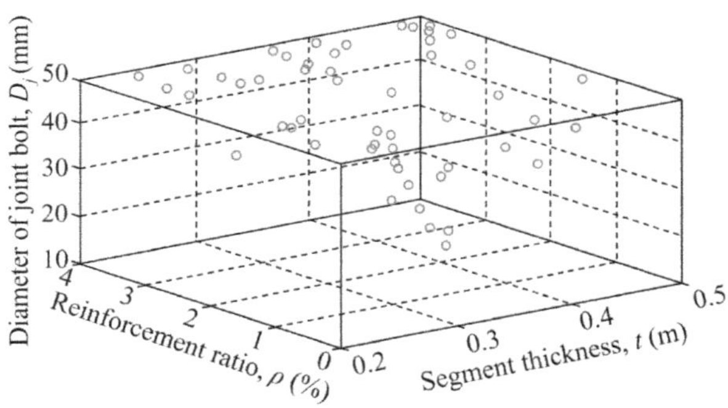

(b) Design parameters of all non-dominated optimal solutions

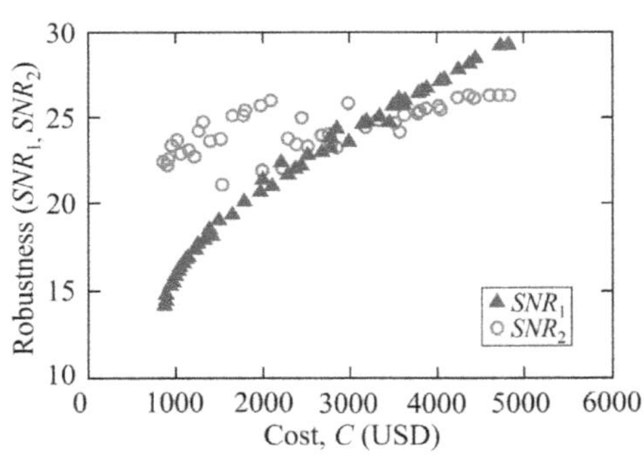

(c) Robustness versus cost of all non-dominated optimal solutions (2-D Pareto front)

FIGURE 5.17 Pareto front obtained using NSGA-II in Design Example 2 (modified from Figure 10 in Gong et al. 2014b, with permission from Elsevier).

As mentioned previously, all designs on the Pareto front, as shown in Figure 5.17(a are non-dominated optimal designs. The most preferred design in the design spac may be selected by the engineer based on a target level of cost or design robust ness. For example, if the target level of robustness is set at SNR = 15, then the leas cost design on the Pareto front could be taken as the most preferred design, whic is defined in this case by the following set of design parameters: $t = 274.0\ mm$ $\rho = 0.72\ \%$, and $D_j = 45.7\ mm$. Of course, the knee point concept can be adopte to identify the most compromised design on the Pareto front.

5.3.3 Design Example 3: Robust Design Optimization of a Longitudinal Tunnel

It should be mentioned that the longitudinal length of a shield tunnel is hundred (or thousands) of meters, while the diameter is usually less than 20 meters. Thus the design of the shield tunnel should be a 3D problem instead of a 2D plane strai problem. Furthermore, the longitudinal variation of the input parameters, induce by the variation of tunnel alignment, spatial variation of soil parameters, and nearb activities, could affect the tunnel performance significantly. The performance of designed tunnel, in terms of the structural safety and serviceability of each segmen ring, might not be correctly reflected in the analysis results of a few "representative" segment rings (or cross-sections), and the design of a shield tunnel cannot be capture by the design of a few tunnel-cross sections, as shown in Design Example 2. The third design example considers the design of a tunnel longitudinal structure, in whic the longitudinal variation of input parameters is explicitly considered. An illustrativ shield tunnel with a longitudinal length of 300 m is analyzed in this example, an the random field theory is adopted to characterize the longitudinal variation of inpu parameters. Note that this tunnel design problem was studied by Gong et al. (2015a)

5.3.3.1 Noise Factors Characterization and Design Space Selection for the Robust Design

In the robust design of tunnel longitudinal structure, the parameters of tunnel seg ment thickness (t), steel reinforcement ratio (ρ), and diameter of steel bolts (i.e., D_c and D_l for the circumferential bolts and longitudinal bolts, respectively) are cate gorized as the easy-to-control design parameters (**d**); and, the parameters of verti cal ground stiffness (k_v), horizontal ground stiffness (k_h), effective cohesion (c) effective friction angle (φ), groundwater table (H_w), and ground surcharge (q_0) are treated as the hard-to-control noise factors ($\boldsymbol{\theta}$).

Here, the tunnel is assumed to be embedded in a homogeneous soil. The station ary random fields are adopted to model the longitudinal variation of noise factors; the statistics of these noise factors are tabulated in Table 5.8. As an example, the mear and standard deviation of noise factors shown in Table 5.8 are estimated based or the available geological and geotechnical data reported in the Shanghai Foundatior Design Code (1999). It is acknowledged that different soil parameters may exhibi different scales of fluctuation in the horizontal direction, typically in the range o 40–60 m (e.g., Phoon and Kulhawy 1999). For simplicity, the scales of fluctuation o

TABLE 5.8
Statistical Characterization of Noise Factors in Design Example 3

Noise factors	Mean (μ)	Coefficient of variation (δ)	Scale of fluctuation (r)	Distribution
Vertical ground stiffness of the ground under the tunnel (k_v, kN/m³)	33,000	0.500	50	Lognormal
Effective cohesion of soil (c, kN/m²)	7.5	0.333	50	Lognormal
Effective friction angle of soil (φ,°)	32.65	0.027	50	Lognormal
Horizontal ground stiffness of soil (k_h, kN/m³)	9,250	0.207	50	Lognormal
Ground water table (H_w, m)	1.25	0.200	50	Lognormal
Ground surcharge (q_0, kN/m²)	10	0.333	50	Lognormal

TABLE 5.9
Design Space of the Robust Design in Design Example 3

Easy-to-control design parameter	Possible value
Segment thickness (t: m)	{0.25, 0.30, 0.35, 0.40, 0.45}
Steel reinforcement ratio of tunnel segment (ρ: %)	{0.5, 1.0, 1.5, 2.0}
Bolt diameter of the circumferential joints (D_c : mm)	{20, 25, 30, 35, 40}
Bolt diameter of the longitudinal joints (D_l : mm)	{20, 25, 30, 35, 40}

all these noise factors are assumed to be 50 m. The data listed in Table 5.8 are typical values that are consistent with the data for the Shanghai Metro. For the design of a real-world shield tunnel, these statistical parameters should be characterized based on site investigation data and local experience. Meanwhile, a discrete design space that consists of 500 discrete candidate designs, as shown in Table 5.9, is adopted in this design example.

Apart from the noise factors and design parameters shown in Tables 5.8 and 5.9, the performance of a longitudinal shield tunnel is also effected by many other parameters. However, the levels of variation of these parameters are relatively small and can generally be ignored. The deterministic parameters involved in this shield tunnel design are tabulated in Table 5.10.

TABLE 5.10
Deterministic parameters in Design Example 3

Category	Parameter	Value
Soil	Unit weight (γ_s, kN/m^3)	18.0
Tunnel geometries	Longitudinal length (L, m)	300.0
	Embedded depth (H, m)	14.0
	Tunnel inner radius (R_{in}, m)	2.75
	With of segment ring (b, m)	1.0
	Position of circumferential joints of half structure (φ_i, °)	8, 73, 138
Tunnel segment	Unit weight of concrete (γ_c, kN/m^3)	25.0
	Elastic modulus of concrete (E_c, kN/m^2)	34.5×10^6
	Compression strength of concrete (f_c, kN/m^2)	39×10^3
	Ultimate plastic strain of concrete (ε_p)	0.0033
Steel reinforcement	Elastic modulus of steel (E_s, kN/m^2)	210×10^6
	Yielding strength of steel (f_y, kN/m^2)	345×10^3
	Thickness of protective concrete cover (a, m)	0.05
Steel bolt	Length of steel bolts (l_b, m)	0.4
	Number of bolts at each circumferential joint	2
	Distance from bolts center to tunnel inside surface (h)	t/3
	Number of steel bolts between adjacent segment rings (or of each longitudinal joint)	17
Water	Unit weight (γ_w: kN/m^3)	9.8

5.3.3.2 Performance Analysis of Tunnel Longitudinal Structure in the Face of Uncertainties

It has long been acknowledged that the longitudinal variation of input parameters must be explicitly considered in the analysis and design of shield tunnels (e.g., ITA 2000; ATRB 2000; Koyama 2003), a convincing framework for shield tunnel performance analysis that considers the longitudinal variation of input parameters is not available. In this design example, the framework proposed by Gong et al. (2015a) is adopted for analyzing the tunnel performance in the face of longitudinal variation of input parameters; and, in reference to Figure 5.18, the performance of a tunnel longitudinal structure can be analyzed with the following steps:

Step 1: Generating input noise factors with Monte Carlo simulation

In this step, the MCS is employed to generate the input noise factors, including the vertical ground stiffness (k_v), horizontal ground stiffness (k_h), effective cohesion (c), effective friction angle (φ), groundwater table (H_w), and ground surcharge (q_0),

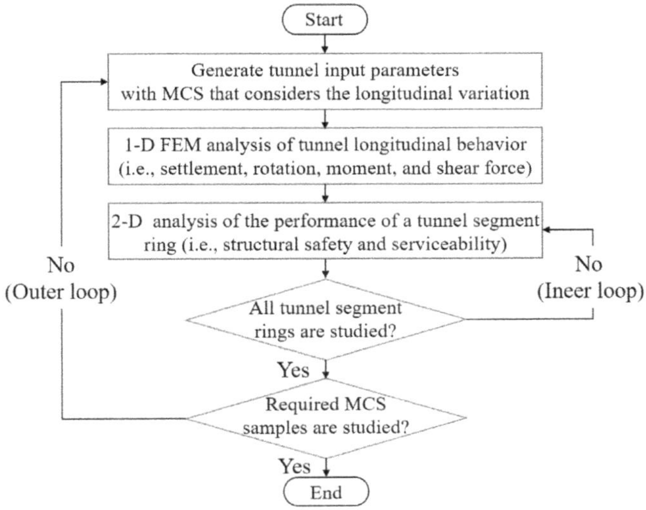

GURE 5.18 Framework for the performance analysis of tunnel longitudinal structure nodified from Figure 1 in Gong et al. 2015a).

ong the tunnel longitudinal direction, with the statistical information shown in able 5.8.

tep 2: Analyzing tunnel longitudinal behavior with 1D FEM solution

ased on the input noise factors generated in **Step 1**, the longitudinal behavior of e shield tunnel, including tunnel settlement (w), longitudinal rotation (θ_L), lon-itudinal bending moment (M_L), and longitudinal shear force (Q_L), are readily nalyzed using the simplified one-dimensional (1D) finite element method (FEM) olution model advanced by Huang et al. (2015). The schematic diagram of this 1D EM solution model is shown in Figure 5.19. Within this 1D FEM solution model, e longitudinal structure of the shield tunnel is modeled as an elastic and continu-us beam, and the effect of tunnel longitudinal joints, which are the joints between egment rings, is simulated by a reduction factor of tunnel longitudinal flexural iffness (e.g., Liao et al. 2008). The soil-structure interaction between the tunnel eam and the ground under the tunnel is modeled using the Winkler model, and the verburden on the tunnel beam is modeled using the pressure load. Here, the verti-al ground stiffness (k_v) of the ground under the tunnel may also be represented by e coefficient of vertical subgrade reaction, which may be determined with either e plate-load test data or estimated from the elastic modulus and Poisson's ratio of e soil determined with laboratory tests (Winkler 1867; Sadrekarimi and Akbarzad 009).

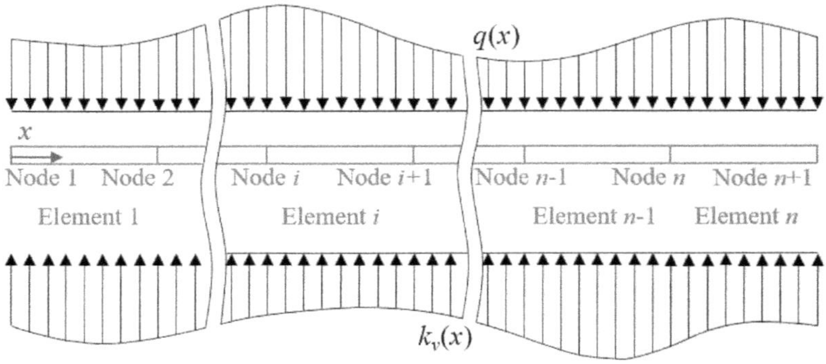

FIGURE 5.19 Schematic diagram of 1D FEM solution of tunnel longitudinal behavi (modified from Figure 2 in Gong et al. 2015a).

Step 3: Analyzing the performance of tunnel cross-sections with 2D analytical solution

Based on the tunnel longitudinal behavior obtained from Step 2 and input paramete generated in Step 1, specifically the horizontal ground stiffness (k_h), effective coh sion (c), effective friction angle (φ), groundwater table (H_w), and ground surchar (q_0) along the tunnel longitudinal direction, the performance (i.e., tunnel structu safety and serviceability) of each segment ring is readily studied with the 2D an lytical solution advanced by Gong et al. (2015a). The schematic diagram of this 2 analytical solution is illustrated in Figure 5.20, in which the shearing effect (e.g., Li et al. 2005) and flattening effect (e.g., Huang et al. 2012) are adopted to consider t impact of tunnel longitudinal behavior on the performance of cross-sections.

The results of this 2D analytical solution are the internal forces (i.e., axial force bending moment M, and shear force Q) and the convergence deformations (i.e., hor zontal convergence deformation Δ_h and vertical convergence deformation Δ_v) of th concerned tunnel cross-section, which are then adopted to assess the performanc of the segment ring. Here, the internal forces are employed to calculate the FS wit respect to the structure safety (ULS performance), denoted as Fs_1, using the meth outlined in Gong et al. (2014c). On the other hand, the convergence deformations a employed to calculate the FS with respect to the serviceability (SLS performance denoted as Fs_2. As specified in the Chinese metro code (MCPRC 2003), the ma imum convergence deformation of the shield tunnel cross-section should be co trolled under 0.4%D to 0.6%D to prevent operational distress, in which D represen the outer diameter of the segment ring. To be consistent with this code, the limitin convergence deformation in this design example is taken as 0.6%D.

Step 4: Repeating the performance analysis in Step 3 for all segment rings

In this step, the performance analysis of the tunnel cross-section presented in Step is repeated for all segment rings of the longitudinal shield tunnel, as shown with th

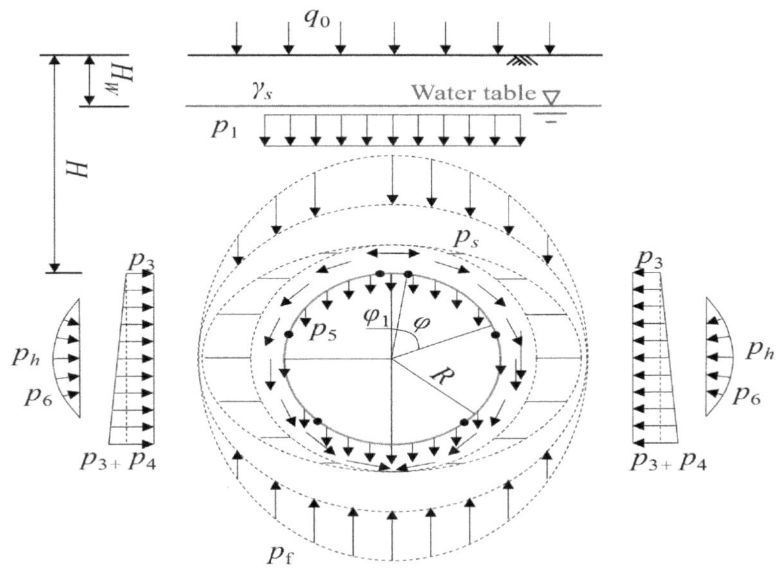

'IGURE 5.20 2D analytical solution of the performance of the tunnel cross-section modified from Figure 3 in Gong et al. 2015a).

nner loop in Figure 5.20. This procedure can yield a series of FSs (with respect to oth structure safety and serviceability), as different segment rings could exhibit dif-erent FSs due to the longitudinal variation of noise factors. Then, the mean values of he FSs, denoted as μ_{Fs1} and μ_{Fs2}, are derived to represent the overall performance of this tunnel longitudinal structure with respect to structure safety and serviceabil-ty, respectively. The standard deviations of the FSs, denoted as σ_{Fs1} and σ_{Fs2}, are also calculated to represent the variation (or degree of uncertainty) of the perfor-nance of the tunnel with respect to structure safety and serviceability, respectively.

'tep 5: Repeating MCS runs to yield a converged solution of tunnel performance

'n this step, the performance analysis procedures of the shield tunnel presented in Steps 1–4 are repeated for a specified number of MCS runs for purposes of deriv-ng a converged solution of tunnel performance, as shown with the outer loop in Figure 5.20. Note that the accuracy of MCS results is affected by the number of MCS samples, and the number of MCS samples required for different targets can be different. In this design example, the convergence criterion of MCS results is determined as follows: if the variation of MCS results caused by the increase of MCS samples is less than 1.0%, the convergence is achieved; otherwise, the con-vergence is not achieved, and the number of MCS samples is not sufficient. The MCS results in this design example are referred to as the statistics of the mean and variation of the tunnel performance (in terms of the statistics of μ_{Fs1}, μ_{Fs2}, σ_{Fs1} and σ_{Fs2}).

5.3.3.3 Formulation of the Design Robustness of the Tunnel Longitudinal Structure

With the longitudinal variation of input parameters (called noise factors in the context of robust design), different segment rings could exhibit different design response (i.e., Fs_1 and Fs_2). Thus, the robustness measures reported in the literature cannot be directly applied to the robust design of a tunnel longitudinal structure. Note that if the reported robustness measures were directly applied to the robust design of a tunnel longitudinal structure, the evaluation of the design robustness of candidate tunnel designs could be computationally prohibitive, and the screening of candidate designs would be a challenge. For example, for a tunnel longitudinal structure that consist of 200 segment rings, 400 robustness indexes need to be computed and further optimized, as both the structure safety and the serviceability of each segment ring must be included. In this design example, the *SNR* of the FS is employed to measure the design robustness of each segment ring.

$$SNR = 10\log_{10}\left(\frac{{\mu_{Fs}}^2}{{\sigma_{Fs}}^2}\right)$$

where μ_{Fs} and σ_{Fs} are the mean and standard deviation of the FS (with respect to the tunnel performance, either ULS or SLS performance), respectively, which can readily be computed from the tunnel analysis framework (see Step 4 in Section 5.3.3.2). Intuitively, a higher *SNR* value signals a lower variability of tunnel performance (i.e., Fs_1 and Fs_2) and, thus, higher robustness. As mentioned above, different MCS samples, in terms of input noise factors (i.e., k_v, k_h, c, φ, H_W and q_0) along tunnel longitudinal direction, can result in different values of the statistics of the design performance (i.e., μ_{Fs1}, μ_{Fs2}, σ_{Fs1}, and σ_{Fs2}). The variation in performance statistics creates further uncertainty in the computed *SNR*. Thus a new robustness index (*R*), based on the statistics of *SNR*, is proposed here to capture the robustness of the tunnel longitudinal structure (against longitudinal variation of noise factors):

$$R = \mu_{SNR} - 2\sigma_{SNR}$$

where μ_{SNR} and σ_{SNR} are the mean and standard deviation of *SNR* with respect to the design performance of interest (either ULS or SLS performance), respectively which can readily be computed using the procedure described in Step 5 in the previous section. As this robustness measure (*R*) is basically the same as *SNR*, greater robustness is achieved with a greater *R* value. Here, the mean minus two standard deviations is adopted to approximate the lower bound of *SNR*. This lower bound of *SNR* is selected so that at least a certain level of *SNR* can be achieved in the robust design optimization of the tunnel longitudinal structure.

5.3.3.4 Robust Design Optimization Setting for the Tunnel Longitudinal Structure

The robust design optimization of the tunnel longitudinal structure is implemented as a multi-objective optimization problem, the optimization setting of which is set up as follows.

$$\begin{aligned}
&\text{Find:} && \text{design parameters } \boldsymbol{d} = \{t, \rho, D_c, D_l\} \\
&\text{Subject to:} && \boldsymbol{d} \in \text{design space } \boldsymbol{DS} \\
& && \text{mean of } \mu_{Fs1} \geq 1.5; \text{ mean of } \mu_{Fs2} \geq 1.5 \\
&\text{Objectives:} && \text{maximizing robustness index } R_1 \\
& && \text{maximizing robustness index } R_2 \\
& && \text{minimizing cost } C(t, \rho, D_c, D_l)
\end{aligned}$$

where the constraint of "$\mu_{Fs1} \geq 1.5$" indicates the ULS requirement (structure safety), and the constraint of "$\mu_{Fs2} \geq 1.5$" indicates the SLS requirement (serviceability). In other words, the target FSs with respect to both the structure safety and the serviceability are set at 1.5.

For illustration purposes, only the material cost of the tunnel longitudinal structure is considered in this design example, which consists of segment concrete cost, steel reinforcement cost, and joint bolt cost. The unit prices of segment concrete (c_c), reinforcement steel bar (c_s), and joint bolt (c_b) are 600 RMB/m^3 (97.77 USD/m^3), 4,000 RMB/10^{3}kg (645.16 USD/10^{3}kg), and 10 RMB/kg (1.61 USD/kg), respectively. In this design example, the cost function $C(t, \rho, D_j)$ in the RGD of the tunnel longitudinal structure is formulated as follows.

$$C(\boldsymbol{d}) = c_c Q_c + c_s Q_s + c_b Q_b$$

where Q_c, Q_s, and Q_b represent the quantity of segment concrete (m^3), reinforcement steel bar (10^{3}kg), and joint bolt (kg) of the shield tunnel longitudinal structure, respectively. As can be seen, the cost estimate in this design example is identical to that in Design Example 2.

5.3.3.5 Robust Design Optimization Results of this Tunnel Longitudinal Structure

With the statistics of noise factors tabulated in Table 5.8, the performance of the example tunnel can readily be analyzed using the performance analysis framework presented in Section 5.3.3.2. Plotted in Figure 5.21 are the noise factors along the tunnel longitudinal direction that are generated with one MCS run. The variance of noise factors in the tunnel longitudinal domain makes determining critical tunnel cross-sections of the shield tunnel quite challenging. Here, the simplified 1D FEM solution model, developed by Huang et al. (2015), is adopted to analyze the longitudinal behavior of this shield tunnel. The results of the 1D FEM analysis, including tunnel settlement (w), longitudinal rotation (θ_L), longitudinal bending moment (M_L),

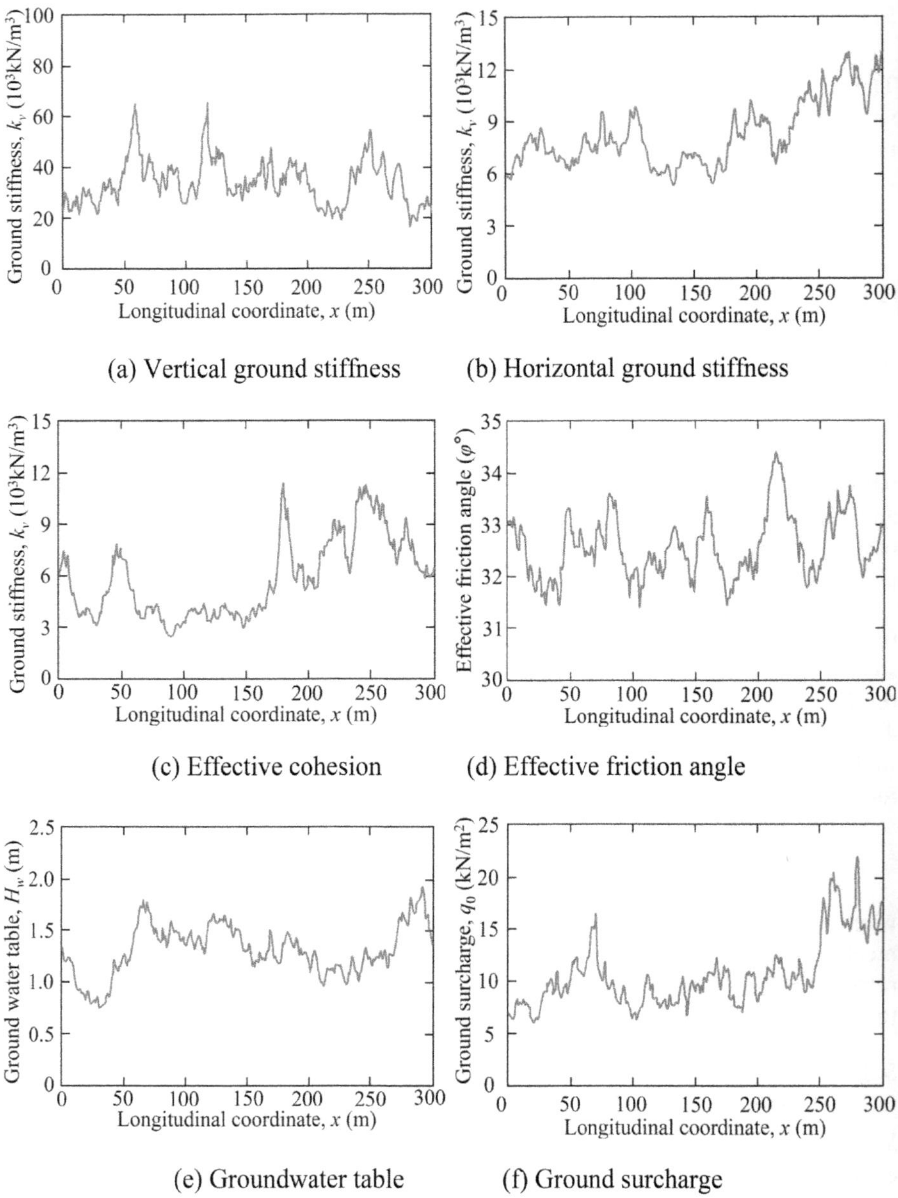

FIGURE 5.21 Illustrative longitudinal variation of noise factors (modified from Figure 5 in Gong et al. 2015a).

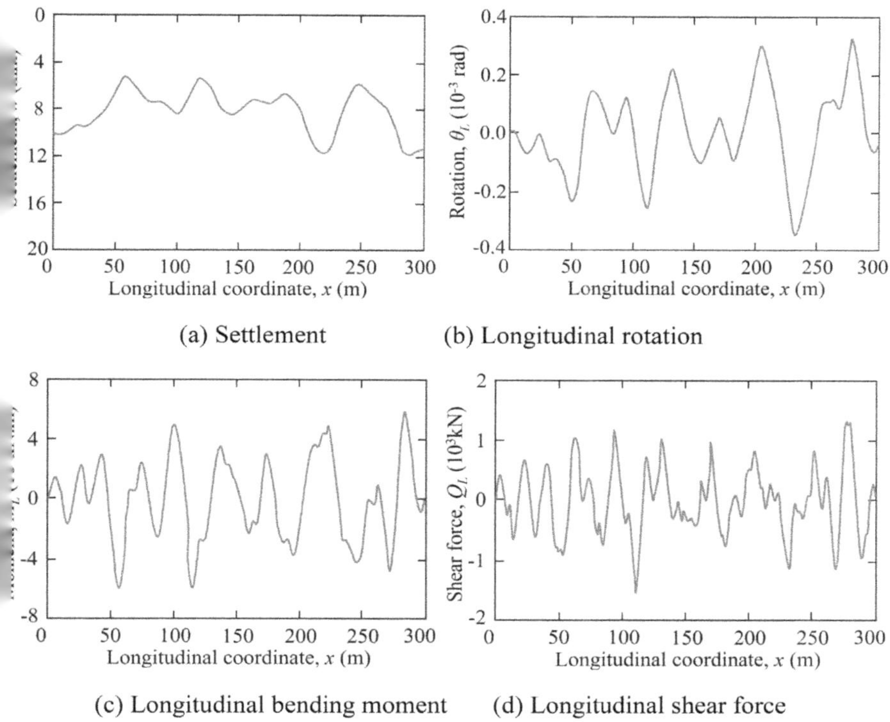

GURE 5.22 Computed tunnel longitudinal behavior, given the noise factors in Figure 5.21 nodified from Figure 6 in Gong et al. 2015a).

nd longitudinal shear force (Q_L), are illustrated in Figure 5.22. As can be expected, n evident longitudinal variation of the tunnel longitudinal behavior exists. It should e noted that the design parameters of the tunnel performance analysis shown in igure 5.22 are $t = 0.35$ m, $\rho = 0.5\%$, $D_c = 30$ mm, and $D_l = 30$, which are compa-tble with the design parameters of typical Shanghai metro tunnels.

Based on the noise factors generated along the longitudinal direction (see igure 5.21) and the computed tunnel longitudinal behavior (see Figure 5.22), the erformance (i.e., structure safety Fs_1 and serviceability Fs_2) of each segment ring f this tunnel can be analyzed using the 2D analytical solution model mentioned in **ection 5.3.3.2** (Gong et al. 2015a). Figure 5.23 shows the resulting longitudinal ariation of the tunnel performance, in terms of Fs_1 and Fs_2 (with respect to the ructure safety and serviceability, respectively) of each segment ring. It is noted hat different performances are exhibited upon different segment rings, which are eflected by the longitudinal variation of the factors of safety (i.e., Fs_1 and Fs_2). hus, the performance of the tunnel longitudinal structure cannot be assessed with he results of a few representative segment rings, as is adopted in the conventional eotechnical practice.

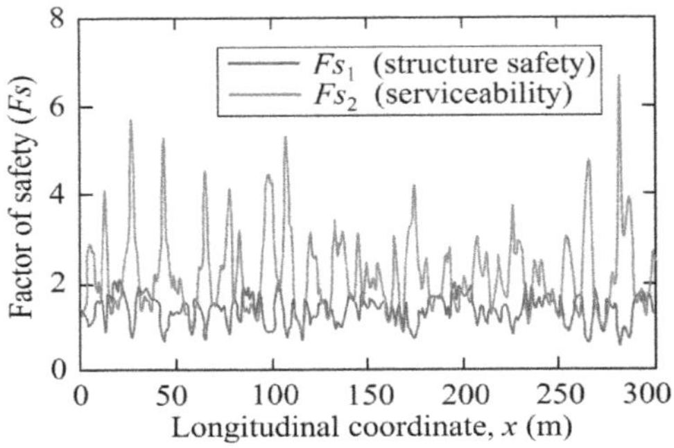

FIGURE 5.23 Assessed tunnel performance given the noise factors shown in Figure 5.: (modified from Figure 7 in Gong et al. 2015a).

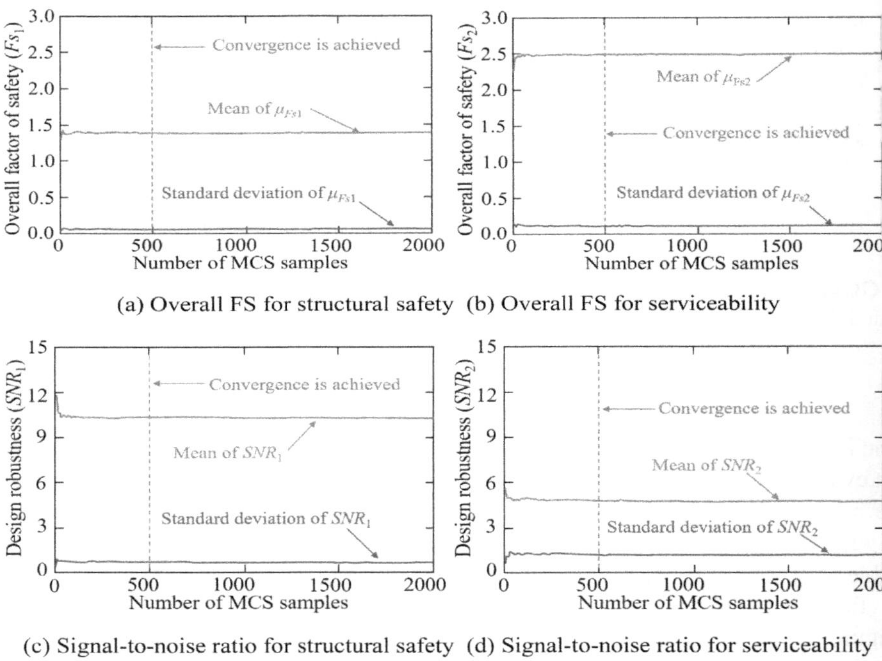

FIGURE 5.24 Convergence of the overall factors of safety and signal-to-noise ratio with tł number of MCS samples (modified from Figure 8 in Gong et al. 2015a).

Note that different MCS runs could generate different longitudinal variations c noise factors, which in turn result in an expected variance in the longitudinal var ation of tunnel performance. For example, the longitudinal variation of tunnel pe formance shown in Figure 5.23 is the analysis result of the longitudinal variation c noise factors shown in Figure 5.21. Therefore, a series of MCS runs must be coı ducted to derive a converged solution of tunnel performance, in terms of the mea

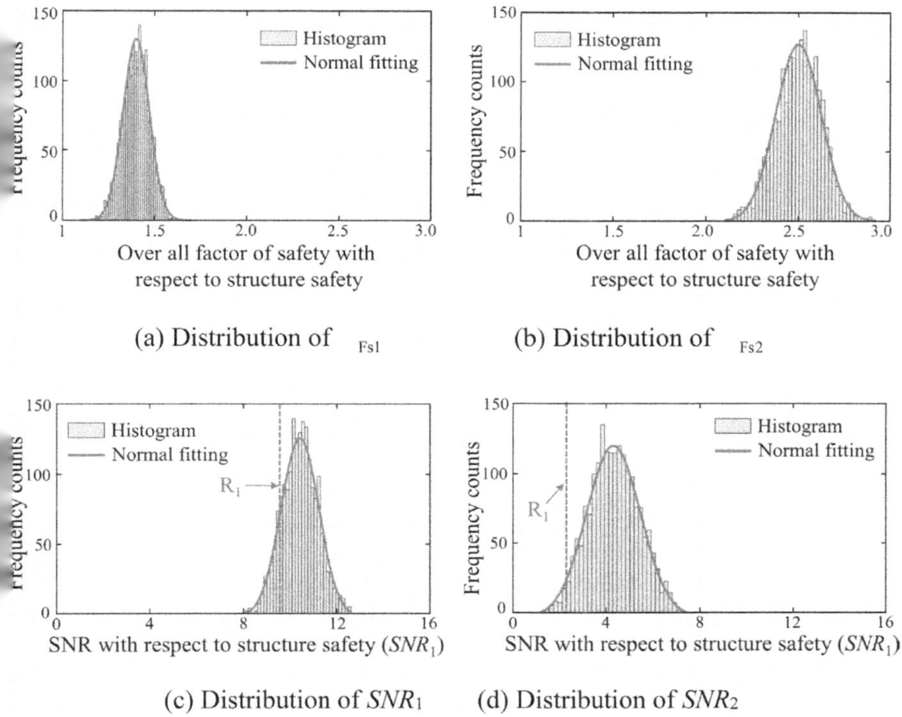

IGURE 5.25 Distributions of the overall factors of safety and *SNR* in 2,000 MCS runs modified from Figure 9 in Gong et al. 2015a).

ind standard deviation of tunnel performance (i.e., μ_{Fs1}, μ_{Fs2}, σ_{Fs1}, and σ_{Fs2}), as per Step 5 of the analysis framework (Section 5.3.3.2). Figures 5.24(a) and (b) show the derived values of the mean and standard deviation of the overall FSs (i.e., μ_{Fs1} and μ_{Fs2}) with different numbers of MCS samples.

It is noted that a converged solution of the mean and standard deviation of the overall FSs may be achieved with 500 MCS samples. Figures 5.24(c) and (d) depict the estimated values of the mean and standard deviation of the SNRs (i.e., SNR_1 and SNR_2) with different numbers of MCS samples. The converged solution of the mean and standard deviation of the SNRs could also be achieved with 500 MCS samples. Thus, the number of MCS samples in this example is set at 500 for sampling the longitudinal variation of noise factors. It is observed that the overall FSs (i.e., μ_{Fs1} and μ_{Fs2}) and SNRs (i.e., SNR_1 and SNR_2) of this tunnel longitudinal structure fit well with normal distributions, as shown in Figure 5.25.

Since a discrete design space (**DS**) shown in Table 5.9 is adopted in this design example, the tunnel performance (i.e., structure safety and serviceability of all segment rings), robustness, and cost of all candidate designs can be assessed, and the robust design optimization is then performed. As shown in Figure 5.26, this optimization yields a series of non-dominated optimal designs, not a single best design.

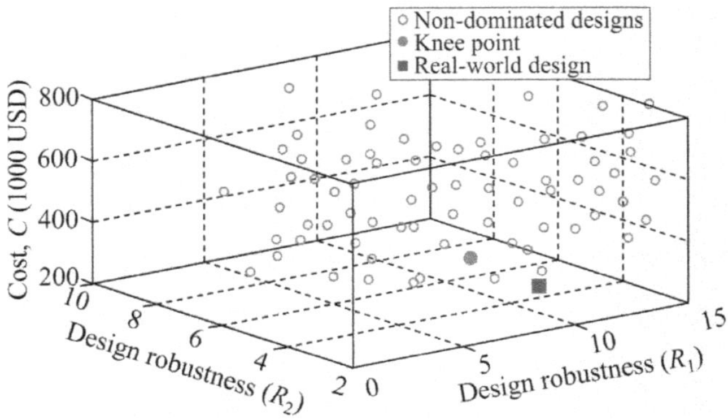

FIGURE 5.26 Robust design optimization results of the tunnel in Design Example 2 (modified from Figure 14 in Gong et al. 2015a).

These non-dominated optimal designs collectively form a 3D Pareto front, as three objectives are involved in this robust design optimization.

In a design decision involving a tradeoff relationship between design robustness and cost efficiency, the determination of an appropriate target level of design robustness or cost is problem-specific and engineer-specific. If no design preference is specified, the knee point on the Pareto front may be taken as the most preferred design (e.g., Juang et al. 2014; Gong et al. 2014b, 2015b), and the marginal utility function approach (e.g., Branke et al. 2004; Gong et al. 2014c) is employed here to locate this knee point. The identified knee point is also plotted in Figure 5.26. Note that on the upper side of this knee point, a slight improvement in the design robustness would significantly increase the cost; on the other side of this knee point, a slight reduction in cost would, however, require a considerable sacrifice in the robustness. Also plotted in Figure 5.26 is the real-world design adopted in Shanghai metro tunnels. It is noted that greater robustness (indicated by a larger R) and greater cost efficiency (indicated by a lower C) can be achieved with this knee point obtained from the robust design. It should be interesting to compare this robust design with the real-world design adopted in Shanghai. A detailed comparison of these two designs is given in Table 5.11.

One of the features shown in Table 5.11 is the reliability-based FI, which is the likelihood that the tunnel could meet the performance requirement (i.e., structure safety and serviceability of all segment rings). This reliability-based FI is defined as follows (e.g., Juang et al. 2014; Juang and Wang 2013).

$$FI_1 = \Pr\left[Fs_1 > 1.0\right]$$

$$FI_2 = \Pr\left[Fs_2 > 1.0\right]$$

where FI_1 and FI_2 denote the reliability-based FIs with respect to the structure safety (ULS performance) and serviceability (SLS performance) of the

TABLE 5.11
Comparison Between the Knee Point in Robust Design and the Actual Design

	Easy-to-control design parameters (d)				Design feasibility		Tunnel performance		Design objectives		
Designs	t (m)	ρ (%)	D_c (MM)	D_l (MM)	FI_1	FI_2	Mean of μ_{Fs1}	Mean of μ_{Fs2}	*Cost, C* (USD 1,000)	*Robustness* R_1	*Robustness* R_2
Knee point	0.35	0.5	20	30	0.9607	0.9033	1.90	1.80	309.57	9.59	4.89
Real-word design	0.35	0.5	30	30	0.7846	0.9695	1.39	2.50	316.67	8.78	2.31

tunnel, respectively; and Pr[$Fs > 1.0$] represents the conditional probability that the computed FS is greater than 1.0 in the face of the unforeseen longitudinal variation of noise factors. Indeed, this reliability-based FI could also be interpreted as the percentage of segment rings that meet the tunnel performance requirement (i.e., $Fs > 1.0$).

Compared to the real-world design in Shanghai, the bolt diameter of the circumferential joints (D_c) of the knee-point design, obtained from the robust design optimization, is significantly reduced from 30 mm to 20 mm. The reliability-based FI with respect to structure safety (FI_1), tunnel performance with respect to structure safety (in terms of the mean of μ_{Fs1}), cost efficiency (C), and robustness with respect to both structure safety (R_1) and serviceability (R_2) are, however, significantly improved. For example, the FI with respect to structure safety (FI_1) is increased from 0.7846 to 0.9607, and the robustness with respect to serviceability (R_2) is increased from 2.31 to 4.89. On the other hand, both the reliability-based FI and the tunnel performance with respect to serviceability (i.e., FI_2 and the mean of μ_{Fs2}) are only degraded slightly. Nevertheless, the tunnel performance requirement can still be satisfied.

It should be noted that the most preferred design in the design space (i.e., knee point on the Pareto front) may vary with the robust design optimization settings (e.g., cost-estimate model, design space, and target factors of safety). Here, the robust design optimization of this shield tunnel is reanalyzed and redesigned with different target factors of safety with respect to the tunnel performance. The optimization results are listed in Table 5.12. The results show that different robust design optimization settings can produce different final designs of this tunnel. For a given cost-estimate model and design space, as the target factors of safety increase, the reliability-based FIs (i.e., FI_1 and FI_2), tunnel performance (i.e., the means of μ_{Fs1} and μ_{Fs2}), and cost (C) of the most preferred designs obtained tend to increase. In contrast, the number of non-dominated optimal designs tends to decrease.

TABLE 5.12
Robust Design Optimization Results of the Example Tunnel with Different Target Factors of Safety

	Easy-to-control design parameters (d)				Design feasibility		Tunnel performance		Number of non-dominated optimal designs	Design objectives		
Target factors of safety	t (m)	ρ (%)	D_c (MM)	D_l (MM)	FI_1	FI_2	Mean of μ_{Fs1}	Mean of μ_{Fs2}		Cost, C (USD 1,000)	Robustness R_1	Robustness R_2
1.5	0.35	0.50	20	30	0.9607	0.9033	1.90	1.80	71	309.57	9.59	4.89
2.0	0.40	0.50	20	30	0.9931	0.9643	2.12	2.10	47	353.18	11.09	4.61
2.5	0.40	1.00	25	35	0.9999	0.9963	3.39	2.53	42	474.85	10.76	4.59
3.0	0.45	1.00	25	35	1.0000	0.9997	3.94	3.04	17	534.22	11.74	4.34

.4 SUMMARY AND REMARKS

ne design situation where only the potential distribution intervals of the geotechnical arameters can be ascertained is common in geotechnical practice. These distribution tervals provide much more knowledge of the uncertainty of the geotechnical parameters than the characteristic or nominal values. Since the potential ranges or variations the noise factors cannot be fully exploited with the robustness measure of gradient-based SI (Chapter 4), the Taguchi SNR-based robustness measures are presented this chapter. The uncertainties of the input geotechnical parameters are alternatively odeled with the fuzzy set theory. For a geotechnical design with fuzzy sets as inputs, e design performance can be studied with the vertex method. The resulting fuzzy esign performance allows for an estimate of the statistics of the design performance, om which the design safety (i.e., mean of design performance and probability of failure) and design robustness (i.e., Taguchi SNR) can be evaluated. With the evaluated esign safety and robustness, the RGD can readily be implemented with the multi-objective optimization setting established in Chapter 3. Three geotechnical design examples are presented to demonstrate the application of this SNR-based RGD approach.

The results of these geotechnical design examples indicate that the Taguchi SNR-ased robustness measures are compatible with the conventional FS-based design and BD. Thus, the SNR-based RGD may also be considered a complementary design pproach to these two conventional design approaches. However, the SNR-based obustness measure captures more knowledge of the uncertainty of the input geotechnical parameters and the design performance than the gradient-based SI. Since e uncertainties of the site-specific geotechnical parameters are characterized by uzzy sets, only limited site investigation effort is required to implement this SNR-ased RGD. Thus, this design approach is suitable in a design situation where only e potential distribution intervals of the geotechnical parameters can be ascertained.

EFERENCES

merican Transportation Research Board (ATRB). (2000) *Design and construction of transportation facilities*. Washington, DC.

ng, A.H., and Tang, W.H. (2004) *Probability concepts in engineering* (2nd edition). John Wiley and Sons, New York.

inder, K., and Heermann, D. (2010) *Monte Carlo simulation in statistical physics: An introduction*. Springer, Berlin, Germany.

ishop, A.W. (1955) The use of the slip circle in the stability analysis of slopes. *Géotechnique*, 5, 7–17.

ranke, J., Deb, K., Dierolf, H., and Osswald, M. (2004) Finding knees in multi-objective optimization. *Parallel Problem Solving from Nature-PPSN*, VIII, 722–731.

hen, W., Allen, J.K., Tsui, K.L., and Mistree, F. (1996) A procedure for robust design: Minimizing variations caused by noise factors and control factors. *Journal of Mechanical Design, Transactions of the ASME*, 118(4), 478–485.

hing, J., Phoon, K.K., and Hu, Y.G. (2009) Efficient evaluation of reliability for slopes with circular slip surfaces using importance sampling. *Journal of Geotechnical and Geoenvironmental Engineering*, 135(6), 768–777.

hristian, J.T. (2004) Geotechnical engineering reliability: How well do we know what we are doing? *Journal of Geotechnical and Geoenvironmental Engineering*, 130(10), 985–1003.

eb, K., and Gupta, S. (2011) Understanding knee points in bicriteria problems and their implications as preferred solution principles. *Engineering Optimization*, 43(11), 1175–1204.

Deb, K., Pratap, A., Agarwal, S., and Meyarivan, T.A.M.T. (2002) A fast and elitist multio jective genetic algorithm: NSGA-II. *IEEE Transactions on Evolutionary Computatio* 6(2), 182–197.

Dodagoudar, G.R., and Venkatachalam, G. (2000) Reliability analysis of slopes using fuz sets theory. *Computers and Geotechnics*, 27(2), 101–115.

Dong, W.M., and Wong, F.S. (1987) Fuzzy weighted averages and implementation of t extension principle. *Fuzzy Sets and Systems*, 21(2), 183–199.

Du, X., and Chen, W. (2000) Towards a better understanding of modeling feasibility robustne in engineering design. *Journal of Mechanical Design*, 122, 385–394.

Duncan, J.M. (2000) Factors of safety and reliability in geotechnical engineering. *Journal Geotechnical and Geoenvironmental Engineering*, 126(4), 307–316.

El-Ramly, H., Morgenstern, N.R., and Cruden, D.M. (2005) Probabilistic assessment of t stability of a cut slope in residual soil. *Géotechnique*, 55(1), 77–84.

Fellenius, W. (1936) Calculation of the stability of earth dams. In *Proceedings of transactio of 2nd congress on large dams*. International Commission on Large Dams (ICOLD 445–462.

Giasi, C.I., Masi, P., and Cherubini, C. (2003) Probabilistic and fuzzy reliability analysis of sample slope near Aliano. *Engineering Geology*, 67(3), 391–402.

Gong, W., Huang, H., Juang, C.H., Atamturktur, S., and Brownlow, A. (2015a) Improv shield tunnel design methodology incorporating design robustness. *Canadian Geotec nical Journal*, 52(10), 1575–1591.

Gong, W., Juang, C.H., Huang, H., Zhang, J., and Luo, Z. (2015b) Improved analytical mod for circumferential behavior of jointed shield tunnels considering the longitudinal diffe ential settlement. *Tunnelling and Underground Space Technology*, 45, 153–165.

Gong, W., Khoshnevisan, S., and Juang, C.H. (2014a) Gradient-based design robustness mea ure for robust geotechnical design. *Canadian Geotechnical Journal*, 51(11), 1331–134

Gong, W., Wang, L., Juang, C.H., Zhang, J., and Huang, H. (2014b) Robust geotechnic design of shield-driven tunnels. *Computers and Geotechnics*, 56, 191–201.

Gong, W., Wang, L., Khoshnevisan, S., Juang, C.H., Huang, H., and Zhang, J. (2014c) Robu geotechnical design of earth slopes using fuzzy sets. *ASCE Journal of Geotechnical ar Geoenvironmental Engineering*, 141(1), 04014084.

Greco, V.R. (1996) Efficient Monte Carlo technique for locating critical slip surface. *Journ of Geotechnical Engineering*, 122(7), 517–525.

Griffiths, D.V., and Lane, P.A. (1999) Slope stability analysis by finite elements. *Géotec nique*, 49(3), 387–403.

Huang, H., Gong, W., Khoshnevisan, S., Juang, C.H., Zhang, D., and Wang, L. (2015) Sin plified procedure for finite element analysis of the longitudinal performance of shie tunnels considering spatial soil variability in longitudinal direction. *Computers and Ge technics*, 64, 132–145.

Huang, X., Huang, H.W., and Zhang, J. (2012) Flattening of jointed shield-driven tunn induced by longitudinal differential settlements. *Tunnelling and Underground Spac Technology*, 31, 20–32.

International Tunnelling Association (ITA). (2000) Guidelines for the design of shield tunn lining. *Tunnelling and Underground Space Technology*, 15(3), 303–331.

Juang, C.H., Jhi, Y.Y., and Lee, D.H. (1998) Stability analysis of existing slopes consideri uncertainty. *Engineering Geology*, 49(2), 111–122.

Juang, C.H., Lee, D.H., and Sheu, C. (1992) Mapping slope failure potential using fuzzy set *Journal of Geotechnical Engineering*, 118(3), 475–94.

Juang, C.H., and Wang, L. (2013) Reliability-based robust geotechnical design of spread four dations using multi-objective genetic algorithm. *Computers and Geotechnics*, 48, 96–10

Juang, C.H., Wang, L., Hsieh, H.S., and Atamturktur, S. (2014) Robust geotechnical design c braced excavations in clays. *Structural Safety*, 49, 37–44.

Kim, J.Y., and Lee, S.R. (1997) An improved search strategy for the critical slip surface usin finite element stress fields. *Computers and Geotechnics*, 21(4), 295–313.

Koyama, Y. (2003) Present status and technology of shield tunneling method in Japan. *Tunnelling and Underground Space Technology*, 18(2), 145–59.

Lee, K.M., Hou, X.Y., Ge, X.W., and Tang, Y. (2001) An analytical solution for jointed shield-driven tunnel lining. *International Journal for Numerical and Analytical Methods in Geomechanics*, 25(4), 365–390.

Liao, S.M., Hou, X.Y., and Peng, F.L. (2005) Longitudinal shear transfer of tunnel and its 1D analytical solution. *Chinese Journal of Rock Mechanics and Engineering*, 24(7), 1110–1116 (in Chinese).

Liao, S.M., Peng, F.L., and Shen, S.L. (2008) Analysis of shearing effect on tunnel induced by load transfer along the longitudinal direction. *Tunnelling and Underground Space Technology*, 23(4), 421–430.

Luo, Z., Atamturktur, S., Juang, C.H., Huang, H., and Lin, P.S. (2011) Probability of serviceability failure in a braced excavation in a spatially random field: Fuzzy finite element approach. *Computers and Geotechnics*, 38(8), 1031–40.

MATLAB version 7.10.0. (2010) Natick, MA: The MathWorks Inc.

Ministry of Construction of the People's Republic of China (MCPRC). (2003) *Code for design of metro* (GB50157–2003). China Building Industry Press, Beijing (in Chinese).

Nagy, Z.K., and Braatz, R.D. (2004) Open-loop and closed-loop robust optimal control of batch processes using distributional and worst-case analysis. *Journal of Process Control*, 14(4), 411–422.

Park, G.J., Lee, T.H., Lee, K.H., and Hwang, K.H. (2006) Robust design: An overview. *AIAA Journal*, 44(1), 181–191.

Park, H.J., Um, J.G., Woo, I., and Kim, J.W. (2012) Application of fuzzy set theory to evaluate the probability of failure in rock slopes. *Engineering Geology*, 125, 92–101.

Phoon, K.K., and Kulhawy, F.H. (1999) Characterization of geotechnical variability. *Canadian Geotechnical Journal*, 36(4), 612–624.

Sadrekarimi, J., and Akbarzad, M. (2009) Comparative study of methods of determination of coefficient of subgrade reaction. *Electronic Journal of Geotechnical Engineering*, 14(1), 45–61.

Shanghai Municipal Construction Committee. (1999) *Foundation design code* (DGJ08–11–1999). Shanghai (in Chinese).

Shrestha, B., and Duckstein, L. (1998) A fuzzy reliability measure for engineering applications. In *Uncertainty modeling and analysis in civil engineering*. Boca Raton: CRC Press, 121–135.

Sonmez, H., Gokceoglu, C., and Ulusay, R. (2003) An application of fuzzy sets to the geological strength index (GSI) system used in rock engineering. *Engineering Applications of Artificial Intelligence*, 16(3), 251–69.

Spencer, E. (1967) A method of analysis of the stability of embankments assuming parallel inter-slice forces. *Géotechnique*, 17(1), 11–26.

Taguchi, G. (1986) *Introduction to quality engineering: Designing quality into products and processes*. Quality Resources, White Plains, New York.

Tsui, K.L. (1992) An overview of Taguchi method and newly developed statistical methods for robust design. *IIE Transactions*, 24(5), 44–57.

Turnbull, W.J., and Hvorslev, M.J. (1967) Special problems in slope stability. *ASCE Journal of Soil Mechanics and Foundations*, 93, 499–528.

Winkler, E. (1867) *Theory of elasticity and strength*. Dominicus Prague, Czechoslovakia.

Wood, A.M.M. (1975) The circular tunnel in the elastic ground. *Géotechnique*, 25(1), 115–27.

Zadeh, L.A. (1965) Fuzzy sets. *Information and Control*, 8(3), 338–353.

Zhang, J., Zhang, L.M., and Tang, W.H. (2011) Reliability-based optimization of geotechnical systems. *Journal of Geotechnical and Geoenvironmental Engineering*, 137(12), 1211–1221.

Zhao, Y.G., and Ono, T. (2000) New point estimates for probability moments. *Journal of Engineering Mechanics*, 126(4), 433–436.

Zhu, S., and Fukushima, M. (2009) Worst-case conditional value-at-risk with application to robust portfolio management. *Operations Research*, 57(5), 1155–1168.

6 Feasibility Index-Based Robust Geotechnical Design

6.1 INTRODUCTION

Traditional geotechnical design often adopts a trial-and-error procedure to identify the acceptable designs and optimize them with regard to cost efficiency, regardless of whether the factor of safety (FS)-based design, reliability-based design (RBD), or load and resistance factor design (LRFD) is adopted. However, the traditional designs focus mainly on the requirements of safety and cost efficiency, while the robustness of the design performance is not explicitly considered. Robust design, as a design approach originated in the field of industrial engineering (Taguchi 1986; Chen et al. 1996), aims to achieve the robustness of the design against hard-to-control and hard-to-characterize noise factors by adjusting easy-to-control design parameters. Due to the limited site-specific data availability and incomplete knowledge of human being, accurate characterization of the statistical information of geotechnical parameters is a challenge. Thus, the effectiveness of the conventional RBD, which explicitly considers the uncertainties of input parameters and solution models, may be degraded. To this end, the feasibility index (FI)-based robust geotechnical design (RGD) could be adopted (Juang et al. 2012, 2013; Juang and Wang 2013; Wang et al. 2013). Note that the noise factors in this RGD framework are the statistical information of geotechnical parameters, and the design parameters are those that the engineer can easily control.

Statistical characterization of a geotechnical parameter typically involves an estimate of the mean, standard deviation (or coefficient of variation, COV), distribution type, and spatial correlation structure. Among these statistical features, the mean can be easily estimated even with limited site-specific data availability. In contrast, the estimate of the standard deviation and the spatial correlation structure requires a large amount of site-specific data (e.g., Phoon et al. 1995; Gong et al. 2017). The distribution type of a geotechnical parameter can be determined utilizing techniques such as the Chi-square or Kolmogorov-Smirnov test (Nikulin 1974) when sufficient site-specific data are available. However, in most geotechnical practices, oftentimes, only limited site investigation efforts can be afforded. As a result, the statistical information of geotechnical parameters in most geotechnical designs cannot be ascertained, although the Bayesian methods (e.g., Gong et al. 2017) may be employed to derive data from similar sites, literature, local experience, and engineering judgment.

The FI-based RGD is complementary to the RBD. The outcome of the RBD depends significantly on the statistical characterization of geotechnical parameters.

DOI: 10.1201/9781003564454-6

When the statistical information of uncertain geotechnical parameters is fully characterized, the RBD can be easily executed by identifying the least-cost design that meets the target reliability index or failure probability requirement. However, in real-world projects, the statistical information of geotechnical parameters is hard to quantify due to incomplete data, measurement errors, and transformation errors. The statistical information of geotechnical parameters may be either overestimated or underestimated, which could result in over-design or under-design. The FI-based RGD aims to achieve robustness against the uncertainties in the statistical information of geotechnical parameters (e.g., statistics and distributions of geotechnical parameters and correlations among geotechnical parameters). Further, it employs the feasibility concept in decision-making. It is well known that the probability of failure in a geotechnical design can be greatly affected by the statistical information of geotechnical parameters. The FI indicates the probability that a geotechnical design is capable of meeting the target reliability index (or failure probability) even with uncertain statistical information on geotechnical parameters.

This chapter focuses on the development of the robustness measure of FI and its application in RGD. This RGD can be applicable to situations where the statistical information of the hard-to-control and hard-to-characterize geotechnical parameters cannot be ascertained due to limited site-specific data. This chapter consists of three parts. First, the uncertainties in the statistical characterization of the geotechnical parameters are discussed, and the establishment of the robustness measure of FI is presented. Next, three geotechnical design examples are presented to demonstrate the FI-based RGD. Finally, the concluding remarks are drawn.

6.2 PRINCIPLE OF FEASIBILITY INDEX-BASED ROBUST GEOTECHNICAL DESIGN

The principle of this FI-based RGD is presented in this section. First, the uncertainties in the statistical characterization of the geotechnical parameters are discussed. Second, the formulation of the robustness measure of FI is outlined. Third, the implementation procedures of the FI-based RGD are presented.

6.2.1 Uncertainties in the Statistical Characterization of Geotechnical Parameters

In most geotechnical projects, geotechnical parameters are often assessed based on a handful of site-specific investigation data. There are considerable and inevitable uncertainties in the statistical characterization of these geotechnical parameters derived from a small sample. The "population" mean of geotechnical parameters can generally be adequately assessed based on the "sample" mean even with a handful of sampling points (e.g., Wu et al. 1989; Gong et al. 2017). However, there could be a considerable error in the evaluation of the COV and spatial correlation structure for the population based on a small sample. Additionally, there exist considerable uncertainties in the model errors for the analytical and numerical models for evaluating the geotechnical performance, as well as the transformation errors to derive the geotechnical parameters based on field testing results.

6.2.1.1 Three-Sigma Bounds for Estimating Uncertainties in the Statistical Characterization

The variability of a geotechnical parameter, in terms of COV, may be estimated with the following approaches (Duncan 2000): 1) direct statistical assessment from site-specific data based on the moment method, 2) estimate based on the three-sigma bounds or its variant, or 3) adoption of published COVs available in the literature. The extent of variation in the COVs of geotechnical parameters may be quantified based on the ranges reported in the literature. Using the design of drilled shafts in loose sand as an example. The COV of the effective friction angle (φ') of loose sand, denoted as $COV[\varphi']$, has a typical range from 0.05 to 0.10 (Amundaray 1994), while the COV of the coefficient of earth pressure at rest (K_0), denoted as $COV\left[K_0\right]$, has a typical range from 0.20 to 0.90 (Phoon et al. 1995). The results of an RBD strongly depend on the estimated COVs of these two geotechnical parameters. In the face of the uncertainties of the statistical characterization, we need to quantify the uncertainties in the COVs (of the uncertain geotechnical parameters) so that we can evaluate the robustness of the design performance of concern (e.g., the estimated failure probability) and then achieve the design that meets the target failure probability requirements even with the uncertainties on the estimated COVs.

Since the typical ranges of COVs of geotechnical parameters are usually available in the literature, the uncertainty in the estimated COVs may be characterized. Using the same drilled shaft in loose sand as an example, based on the typical range of $COV[\varphi']$ from 0.05 to 0.10, the mean and standard deviation of $COV[\varphi']$, denoted as $\mu_{COV[\varphi']}$ and $\sigma_{COV[\varphi']}$, respectively, may be readily estimated using the "two-sigma bounds" method. Here, we treat 0.05 as the lowest conceivable value (LCV) and 0.10 as the highest conceivable value (HCV). From the two-sigma bounds and the reported range, it is determined that $\mu_{COV[\varphi']} \approx 0.07$ and $\sigma_{COV[\varphi']} \approx (0.10-0.05)/4 = 0.0125$. It is noted that the "two-sigma bounds" method is handy for cases with limited site-specific data, while the typical range of COV is available in the literature.

Following a similar procedure for this drilled shaft problem, based on the range of $COV\left[K_0\right]$ from 0.20 to 0.90 reported in the literature (e.g., Phoon et al. 1995), the mean and standard deviation of $COV\left[K_0\right]$, denoted as $\mu_{COV[K_0]}$ and $\sigma_{COV[K_0]}$, respectively, may be calculated as $\mu_{COV[K_0]} \approx 0.50$ and $\sigma_{COV[K_0]} \approx (0.9–0.2)/4 = 0.175$. In addition, the uncertainties in the correlation between these two geotechnical parameters may also be considered. For example, the parameters φ' and K_0 of loose sands are typically negatively correlated, and the correlation coefficient, denoted as ρ_{φ',K_0}, is generally in the range from -0.6 to -0.9 (Mayne and Kulhawy 1982; Juang et al. 2013). Following the "two-sigma bounds" method, the mean and standard deviation of the correlation coefficient ρ_{φ',K_0}, denoted as $\mu_{\rho_{\varphi',K_0}}$ and $\sigma_{\rho_{\varphi',K_0}}$, respectively, are determined to be −0.75 and 0.075, respectively. Based on the uncertainty characterization of the COVs of the geotechnical parameters φ' and K_0 and their correlation coefficient ρ_{φ',K_0}, probabilistic analyses can be performed in the RBD framework to achieve the design robustness. **Example 1** demonstrates the characterization of the uncertainty of COVs of soil parameters using the three-sigma bounds method.

Example 1

The first example focuses on the design of a spread foundation that rests on a stiff till. For the geotechnical design of this spread foundation, considering the ultimate limit state (ULS) and serviceability limit state (SLS) requirements, three soil parameters, including the effective friction angle (φ'), undrained shear strength (c_u), and coefficient of volume compressibility (m_v), are taken as the main parameters of concern in the RGD. It should be noted that the first soil parameter is the key parameter for the ULS bearing capacity evaluation, and the latter two are key parameters for the SLS failure evaluation (Wang 2011). Because of the budget constraints, only a limited effort could be made to conduct a site investigation to characterize these geotechnical parameters. Generally, the mean values for soil parameters can be adequately estimated based on the interpretation of testing data. However, the site-specific coefficients of variation (COVs) of these three soil parameters are usually unavailable. In this regard, they have to be estimated based on the published literature with engineering judgment. Here, the COV of effective friction angle (φ'), denoted as COV[φ'], ranges from 4% to 11%; the COV of undrained shear strength *(cu)*, denoted as $\mathrm{COV}\left[c_u\right]$, ranges from 20% to 40%; and the COV of volume compressibility coefficient (m_v), denoted as $\mathrm{COV}\left[m_v\right]$, ranges from 20% to 40%. The problem here is to evaluate the mean and standard deviation of the statistics COV[φ'], $\mathrm{COV}\left[c_u\right]$, and $\mathrm{COV}\left[m_v\right]$ based on the three-sigma bounds method.

Solution:

According to the information provided in the problem, the ranges of COVs of the uncertain soil parameters (i.e., effective friction angle φ', undrained shear strength c_u, and coefficient of volume compressibility m_v) are listed as follows.

$$\mathrm{COV}\left[\varphi'\right] = 4\%-11\%,\ \mathrm{COV}\left[c_u\right] = 20\%-40\%,\ \mathrm{COV}\left[m_v\right] = 20\%-40\%.$$

Step 1: Derive the formulas for the three-sigma bounds method

The "three-sigma bounds" method assumes that the LCV and HCV of a parameter correspond to three standard deviations below and above the mean. Under the three-sigma rule, we treat the bounds as:

$$\mathrm{LCV} \approx \mu - 3\sigma,\ \mathrm{HCV} \approx \mu + 3\sigma.$$

Then, the mean (μ) and standard deviation (σ) can be estimated as follows.

$$\mu \approx \frac{1}{2}(\mathrm{LCV} + \mathrm{HCV}),\ \sigma \approx \frac{1}{6}(\mathrm{HCV} - \mathrm{LCV})$$

Step 2: Calculate statistics of the COV of effective friction angle (*COV*[φ'])

Range: 4%–11%. LCV = 4%, HCV = 11%.

Mean of COV for effective friction angle (COV[φ']) is estimated below.

$$\mu_{COV[\varphi']} \approx \frac{4\% + 11\%}{2} = 7.5\%$$

Standard deviation of COV for effective friction angle (COV[φ']) is estimated below.

$$\sigma_{COV[\varphi']} \approx \frac{11\% - 4\%}{6} = 1.17\%$$

Step 3: Calculate statistics of the COV of undrained shear strength $\boldsymbol{COV}[\boldsymbol{cu}]$)

Range: 20%–40%. LCV = 20%, HCV = 40%.

Mean of COV for undrained shear strength (COV$[c_u]$) is estimated below.

$$\mu_{COV[c_u]} \approx \frac{20\% + 40\%}{2} = 30\%$$

Standard deviation of COV for undrained shear strength (COV$[c_u]$) is estimated below.

$$\sigma_{COV[c_u]} \approx \frac{40\% - 20\%}{6} = 3.33\%$$

Step 4: Calculate statistics of the COV of volume compressibility ($\boldsymbol{COV}[\boldsymbol{M}]$)

Range: 20%–40%. LCV = 20%, HCV = 40%.

Mean of COV for volume compressibility (COV$[m_v]$) is estimated below.

$$\mu_{COV[m_v]} \approx \frac{20\% + 40\%}{2} = 30\%$$

Standard deviation of COV for volume compressibility (COV$[m_v]$) is estimated below.

$$\sigma_{COV[m_v]} \approx \frac{40\% - 20\%}{6} = 3.33\%$$

6.2.1.2 Bootstrapping for Estimating Uncertainties in the Statistical Characterization

In addition to the three-sigma bounds estimation, the non-parametric bootstra method (Amundaray 1994; Luo et al. 2012b) may also be adopted to evaluate th mean and standard deviation of the COV of a geotechnical parameter by resamplin a large number of smaller samples through a statistical manner. Bootstrapping is re-sampling technique that yields an estimate of the mean and standard deviation c the sample statistics. In reference to Figure 6.1, the procedures for implementing this bootstrapping technique can be summarized as follows (Bourdeau and Amundaray 2005).

Step 1: A large number (N) of re-samples Y_j^* ($j = 1, k$) are generated from the original sample Y (with k data points) using "random sampling with replacement." Here, each data point (i.e., $y_{j,1}^*$) of Y_j^* can be assigned with the value of any data point of Y.

Step 2: For each re-sample Y_j^*, the statistics, such as mean and standard deviation of the interested statistics X_i (e.g., mean and standard deviation), are calculated.

Step 3: Step 2 is repeated N times. With the N set of statistics X_i, the mean (μ_{X_i}) and standard deviation (σ_{X_i}) of statistics X_i can be evaluated.

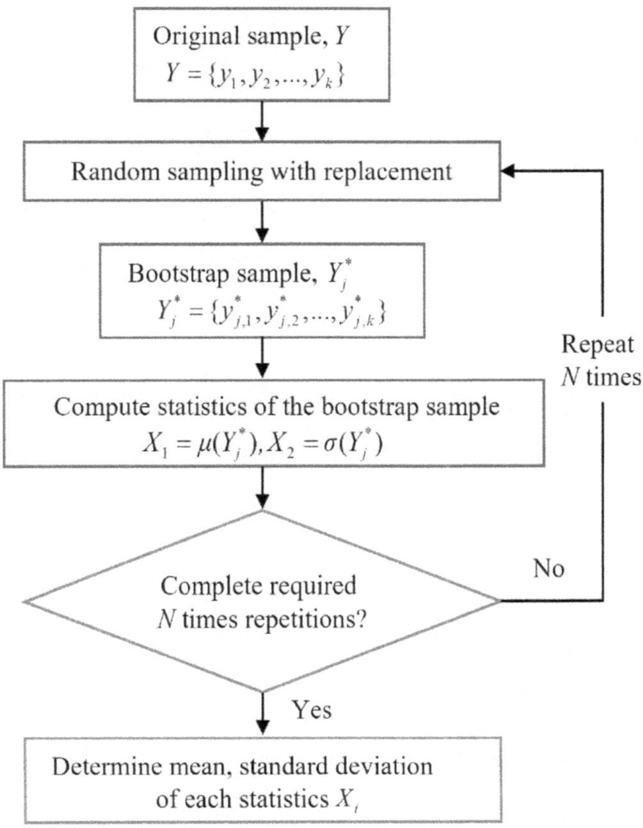

FIGURE 6.1 Flowchart of bootstrap procedures for characterizing uncertainty in sample statistics.

Example 2 demonstrates an exercise using the bootstrapping method for uncertainty characterization of sample statistics of the effective friction angle for the shallow foundation design. In addition to the uncertainty characterization of statistics of geotechnical parameters, the bootstrapping technique may also be employed for uncertainty quantification of the statistics of model error. Using the design of a shallow foundation as an example, a multiplicative model may be constructed to describe the model uncertainty for the ULS bearing capacity using a model bias factor (BF_Q):

$$BF_Q = \frac{\text{Observed capacity based on field load test}}{\text{Predicted model capacity}} = \frac{Q_o}{Q_P} \quad (6.1)$$

For the ULS bearing capacity of a shallow foundation, the predicted bearing capacity (Q_P) is one calculated by an analytical or empirical solution model, while the observed bearing capacity (Q_O) is the failure load interpreted from full-scale field load tests. Based on the database of field load tests compiled by Akbas and Kulhawy (2009b), the uncertainty in the mean (μ_{BF}) and standard deviation (σ_{BF}) of the model bias factor for the ULS bearing capacity may be estimated with the bootstrapping method. The results are shown in Table 6.1.

For the SLS failure, the uncertainty of the solution model is reflected in parameters a and b, in addition to the bias factor BF_Q. Here, the mean (μ_a) and standard deviation (σ_a) of parameter a, the mean (μ_b) and standard deviation (σ_b) of parameter b, and the correlation coefficient (ρ_{ab}) between parameters a and b are assessed using the database compiled by Akbas and Kulhawy (2009a). To this end, the bootstrapping method may be employed to evaluate the possible variation in these statistical parameters, and the characterization results are tabulated in Table 6.2.

TABLE 6.1
Sample Statistics of Model Bias Factor BF_Q for the ULS Failure Evaluation of Shallow Foundation by the Bootstrapping Method

Uncertain variables	μ_{BF}	σ_{BF}
Mean	1.010	0.203
Std. dev.	0.033	0.034

TABLE 6.2
Results from the Bootstrapping Method for Estimating Uncertainty in Statistics of Model Parameters a and b for the SLS Failure Evaluation of Shallow Foundation

Uncertain variables	μ_a	μ_b	σ_a	σ_b	ρ_{ab}
Mean	0.6992	1.7675	0.1549	0.9416	−0.7177
Std. dev.	0.0139	0.0845	0.0125	0.0794	0.0472

Example 2

This example addresses the design of a shallow foundation considering the uncertainty characterization for the statistics of the effective friction angle (φ') of sandy soil at a site. Ten sets of effective friction angles (φ') were obtained from the site investigation, and the data obtained are listed in Table 6.3. We are interested in evaluating the uncertainties of the mean and standard deviation of the effective friction angle (φ') derived from the small sample of data. The problem here is to characterize the uncertainties in the mean and standard deviation of the effective friction angle (φ') using the bootstrapping method.

Solution:

Because the available dataset is limited, the sample mean ($\overline{\varphi}'$) and sample standard deviation ($S_{\varphi'}$) of the effective friction angle (φ') cannot be treated as deterministic values but as random estimators that are affected by sampling uncertainty. To quantify this uncertainty, the bootstrap resampling method is applied. By repeatedly drawing resamples (of the same size) with replacement from the original dataset, we can obtain j^{th} bootstrap sample ($\varphi'^{*(j)}$) and calculate the corresponding mean ($\overline{\varphi}'^{*(j)}$) and standard deviation ($s_{\varphi'}^{*(j)}$) of this bootstrap sample. With a sufficiently large number of repetitions (N), stable sampling distributions of the mean $\overline{\varphi}'^{*(j)}$ and standard deviation $s_{\varphi'}^{*(j)}$ can be derived. Based on the distributions, the mean and standard deviation of the bootstrapped means (denoted as $\mu_{\overline{\varphi}'}$ and $\sigma_{\overline{\varphi}'}$) and those of the bootstrapped standard deviations (denoted as $\mu_{s_{\varphi'}}$ and $\sigma_{s_{\varphi'}}$) can be evaluated to capture the uncertainty in the sample mean ($\overline{\varphi}'$) and sample standard deviation ($S_{\varphi'}$) of the original data of the effective friction angle, respectively.

TABLE 6.3
Triaxial Test Results of Effective friction Angle

Test No.	Effective friction angle, φ'(°)
1	31.5
2	32.5
3	33.0
4	33.5
5	33.5
6	34.5
7	35.0
8	36.0
9	37.0
10	37.5

Source: Orr and Farrell, 1999.

In this example, the original dataset of the effective friction angle is $\boldsymbol{\varphi}'$ = {31.5°, 32.5°, 33.0°, 33.5°, 33.5°, 34.5°, 35.0°, 36.0°, 37.0°, 37.5°}, and the size of the sample is $k = 10$. With the aid of the bootstrapping method, we need to obtain: 1) mean ($\mu_{\bar{\varphi}'}$) and standard deviation ($\sigma_{\bar{\varphi}'}$) of the bootstrapped means; and 2) mean ($\mu_{s_{\varphi'}}$) and standard deviation ($\sigma_{s_{\varphi'}}$) of the bootstrapped standard deviations. Here, the number of bootstrap replications is set as 10,000 (i.e., N = 10,000) to ensure the convergence of mean and standard deviation of the bootstrapped means and standard deviations. To make the results reproducible, a random number generator seed is taken (e.g., random seed = 2025). This ensures that repeated runs of the bootstrap procedure will give identical outputs.

Step 1: Apply the bootstrap resampling scheme

For each bootstrap replication $j = 1, 2, 3, \ldots, N$, the following procedures are conducted: 1) draw k indices ($I_1^{*(j)}, I_2^{*(j)}, I_3^{*(j)}, \ldots, I_k^{*(j)}$) and these indices follow an independent and identically distributed (i.i.d.) uniform distribution over the set {1, 2, 3, . . . , k}; and 2) form the bootstrap sample $\boldsymbol{\varphi}'^{*(j)} = \{\varphi'_{I_1^{*(j)}}, \varphi'_{I_2^{*(j)}}, \varphi'_{I_3^{*(j)}}, \ldots, \varphi'_{I_k^{*(j)}}\}$, where $\varphi_i'^{*(j)}$ is the ith resampled value of the original dataset of the effective friction angle in the jth bootstrap replication, the subscript $I_i^{*(j)}$ is the ith index randomly drawn from an independent and identically distributed (i.i.d.) uniform distribution over the set {1, 2, 3, . . . , k} in the jth bootstrap replication. It should be noted that each bootstrap sample has the same size k as the original data but represents one possible realization of sampling variability.

Step 2: Compute the bootstrap statistics per replication

For j^{th} bootstrap sample ($\boldsymbol{\varphi}'^{*(j)}$), the mean ($\bar{\varphi}'^{*(j)}$) and standard deviation ($s_{\varphi'}^{*(j)}$) of this bootstrap sample can be estimated as follows.

$$\bar{\varphi}'^{*(j)} = \frac{1}{k}\sum_{i=1}^{k}\varphi_i'^{*(j)}, \quad s_{\varphi'}^{*(j)} = \sqrt{\frac{1}{k-1}\sum_{i=1}^{k}(\varphi_i'^{*(j)} - \bar{\varphi}'^{*(j)})^2}$$

This step yields two arrays with N replications, in terms of $\{\bar{\varphi}'^{*(j)}\}_{j=1}^{N} = \left\{\bar{\varphi}'^{*(1)}, \bar{\varphi}'^{*(2)}, \bar{\varphi}'^{*(3)} \ldots, \bar{\varphi}'^{*(N)}\right)$ and $\{s_{\varphi'}^{*(j)}\}_{j=1}^{N} = \left\{s_{\varphi'}^{*(1)}, s_{\varphi'}^{*(2)}, s_{\varphi'}^{*(3)}\right\}$, which represent the empirical sampling distributions of the estimators $\bar{\varphi}'$ and $s_{\varphi'}$.

Step 3: Summarize the distributions of bootstrap statistics

The overall mean and standard deviation of these bootstrap distributions are obtained as follows:

$$\mu_{\bar{\varphi}'} = \frac{1}{N}\sum_{j=1}^{N}\bar{\varphi}'^{*(j)}, \quad \sigma_{\bar{\varphi}'} = \sqrt{\frac{1}{N-1}\sum_{j=1}^{N}(\bar{\varphi}'^{*(j)} - \hat{\mu}_{\bar{\varphi}'})^2}$$

$$\mu_{s_{\varphi'}} = \frac{1}{N}\sum_{j=1}^{N} s_{\varphi'}^{*(j)}, \quad \sigma_{s_{\varphi'}} = \sqrt{\frac{1}{N-1}\sum_{j=1}^{N}(s_{\varphi'}^{*(j)} - \hat{\mu}_{s_{\varphi'}})^2}$$

where $\mu_{\bar{\varphi}'}$ represents the mean of the bootstrapped means; $\sigma_{\bar{\varphi}'}$ represents the standard deviation of the bootstrapped means; $\mu_{s_{\varphi'}}$ represents the mean of the bootstrapped standard deviations; and, $\sigma_{s_{\varphi'}}$ represents the standard deviation of the bootstrapped standard deviations.

Illustrated in Figure 6.2 are the histograms for the bootstrapped means $\{\bar{\varphi}'^{*(j)}\}_{j=1}^{N} = \{\bar{\varphi}'^{*(1)}, \bar{\varphi}'^{*(2)}, \bar{\varphi}'^{*(3)}, \ldots, \bar{\varphi}'^{*(N)}\}$ and the bootstrapped standard deviations $\{s_{\varphi'}^{*(j)}\}_{j=1}^{N} = \{s_{\varphi'}^{*(1)}, s_{\varphi'}^{*(2)}, s_{\varphi'}^{*(3)}, \ldots, s_{\varphi'}^{*(N)}\}$. In this example, the bootstrapped mean ($\bar{\varphi}'$) ranges from 32° to 37° while the bootstrapped standard deviation ($s_{\varphi'}^{*}$) ranges from 0.5° to 3°.

Based on the bootstrap samples generated, these statistics can readily be estimated: $\mu_{\bar{\varphi}'}$ = 34.4055°, $\sigma_{\bar{\varphi}'}$ = 0.5893°, $\mu_{s_{\varphi'}}$ =1.8387°, and $\sigma_{s_{\varphi'}}$ = 0.3329°.

The code for implementing the bootstrap method in this example is provided below.

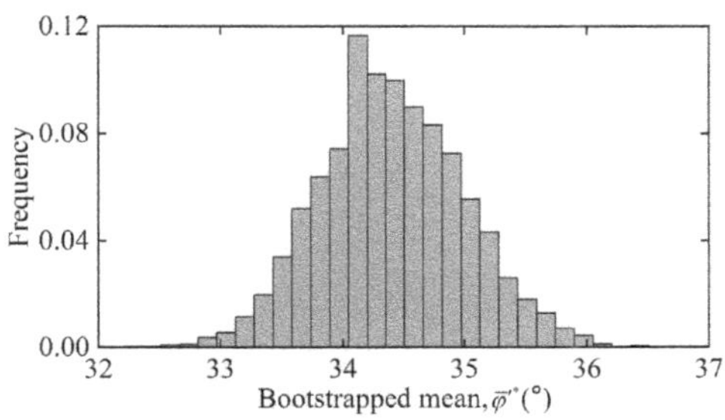

(a) Histogram of bootstrapped mean ($\bar{\varphi}'^{*}$)

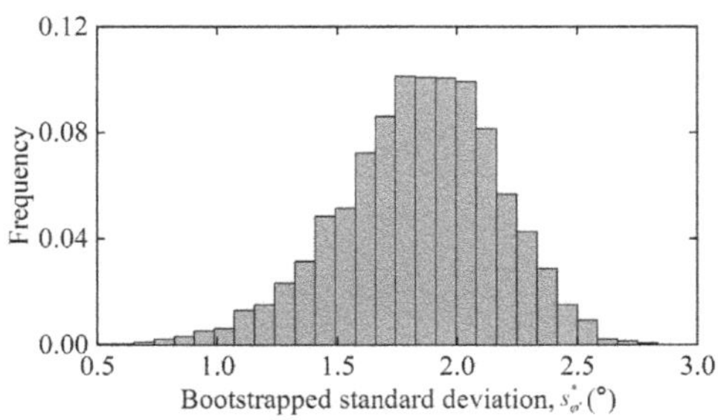

(b) Histogram of bootstrapped standard deviation ($s_{\varphi'}^{*}$)

FIGURE 6.2 Distributions of bootstrap statistics of the effective friction angle (φ') in Example 2.

```
function Bootstrapping
clc;
clear;
% Original data (effective friction angles)
data = [31.5, 32.5, 33.0, 33.5, 33.5, 34.5, 35.0, 36.0, 37.0, 37.5];
% Number of re-samples
N = 10000;
% Number of data points in the original sample
k = length(data);
% Initialize arrays to store bootstrapped statistics
boot_means = zeros(1, N);
boot_stds = zeros(1, N);
rng(2025,'twister'); % make results reproducible (optional)
% Step 1: Generate N re-samples and compute statistics
for j = 1:N
% Generate a bootstrap sample using random sampling with replacement
bootstrap_sample = data(randi(k, 1, k));
% Step 2: Compute the mean and standard deviation of the bootstrap sample
boot_means(j) = mean(bootstrap_sample);
boot_stds(j) = std(bootstrap_sample);
end
% Step 3: Calculate the mean and standard deviation of the bootstrapped statistics
mean_of_means = mean(boot_means);
std_of_means = std(boot_means);
mean_of_stds = mean(boot_stds);
std_of_stds = std(boot_stds);
% Display the results
disp('Bootstrapping Results:');
fprintf('Mean of bootstrapped means: %.4f\n', mean_of_means);
fprintf('Standard deviation of bootstrapped means: %.4f\n', std_of_means);
fprintf('Mean of bootstrapped standard deviations: %.4f\n', mean_of_stds);
fprintf('Standard deviation of bootstrapped standard deviations: %.4f\n', std_of_stds
% Plot the distribution of bootstrapped statistics
figure;
histogram(boot_means, 30, 'Normalization', 'probability');
xlabel('Bootstrapped mean, $\bar{\phi}"^{*}$', 'Interpreter', 'latex', 'FontSize', 12);
ylabel('Frequency');
xlim([32,37]);
set(gca,'FontSize',12,'FontName','Times New Roman');
figure;
histogram(boot_stds, 30, 'Normalization', 'probability');
xlabel('Bootstrapped standard deviation, $s_{\phi"}^{*}$','Interpreter', 'latex');
ylabel('Frequency');
xlim([0.5,3.0]);
xticklabels(arrayfun(@(x) sprintf('%.1f', x), 0.5:0.5:3.0, 'UniformOutput', false));
set(gca,'FontSize',12,'FontName','Times New Roman');
end
```

Example 3

This example focuses on the pad foundation design with available CPT data. The soil consists of a very dense fine glacial outwash sand with a mean particle size of 0.14 mm. The soil has a bulk weight density of 20 kN/m³. The groundwater is assumed at a great depth. Four CPT tests were carried out at the site with respect to the center of the proposed foundation (Hara et al. 2011). The tip resistance (q_c) profiles of the four CPT tests are shown in Figure 6.3. The effective friction angle (φ') in this example can be obtained by transformation from CPT q_c values following the empirical relationship by Kulhawy and Mayne (1990).

$$\varphi' = 17.6 + 11.0\log\left[\frac{(q_c / p_a)}{(\sigma'_{v0} / p_a)^{0.5}}\right]$$

where p_a is the atmospheric pressure and σ_{v0}' is the effective overburden stress. We focus on the uncertainty due to the small sample size while the transformation uncertainty caused by the Kulhawy and Mayne (1990) relationship is ignored.

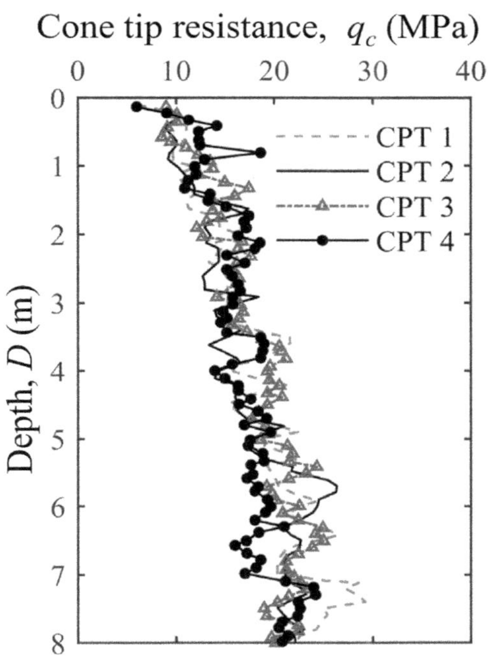

FIGURE 6.3 Four sets of CPT results obtained at the foundation design site (after Hara et al. 2011) (? 326241570.pdf commercial remixing and sharing from this source is allowed through CC BY 4.0 license).

Please use the bootstrapping method to estimate the uncertainties in the mean and standard deviation of effective friction angle (φ').

Solution:

Step 1: Preprocessing of the CPT data

Field data were obtained from four cone penetration tests (CPTs) performed near the proposed foundation location. The data are first extracted from the four CPTs and stored in a text file named *CPT_data.txt* with two columns:

Column 1: D_a (depth in m); Column 2: $q_{c,i}$(cone tip resistance in MPa).

The dataset constructed consists of k sampling points (D_i, $q_{c,i}$), $i = 1, 2, 3, \ldots, k$, combined from the four CPT profiles, where k represents the size of the dataset. The effective overburden stress ($\sigma'_{v0,i}$) at each depth D_i is calculated as follows.

$$\sigma'_{v0,i} = \gamma D_i$$

where γ is the bulk unit weight of the soil and the groundwater table is assumed to be deep enough that pore pressure can be neglected. The effective friction angle (φ'_i) at each depth is then estimated using the empirical relationship proposed by Kulhawy and Mayne (1990):

$$\varphi'_i = 17.6 + 11.0\log[\frac{q_{c,i} / p_a}{(\sigma'_{v0,i} / p_a)^{0.5}}]$$

where p_a is the atmospheric pressure, which is assumed as 100 kPa. Each measured data pair (D_i, $q_{c,i}$) produces a corresponding effective friction angle (φ'_i), resulting in a dataset $\boldsymbol{\varphi}' = \{\varphi_1', \varphi_2', \varphi_3', \ldots, \varphi_k'\}$. This dataset forms the basis for the subsequent bootstrap analysis.

Step 2: Estimate bootstrap statistics per replication through resampling scheme

In this example, the bootstrap method is adopted to characterize the uncertainty in the sample mean ($\overline{\varphi}'$) and sample standard deviation ($s_{\varphi'}$) of the dataset $\boldsymbol{\varphi}' = \{\varphi_1', \varphi_2', \varphi_3', \ldots, \varphi_k'\}$. Specially, the mean and standard deviation of the bootstrapped means (denoted as $\mu_{\overline{\varphi}'}$ and $\sigma_{\overline{\varphi}'}$) and those of the bootstrapped standard deviations (denoted as $\mu_{s_{\varphi'}}$ and $\sigma_{s_{\varphi'}}$) are estimated to capture the uncertainty in the sample mean ($\overline{\varphi}'$) and that of the sample standard deviation ($s_{\varphi'}$) of the dataset $\boldsymbol{\varphi}' = \{\varphi_1', \varphi_2', \varphi_3', \ldots, \varphi_k'\}$, respectively. The bootstrapped mean and bootstrapped standard deviation are the mean and standard deviation of the bootstrap samples generated from the bootstrap method per replication, respectively. The procedures for applying the bootstrap method are available in Example 2, and the code for implementing the bootstrap method in this example is provided below.

```
function Example3
clc;
clear;
CPT_data=load('CPT_data.txt');
% Calculate the effective friction angle (phi) by equation from Kulhawy and Mayne
(1990)
%Assume pa is 100 kPa. The effective overburden stress σ'v0 = γ·D
phi=17.6+11*log10((CPT_data(:,2).*1000./100)./(20*CPT_data(:,1)/100).^0.5);
data=phi;
N = 10000;
% Number of data points in the original sample
k = length(data);
% Initialize arrays to store bootstrapped statistics
boot_means = zeros(1, N);
boot_stds = zeros(1, N);
rng(2025,'twister'); % make results reproducible (optional)
% Step 1: Generate N re-samples and compute statistics
for j = 1:N
% Generate a bootstrap sample using random sampling with replacement
bootstrap_sample = data(randi(k, 1, k));
% Step 2: Compute the mean and standard deviation of the bootstrap sample
boot_means(j) = mean(bootstrap_sample);
boot_stds(j) = std(bootstrap_sample);
end
% Step 3: Calculate the mean and standard deviation of the bootstrapped statistics
mean_of_means = mean(boot_means);
std_of_means = std(boot_means);
mean_of_stds = mean(boot_stds);
std_of_stds = std(boot_stds);
% Display the results
disp('Bootstrapping Results:');
fprintf('Mean of bootstrapped means: %.4f\n', mean_of_means);
fprintf('Standard deviation of bootstrapped means: %.4f\n', std_of_means);
fprintf('Mean of bootstrapped standard deviations: %.4f\n', mean_of_stds);
fprintf('Standard deviation of bootstrapped standard deviations: %.4f\n', std_of_stds);
% Plot the distribution of bootstrapped statistics
figure;
histogram(boot_means, 30, 'Normalization', 'probability');
xlabel('Bootstrapped mean, $\bar{\phi}''^{*}$', 'Interpreter', 'latex');
ylabel('Frequency');
xlim([42.9,43.5]);
xticklabels(arrayfun(@(x) sprintf('%.1f', x), 42.9:0.1:43.5, 'UniformOutput', false));
set(gca,'FontSize',12,'FontName','Times New Roman');
figure;
histogram(boot_stds, 30, 'Normalization', 'probability');
xlabel('Bootstrapped standard deviation, $s_{\phi''}^{*}$','Interpreter', 'latex');
ylabel('Frequency');
xlim([0.8,1.4]);
xticklabels(arrayfun(@(x) sprintf('%.1f', x), 0.8:0.1:1.4, 'UniformOutput', false));
set(gca,'FontSize',12,'FontName','Times New Roman');
end
```

Note that the number of bootstrap replications in this example is set as 10,000 (i.e., N = 10,000) to obtain the stable distributions of the bootstrapped means and standard deviations, and the random number generator seed is set as 2025 to make the results reproducible. Illustrated in Figure 6.4 are the histograms for the bootstrapped means $\{\bar{\varphi}'^{*(j)}\}_{j=1}^{N} = \{\bar{\varphi}'^{*(1)}, \bar{\varphi}'^{*(2)}, \bar{\varphi}'^{*(3)}, \ldots, \bar{\varphi}'^{*(N)}\}$ and the bootstrapped standard deviations $\{s_{\varphi'}^{*(j)}\}_{j=1}^{N} = \{s_{\varphi'}^{*(1)}, s_{\varphi'}^{*(2)}, s_{\varphi'}^{*(3)}, \ldots, s_{\varphi'}^{*(N)}\}$. In this example, the bootstrapped mean ($\bar{\varphi}'$) ranges from 42.9° to 43.5° while the bootstrapped standard deviation ($s_{\varphi'}^{*}$) ranges from 0.8° to 1.4°.

Based on the bootstrap samples generated, the mean ($\mu_{\bar{\varphi}'}$) and standard deviation ($\sigma_{\bar{\varphi}'}$) of the bootstrapped means ($\bar{\varphi}'$) can be estimated as: $\mu_{\bar{\varphi}'}$ = 43.2315° and $\sigma_{\bar{\varphi}'}$ = 0.0677°; and, the mean ($\mu_{s_{\varphi'}}$) and standard deviation ($\sigma_{s_{\varphi'}}$) of the bootstrapped standard deviations ($s_{\varphi'}^{*}$) can be estimated as: $\mu_{s_{\varphi'}}$ = 1.1353° and $\sigma_{s_{\varphi'}}$ = 0.0703°.

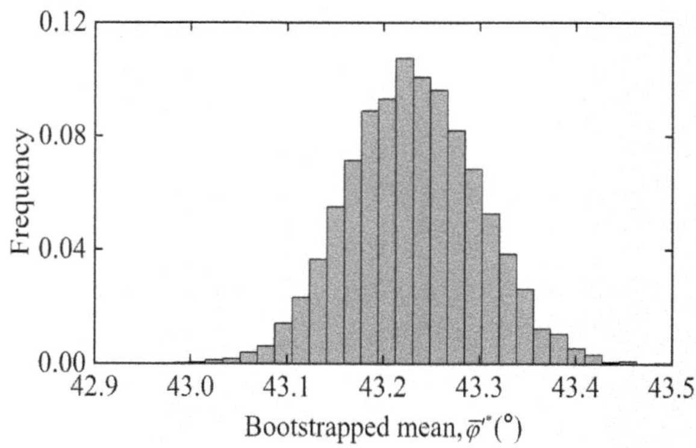

(a) Histogram of bootstrapped mean ($\bar{\varphi}'^*$)

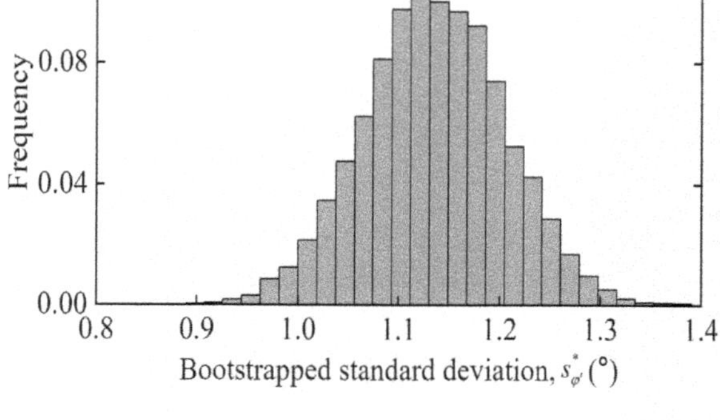

(b) Histogram of bootstrapped standard deviation ($s_{\varphi'}^*$)

FIGURE 6.4 Distributions of bootstrap statistics of the effective friction angle (φ') in Example 3.

.2.2 Formulation of Robustness Measure of Feasibility Index

easibility robustness is defined to ensure that the design can remain feasible relative the safety constraint for a certain probability even though it undergoes variations Parkinson et al. 1993). The safety constraint requires that the failure probability (P_f) f the critical limit state is less than the target failure probability (P_{fT}), expressed as $P_f \le P_{fT}$. In the face of the uncertainty in the statistical information of noise factors .g., geotechnical parameters), the evaluated failure probability (P_f) is a random ariable, while the target failure probability (P_{fT}) is a pre-defined fixed value. In eference to Juang et al. (2012, 2013) and Juang and Wang (2013), the FI can be efined as follows.

$$FI = Pr\left[P_f < P_{fT}\right] = Pr[\beta > \beta_T] \ge P_0 \tag{6.2}$$

here Pr[·] is the conditional probability; β is the reliability index of the design erformance; β_T is the target reliability index in the geotechnical design; P_0 is an cceptable probability pre-specified by the engineer. When the probability distribu-on of the reliability index (β) is assumed to follow a normal distribution (e.g., Most nd Knabe 2010; Luo et al. 2012a), the FI can be computed as follows.

$$\mathrm{FI} = \Phi\left(\frac{\mu_\beta - \beta_T}{\sigma_\beta}\right) \ge P_0 \tag{6.3}$$

here $\Phi(\cdot)$ is the cumulative distribution function (CDF) of the standard normal vari-ble; and μ_β and σ_β are the mean and standard deviation of the reliability index (β), espectively.

Note that the FI can also be evaluated as follows:

$$\beta_\beta = \frac{\mu_\beta - \beta_T}{\sigma_\beta} \tag{6.4}$$

The FI can aid in the design decision-making. Note that there is usually a trade-ff scenario in most geotechnical designs: a higher feasibility robustness (higher I and β_β) is generally associated with a higher cost. Thus, feasibility robustness ffers a user-friendly index to select the most cost-efficient design on the Pareto front ased on the target feasibility robustness level (*FIT* or β_β^T). For example, for a tar-et feasibility robustness of $\beta_\beta^T = 1$ (corresponding to a confidence probability of $P_0 = 84.13\%$), the most cost-efficient design is readily identified. Similarly, if the arget feasibility robustness is set at $\beta_\beta^T = 2$, the failure probability or reliability dex (of the design) would correspond to a confidence probability of $P_0 = 97.72\%$. he most cost-efficient designs based on different target feasibility robustness levels an be easily identified to aid in the final design decision.

.2.3 Implementation Procedures of the FI-Based RGD

s shown in Figure 6.5, the procedures for implementing the FI-based RGD are utlined in the following.

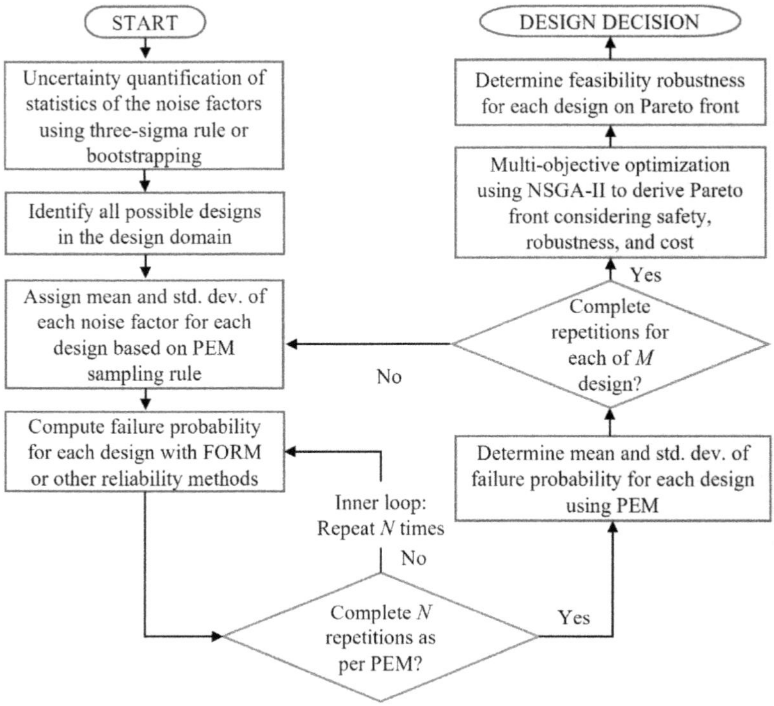

FIGURE 6.5 Flowchart illustrating FI-based RGD methodology (modified from Juang et a 2012).

Step 1: Characterize the uncertainty in the statistics of noise factors and identif all possible designs in the design space. Here, the noise factors can include both th geotechnical parameters and model parameters. The uncertainty in the statistics o each of the noise factors can be estimated with either the three-sigma bounds or th bootstrapping method.

The design space (**DS**) can be specified based on the typical ranges in the eng neering practice. Let's use the design of a square shallow foundation as an exampl in which the design parameters include the width (B) and the embedment depth (L of the footing. The width of footing (B) typically ranges from 1.0 m to 5.0 m (Akba and Kulhawy 2011), while the embedment depth (D) typically ranges from 1.0 r to 2.0 m based on the load level and environmental protection requirement (e.g Coduto 2010; Wang and Kulhawy 2008). The ratio of embedment depth to widt (D/B) is generally less than 4 for shallow foundations. Further, all possible design are restricted to the discrete design space for construction convenience. For exampl the dimensions of the foundation are generally rounded to the nearest 0.1 m (Wan 2011). Based on all the geometric constraints, a total number of 450 designs exist i the discrete design space (M = 450 in Figure 6.5).

Step 2: Evaluate the mean and standard deviation (std. dev.) of the failur probability (P_f) for each candidate design in the design space (**DS**). For eac candidate design, this step involves the inner loop in Figure 6.5 to assess the failure probability for a given set of statistics of noise factors and an outer loop in Figure 6.5 to assess the mean and standard deviation of the failure probability considering the uncertainties in the statistics of noise factors. In this FI-based RGD framework, the failure probability based on the critical limit state, either the ULS or SLS, can be adopted as a measure of design performance. A design is considered robust if its failure probability is insensitive to the uncertainty characterization of noise factors. In the outer loop, for each design of the M possible designs in the design space, the mean and standard deviation (also a measure of robustness) of the failure probability are assessed with the point estimate method (PEM; Zhao and Ono 2000).

When the statistics of the input parameters (e.g., geotechnical parameters and model parameters) are fixed values, the traditional reliability methods such as the first-order reliability method (FORM), Monte Carlo simulations (MCS), or system reliability approach can be used to evaluate the failure probability. With the uncertainties in the statistics of input parameters, the resulting failure probability is no longer a fixed value. In this scenario, the PEM may be employed to evaluate failure probability at each of a set of statistics of input parameters. As an example, let's use the seven-point estimate method to compute the failure probability. In this case, we need to repeat the calculations for a total of $N = 7 \times m$ times, where m is the number of input parameters and the multiplier "7" is the number of sampling points for each input parameter (Zhao and Ono 2000). As depicted in the inner loop of Figure 6.5, the resulting N failure probabilities can be employed to evaluate the mean and standard deviation of the failure probability based on the PEM.

Step 3: Conduct a multi-objective optimization utilizing the non-dominated sorting genetic algorithm to build a Pareto front, followed by the evaluation of feasibility robustness to select the best design. In this step, a multi-objective optimization is performed while simultaneously considering design safety, robustness, and cost. The robustness of each candidate design can be assessed with the standard deviation of the failure probability as a measure of robustness (see **Step 2**). Robustness and cost are treated as two design objectives, while safety constraints are applied in the optimization based on the target failure probability requirements. In this step, the Non-dominated Sorting Genetic Algorithm version II (NSGA-II), developed by Deb et al. (2002), or other similar multi-objective optimization algorithms, can be employed to establish the Pareto front. The designs on the Pareto front represent the best compromising designs with respect to both cost efficiency and design robustness.

It should be noted that if the uncertainty in the probability distribution (normal versus lognormal) for the noise factors is considered, the optimization can be formulated as a three-objective optimization problem and treat both the robustness measure in terms of normal distribution and lognormal distribution as objectives (Juang and Wang 2013). The designs on the Pareto front are based on the non-dominated rank and crowding distance to ensure the designs are more uniformly spread out in the solution space. All the designs on the Pareto front can satisfy the safety constraints while optimizing to both robustness and cost efficiency. Each point on the Pareto front represents a unique design with a set of design parameters. Although the Pareto front provides a trade-off relationship between cost efficiency and robustness, the feasibility robustness concept can be used as a more user-friendly index for design robustness, which can result in a more informed design decision.

6.3 DESIGN EXAMPLES FOR FEASIBILITY INDEX-BASED ROBUST GEOTECHNICAL DESIGN

Three geotechnical designs are undertaken in this section to demonstrate the application and effectiveness of the FI-based RGD.

6.3.1 Design Example 1: Design of a Square Footing in Cohesionless Soil

The first example concerns the design of a square footing in cohesionless soil under compressive loads, which was studied by Juang et al. (2012). As shown in Figure 6.6, the footing is designed to support vertical central loads with a permanent load component of G = 2,000 kN and a transient load component of Q = 1,000 kN. Both permanent load G and transient load Q follow lognormal distributions with a COV of 10% and 18%, respectively (Zhang et al. 2011). The soil supporting the footing is dry sand with a deterministic unit weight of γ = 18.5 kN/m^3 (Orr and Farrel 1999). Triaxial tests were conducted to evaluate the effective friction angle of the sand (ϕ'), and ten data points obtained are listed in Table 6.3. The groundwater level is assumed to be very deep, which has negligible effects on the foundation design. The SLS requires a maximum allowable settlement of 25 mm.

6.3.1.1 Deterministic Solution Models for Evaluating the ULS and SLS Bearing Capacities

The ULS bearing capacity of footings in cohesionless soil under compressive loads can be estimated following the procedures by Vesić (1975) and Kulhawy et al. (1983). The ULS bearing capacity (R_{ULS}) of a footing with width B, length L, and embedment depth D can be determined as follows (Akbas and Kulhawy 2009b).

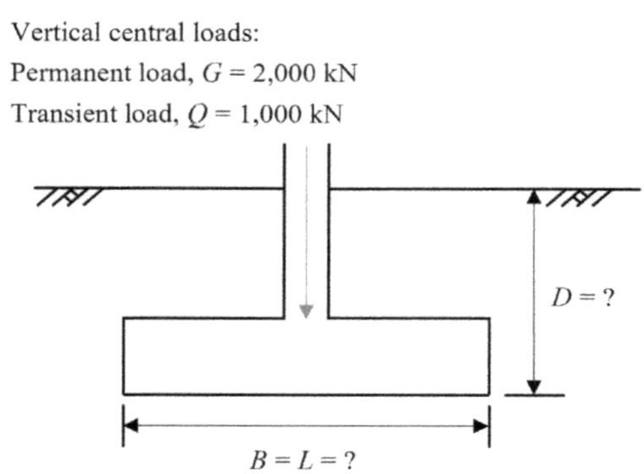

FIGURE 6.6 Schematic plot for the square footing design in Design Example 1 (modified from Figure 1 in Juang et al. 2012, with permission from *Journal of GeoEngineering*).

$$R_{ULS} = [(1/2)B\gamma' N_\gamma \xi_{\gamma s}\xi_{\gamma d}\xi_{\gamma r} + q' N_q \xi_{qs}\xi_{qd}\xi_{qr}](BL)$$

where γ' is the effective unit weight of soil below foundation; q' is the effective overburden stress at foundation level; N_γ and N_q are bearing capacity factors defined by Vesić (1975); $\xi_{\gamma s}$ and ξ_{qs} are shape correction factors; $\xi_{\gamma d}$ and ξ_{qd} are depth correction factors; and, $\xi_{\gamma r}$ and ξ_{qr} are rigidity correction factors. These correction factors are defined in Kulhawy et al. (1983). The ULS failure is said to happen if the bearing capacity calculated (R_{ULS}) is less than the applied loading (G+Q). The SLS bearing capacity (R_{SLS}) is formulated based on the normalized load-settlement behavior using the following equation (Akbas and Kulhawy 2011).

$$R_{SLS} = \frac{R_{ULS}(s_t / B)}{a(s_t / B) + b}$$

where s_t is the allowable settlement of 25 mm, and the coefficients a and b are model parameters of a hyperbolic model that fit the normalized load-settlement curve obtained from the full-scale field tests (Akbas and Kulhawy 2009a). The SLS failure is deemed to occur if the SLS bearing capacity based on the allowable settlement limit (R_{SLS}) is less than the applied loading (G+Q).

6.3.1.2 Robust Design Optimization Results of the Square Footing Design Problem

In this design example, both soil parameters and model parameters are treated as noise factors. Based on field load testing results and bootstrapping technique, the uncertainty characterization for the ULS model bias factor BF_Q can readily be conducted, and the results are tabulated in Table 6.1. Similarly, the uncertainty in the mean and standard deviation of two model parameters for the SLS capacity (i.e., a and b) can be quantified, and their statistics are listed in Table 6.2. The uncertainties in the mean and standard deviation of the effective friction angle of the sand (ϕ') are tabulated in Table 6.3. All these uncertainties are considered to evaluate the standard deviation of failure probability and associated feasibility robustness. The design parameters in this example are chosen to reflect both construction convenience and practical limitations related to equipment and local practice. The foundation width B is discretized from 1.0 m to 5.0 m in increments of 0.1 m., which covers typical ranges for spread foundation construction while maintaining a manageable number of alternatives. The foundation depth D is varied between 1.0 m and 2.0 m with an increment of 1.0 m, consistent with common excavation depths and avoiding excessive embedment. To eliminate excessively conservative alternatives, designs associated with a mean failure probability lower than 10^{-8} are excluded. The resulting discrete design space provides a rational basis for subsequent reliability and robustness evaluations.

Note that ULS is the critical limit state in this case; only the ULS failure probability is considered in this design example. The reliability requirement for the ULS requirement defined in Eurocode 7 is a reliability index $\beta_T^{ULS} = 3.8$ (corresponding to the target failure probability $p_T^{ULS} = 7.2 \times 10^{-5}$). Based on the inner loop of the RGD flowchart shown in Figure 6.5, the ULS failure probability for the given set of statistics is evaluated using the first-order reliability method (FORM). In the outer loop, the mean and standard deviation of the ULS failure probability, denoted as μ_p^{ULS} and σ_p^{ULS}, respectively, are evaluated for all designs in the design space using the PEM. Figure 6.7 depicts the mean of ULS failure probability (μ_p^{ULS}) for selected designs with D = 1.0 m, 1.5 m, and 2.0 m. Figure 6.8 shows the standard deviation of ULS failure probability (σ_p^{ULS}) of selected satisfactory designs with D = 1.0 m, 1.5 m, and 2.0 m.

In addition to design robustness evaluation (in terms of the standard deviation of failure probability), the total cost for a footing can be summarized as the cost of five individual tasks based on Wang and Kulhawy (2008).

$$Z = Q_e c_e + Q_f c_f + Q_c c_c + Q_r c_r + Q_b c_b$$

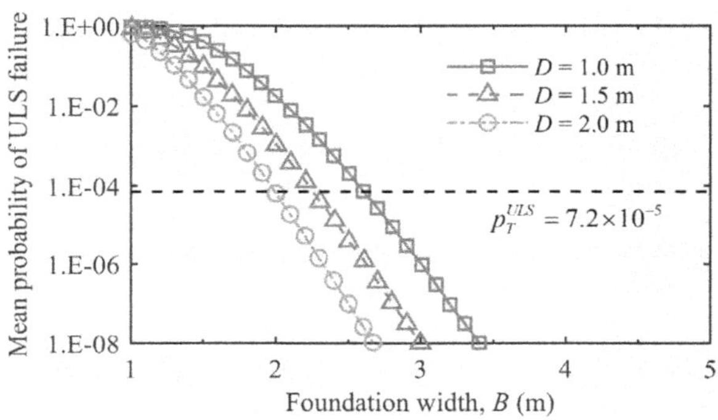

FIGURE 6.7 Mean ULS failure probabilities of selected designs in Design Example 1 (modified from Figure 7 in Juang et al. 2012, with permission from *Journal of GeoEngineering*).

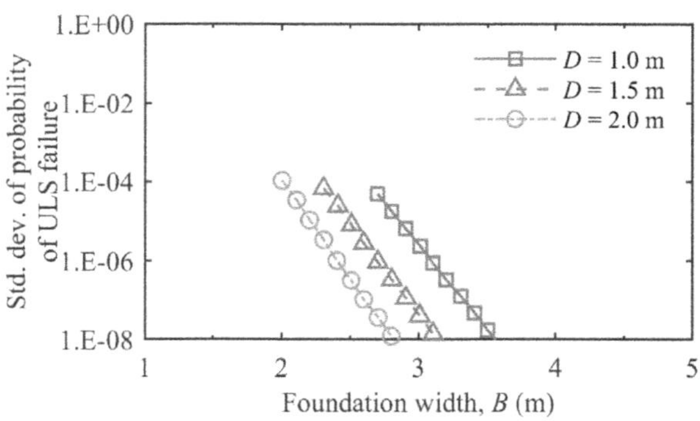

FIGURE 6.8 The standard deviation of ULS failure probabilities of selected satisfactory designs in Design Example 1 (modified from Figure 8 in Juang et al. 2012, with permission from *Journal of GeoEngineering*).

here Q_e, Q_f, Q_c, Q_r, *and* Q_b are the quantities for excavation, formwork, conete, reinforcement, and compacted backfill, respectively, which are functions of undation geometry including width *B*, length *L*, and embedment depth *D*; and c_e, $_f$, c_c, c_r, c_b are the unit prices for excavation, formwork, concrete, reinforcement, nd compacted backfill, respectively, which are documented in Wang and Kulhawy 008). Based on the evaluation of design robustness and cost for all designs in the screte design space, a bi-objective optimization can be performed by using the rget failure probability as a constraint while optimizing the design with respect to oth design robustness and cost, which can be set up as follows.

Find→d = [*B*, *D*]
Subject to: $B \in \{1.0\text{m}, 1.1\text{m}, 1.2\text{m}, \ldots, 5.0\text{m}\}$
$D \in \{1.0\text{m}, 1.1\text{m}, 1.2\text{m}, \ldots, 2.0\text{m}\}$
Safety requirement $\mu_p^{ULS} < p_T^{ULS} = 7.2\times10^{-5}$
Objectives: Minimizing standard deviation of ULS failure probability (σ_p)
Minimizing the cost of the footing (*Z*)

With the NSGA-II optimization (Deb et al. 2002), a converged Pareto front is btained, which is defined by 62 "unique" designs. Each design is represented as ne point, as depicted in Figure 6.9. A trade-off relationship between cost and design obustness is clearly observed, which means higher robustness can only be achieved t the expense of more cost. Furthermore, the feasibility robustness β_β for each of ne 62 designs on the Pareto front can be evaluated, and the correlation between β_β nd the cost for the 62 designs is depicted in Figure 6.10. The results show that a esign with higher feasibility robustness (higher β_β) can be achieved only with a igher cost. However, it is found that in the lower cost range, the curve is relatively at, which means a slight decrease in cost can bring a considerable reduction in feaibility robustness. In the higher cost range, however, a huge cost increase is needed

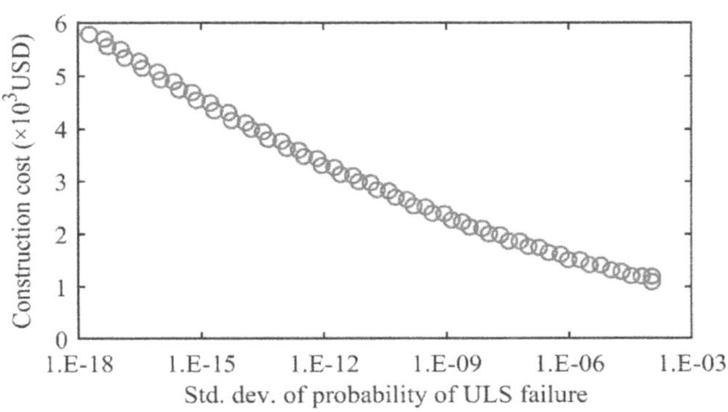

IGURE 6.9 Pareto front for the footing design obtained by NSGA-II in Design xample 1 (modified from Figure 10 in Juang et al. 2012, with permission from *Journal of eoEngineering*).

to achieve little gain in the improvement of feasibility robustness. Thus, feasibili robustness can be used as a user-friendly decision aid tool to finalize the optim design selection.

Once a target feasibility robustness level (β_β^T) is specified, the least-cost desig can readily be identified on the Pareto front. The least cost designs for different ta get feasibility robustness levels are listed in Table 6.4. For example, with the targ feasibility robustness set at $\beta_\beta^T = 1$, which corresponds to a confidence probabili of $P_0 = 84.13\%$, the least-cost design is $B = 2.1$ m and $D = 1.9$ m with a total co of USD 1,200. The target feasibility robustness $\beta_\beta^T = 3$ corresponds to a confiden probability of $P_0 = 99.87\%$, and the least-cost design is $B = 2.6$ m and $D = 2.0$ r with a total cost of USD 1,764. From there, the feasibility robustness provides a easy-to-use quantitative measure for a more informed design decision by explicit considering the design robustness.

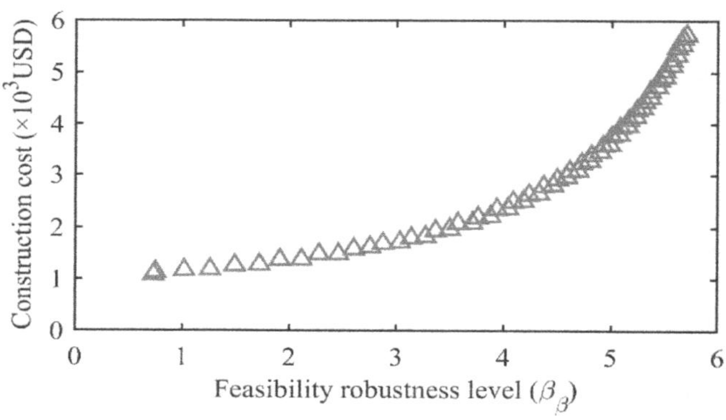

FIGURE 6.10 Cost versus feasibility robustness for all designs on Pareto front in Desig Example 1 (modified from Figure 11 in Juang et al. 2012, with permission from *Journal GeoEngineering*).

TABLE 6.4
Selected Final Designs at Various Feasibility Robustness Levels in Design Example 1

β_β	P_0	$B(m)$	D (m)	Cost (USD)
1	84.13%	2.1	1.9	1,200
2	97.72%	2.3	2.0	1,424
3	99.87%	2.6	2.0	1,764
4	99.997%	3.1	2.0	2,410

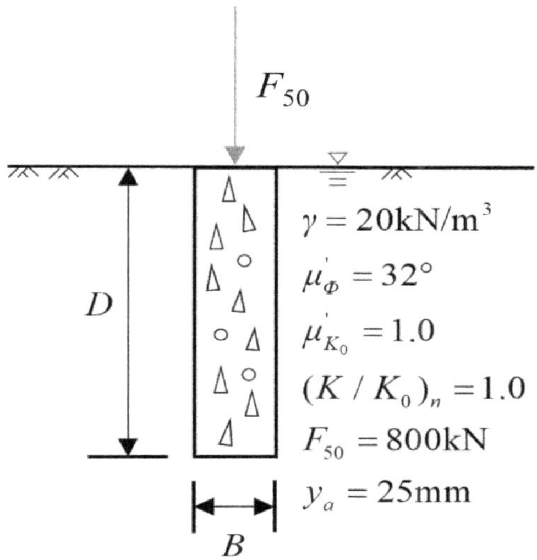

FIGURE 6.11 Schematic diagram of the design problem of a drilled shaft in loose sand under compression load in Design Example 2 (modified from Figure 1 in Juang et al. 2013, with permission from American Society of Civil Engineers).

6.3.2 Design Example 2: Design of a Drilled Shaft in Loose Sand

The second example is for a drilled shaft design in sand, which was studied by Juang et al. (2013). Figure 6.11 shows the schematic diagram of the design problem of a drilled shaft in loose sand subjected to a compression load. The water table in this example is located at the ground surface. The geometry of the drilled shaft, including diameter B and embedment depth D, are design parameters to optimize in the RGD framework. For both ULS and SLS designs, the vertical compression load F is set as the 50-year return period load $F_{50} = 800\ kN$ in this example. The drilled shaft is said to fail if F_{50} exceeds either ULS or SLS capacities.

6.3.2.1 Deterministic Solution Models for Evaluating the ULS and SLS Bearing Capacities

For the design of drilled shafts, both ULS and SLS bearing capacities need to be considered. The ULS bearing capacity, denoted as Q_{ULS}, can be calculated using the following equation (Kulhawy 1991).

$$Q_{ULS} = Q_{side} + Q_{tip} - W$$

where Q_{side}, Q_{tip}, and W represent friction resistance, tip resistance, and effective shaft weight, respectively. For sandy soil, the cohesion term is neglected. The terms Q_{side} and Q_{tip} in the above equation can be calculated as follows (Kulhawy 1991).

$$Q_{side} = \pi BD\left(K / K_0\right)_n K_0 \sigma'_{vm} \tan\varphi$$

$$Q_{tip} = 0.25\pi B^2 \left[0.5B(\gamma - \gamma_w) N_\gamma \xi_{\gamma s} \xi_{\gamma d} \xi_{\gamma r} + (\gamma - \gamma_w) D N_q \xi_{qs} \xi_{qd} \xi_{qr}\right]$$

where $(K/K_0)_n$ is the nominal operative in-situ horizontal stress coefficient ratio; σ'_{vm} is the mean vertical effective stress along the depth of drilled shaft; φ' is the effective friction angle; N_γ and N_q are the bearing capacity factors defined by Vesic (1975); $\xi_{\gamma s}$ and ξ_{qs} are the shape correction factors; $\xi_{\gamma d}$ and ξ_{qd} are the depth correction factors; and, $\xi_{\gamma r}$ and ξ_{qr} are the rigidity correction factors for respective bearing capacity factors. Formulations for calculating these correction factors are detailed in Kulhawy (1991). The SLS bearing capacity, denoted as Q_{SLS}, can be calculated using the following equation (Phoon et al. 1995).

$$Q_{SLS} = 0.625a\left(\frac{y_a}{B}\right)^b Q_{ULS}$$

where a = 4.0 and b = 0.4 are the curve-fitted parameters of the empirical model describing the load-displacement behavior, and y_a is the allowable displacement (25 mm in this case).

6.3.2.2 Robust Design Optimization Results of the Drilled Shaft Design Problem

In this design example, the geometry of the shaft, including diameter (B) and depth (D), is treated as the design parameters, and the soil parameters φ' and K_0 are treated as the noise factors. The uncertainty characterization of the COV and correlation coefficient statistics of the noise factors are conducted based on the three-sigma bounds method guided by literature data and experience. The mean and standard deviation of the COV of the effective friction angle (φ') are estimated as $\mu_{COV[\varphi']} \approx 0.07$ and $\sigma_{COV[\varphi']} \approx 0.0125$. Similarly, the mean and standard deviation of the COV of the in-situ horizontal stress coefficient (K_0) are estimated as $\mu_{COV[K_0]} \approx 0.50$ and $\sigma_{COV[K_0]} \approx 0.175$. The mean and standard deviation of the correlation coefficient between these two noise factors (ρ_{φ',K_0}) are calculated as $\mu_{\rho_{\varphi',K_0}} = -0.75$ and $\sigma_{\rho_{\varphi',K_0}} = 0.075$.

The design parameters in this example are restricted to those in the discrete design space due to the construction convenience and limitations on equipment and local practice. For instance, in this case, three discrete values of shaft diameter B (i.e., 0.9 m, 1.2 m, and 1.5 m) are considered here. The depth D is to be taken from the range of 2 m to 8 m with an increment of 0.2 m. Thus, a number of designs (M = 93) in the design space is specified.

The cost of a drilled shaft is a function of the shaft geometry (i.e., diameter and depth), which may be estimated from published literature, such as the Means Building Construction Cost Data (e.g., R.S. Means Co. 2007). For demonstration purposes, the U.S. national average unit costs for constructing drilled shafts of unit depth (0.3 m) are USD 77.5, 116, and 157 for the shaft diameters of 0.9 m, 1.2 m, and 1.5 m, respectively (Wang et al. 2011).

Based on the specified design space and uncertainty quantification of the statistics of key soil parameters (noise factors), the mean and standard deviation of ULS

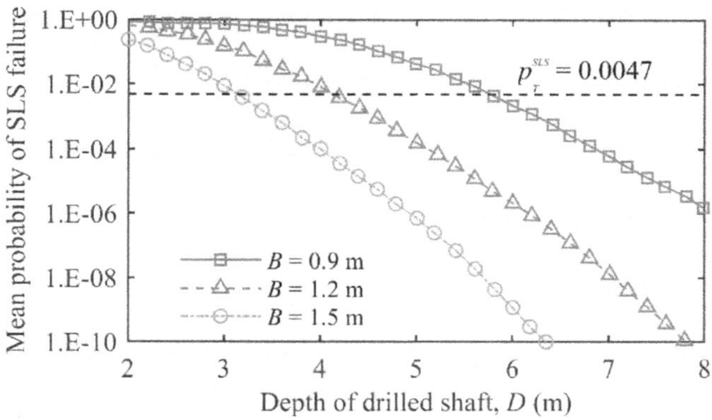

FIGURE 6.12 Mean of SLS failure probability for the drilled shaft design in Design Example 2 (modified from Figure 5 in Juang et al. 2013, with permission from American Society of Civil Engineers).

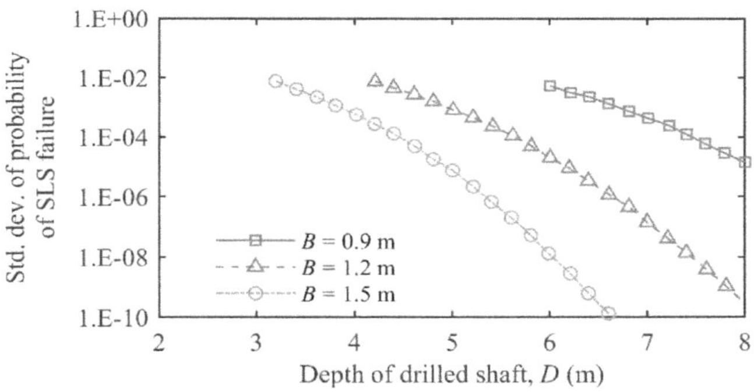

FIGURE 6.13 The standard deviation of SLS failure probability for all satisfactory designs in Design Example 2 (modified from Figure 6 in Juang et al. 2013, with permission from American Society of Civil Engineers).

and SLS failure probabilities can be evaluated. The target ULS failure probability requirement $p_T^{ULS} = 0.00069$ is based on a reliability index of 3.2, while the target SLS failure probability requirement $p_T^{SLS} = 0.0047$ is based on a reliability index of 2.6 (Wang et al. 2011). It was found that the SLS controls the design of this drilled shaft. Thus, SLS is considered the critical limit state, and the SLS failure probability is adopted in the analysis afterward. Figure 6.12 depicts the mean of SLS failure probability for all possible designs in the design space against the target SLS failure probability requirement. Figure 6.13 shows the standard deviation of SLS failure probability for all satisfactory designs, which is adopted for the assessment of design robustness.

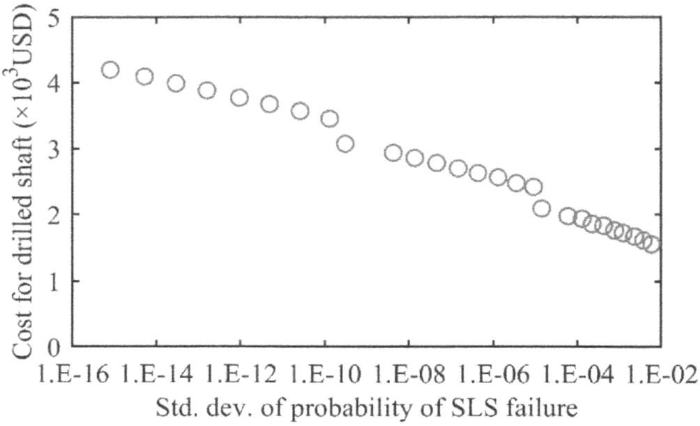

FIGURE 6.14 Pareto front obtained from the two-objective non-dominated sorting in Design Example 2 (modified from Figure 8 in Juang et al. 2013, with permission from American Society of Civil Engineers).

A robust design optimization can be performed to select the designs on the Pareto front. For this drilled shaft design problem, the multi-objective optimization is set up as follows.

Find→d = [B, D]
Subject to: $B \in \{0.9\text{m}, 1.2\text{m}, 1.5\text{m}\}$
$D \in \{2\text{m}, 2.2\text{m}, 2.4\text{m}, \ldots, 8\text{m}\}$
ULS requirement $\mu_p^{ULS} < p_T^{ULS} = 0.00069$
SLS requirement $\mu_p^{SLS} < p_T^{SLS} = 0.0047$
Objectives: Minimizing standard deviation of SLS failure probability (σ_p)
Minimizing the cost of the drilled shaft

Among the 93 possible designs in the specified design space, 27 designs are deemed non-dominated, and they collectively define the Pareto front. All of them are satisfactory with respect to safety constraints while optimized for both design robustness and cost. Figure 6.14 shows a clear trade-off relationship between cost and design robustness (in terms of standard deviation of SLS failure probability). Furthermore, the feasibility robustness can be evaluated for each design on the Pareto front. Here, the feasibility robustness is evaluated against the SLS target failure probability requirement $p_f^{SLS} < p_T^{SLS} = 0.0047$. The uncertainties in the statistics of key geotechnical parameters, in terms of $COV[\phi']$, $COV[K_0]$, and ρ_{φ',K_0}, are considered. The feasibility robustness is assessed with the following constraint.

$$\Pr[(p_f^{SLS} - p_T^{SLS}) < 0] \geq P_0$$

where P_0 is a confidence probability that the SLS requirement is satisfactory. It is observed that the histogram of the reliability index with respect to the SLS quirement, β*SLS*, generally fits well with a lognormal distribution for this design ase. An equivalent counterpart, $\Pr[(\beta^{SLS}-\beta_T^{SLS})>0]$, can be formulated below to sess the feasibility robustness.

$$\mathrm{FI}=\Pr[(\beta^{ULS}-\beta_T^{SLS})>0]=\Phi(\beta_\beta)\geq P_0$$

here β_T^{SLS} is the target reliability index with respect to the SLS requirement, the alue of which is 2.6 (corresponding to the target failure probability of p_T^{SLS} = 0047); $\Phi(\cdot)$ is the CDF of the standard normal variable; and, β_β is defined as llows.

$$\beta_\beta=\frac{\ln\left[\mu_\beta\Big/\sqrt{1+\left(\frac{\sigma_\beta}{\mu_\beta}\right)^2}\right]-\ln\left(\beta_T^{SLS}\right)}{\sqrt{\ln\left[1+\left(\frac{\sigma_\beta}{\mu_\beta}\right)^2\right]}}$$

here μ_β and σ_β are the mean and standard deviation of the reliability index with spect to the SLS requirement, respectively. If the acceptable level of the feasi-ility probability is set as P_0 = 97.72%, then the required β_β value will be 2. In ther words, if the β_β computed based on μ_β and σ_β is equal to 2, then there is a asibility probability of 97.72% that the SLS requirement ($p_f^{SLS}<p_T^{SLS}=0.0047$ or quivalently $\beta^{SLS}>\beta_T^{SLS}=2.6$) is satisfied.

The FI and reliability index β_β are then calculated for each of the 27 designs n the Pareto front. Plotted in Figure 6.15 is the correlation between the feasibility obustness index and the cost. A higher cost is needed to achieve a design with higher easibility robustness. The final design may be selected based on the target feasibility

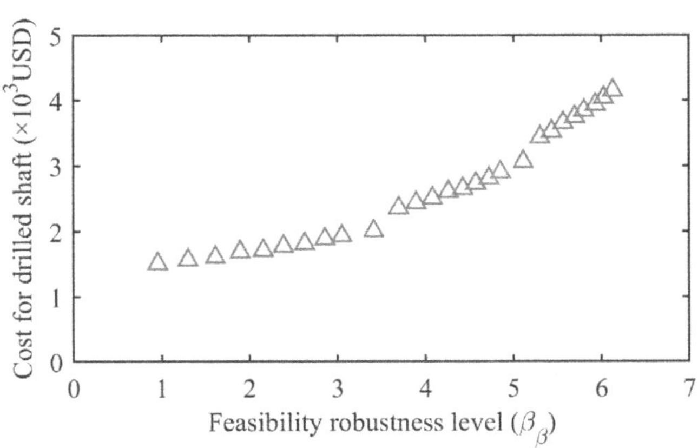

IGURE 6.15 Cost versus feasibility robustness for all designs on the Pareto front in Design Example 2 (modified from Figure 10 in Juang et al. 2013, with permission from American Society of Civil Engineers).

TABLE 6.5
Final Designs Selected Based on Different Feasibility Robustness Levels in Design Example 2

β_β	P_0	B (m)	D (m)	Cost (USD)
1	84.13%	0.9	6.2	1,602
2	97.72%	0.9	6.8	1,757
3	99.87%	0.9	7.6	1,963
4	99.997%	1.2	6.6	2,552

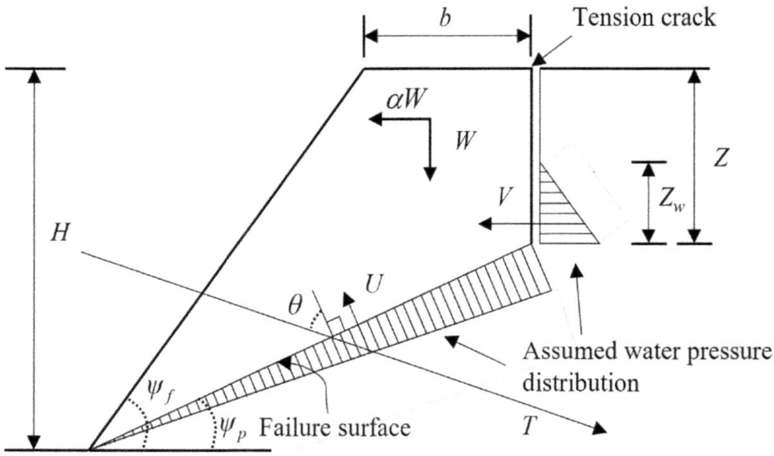

FIGURE 6.16 Geometry for the Sau Mau Ping rock slope studied in Design Example 3 (modifie from Figure 1 in Wang et al. 2015, with permission from American Society of Civil Engineers).

robustness level by choosing the least-cost designs within this robustness level o the Pareto front. The final designs selected based on different feasibility robustnes levels are tabulated in Table 6.5. For example, for a feasibility robustness level c $\beta_\beta = 2$, which corresponds to a confidence probability of $P_0 = 97.72\%$, the chose final design $B = 0.9$ m and $D = 6.8$ m will be the most cost-efficient design within thi robustness level. Thus, the feasibility robustness can provide a quantitative measur to choose the best design based on the target robustness level.

6.3.3 Design Example 3: Design of Rock Bolts for Slope Stabilization

This example focuses on the design of rock bolts to stabilize a real-world slope i Hong Kong, which was examined by Wang et al. (2015). The rock slope of concer is the Sau Mau Ping slope in Hong Kong. This design example focuses on the desig of a rock bolt system for the slope reinforcement. The schematic layout of the roc slope is shown in Figure 6.16. The geotechnical and geometric parameters for th

TABLE 6.6
Geotechnical and Geometric Parameters of the Sau Mau Ping Slope Studied in Design Example 3

Parameter	Value
Unit weight of rock, γ (ton/m^3)	2.6
Unit weight of water, γ_w (ton/m^3)	1
Height of overall slope, H (m)	60
Angle of failure surface, ψ_p (°)	35
Overall slope angle, ψ_f (°)	50

Source: Hoek, 2006.

rock slope are tabulated in Table 6.6. The design parameters include the design force of rock bolts T (ton/m) and the inclination angle of rock bolt θ (°). The design case is simplified as a two-dimensional (2D) problem, and the force of rock bolts refers to that applied to a one-meter-thick slice of the slope.

6.3.3.1 Deterministic Solution Models for Stability Analysis of the Rock Slope

A 2D limit equilibrium model by Hoek and Bray (1981) is used in the stability analysis of the rock slope example. The *FS* of the rock slope is evaluated as follows (Hoek and Bray 1981; Hoek 2006).

$$FS = \frac{cA + [W(\cos\psi_p - \alpha\sin\psi_p) - U - V\sin\psi_p + T\cos\theta]\tan\varphi}{W(\sin\psi_p + \alpha\cos\psi_p) + V\cos\psi_p - T\sin\theta}$$

where c is the cohesion of the rock discontinuity (ton/m^2); A is the base area of the wedge (m); W is the weight of the rock wedge above the failure surface (ton); ψ_p is the angle of the failure surface (°); α is the ratio of horizontal earthquake acceleration to gravitational acceleration; U is the uplift force due to water pressure on the failure surface (ton); V is the horizontal force induced by the water in tension crack (ton); θ is the inclination angle of rock bolt, anti-clockwise from normal vector of the failure surface (°); ϕ is the friction angle of the rock discontinuity (°); and, T is the applied force of rock bolts (ton). The slope failure is said to occur if the computed *FS* is less than 1.0. The formulations for calculating the terms in the above equation are formulated as follows:

$$A = (H - z)/\sin\psi_p$$

$$z = H - (b + H\cot\psi_f)\tan\psi_p$$

$$W = 0.5\gamma H^2[(1 - (\frac{z}{H})^2)\cot\psi_p - \cot\psi_f]$$

$$U = 0.5\gamma_w z_w A$$

$$V = 0.5\gamma_w z_w^2$$

$$z_w = i_w \cdot z$$

where H is the height of the rock slope (m); z is the depth of tension crack (m); b is the horizontal distance of tension crack behind the crest (m); i_w is the ratio of depth of water filled in the tension crack to the depth of tension crack; z_w is depth of water in the tension crack (m); ψ_f is the overall slope angle (°); γ is the unit weight of rock (kN/m^3); and, γ_w is the unit weight of water (kN/m^3).

6.3.3.2 Robust Design Optimization Results of the Rock Bolts Design Problem

This design example aims to optimize two design parameters, in terms of the applied force of rock bolts (T) and inclination angle of rock bolt (θ), so that the rock slope performance (in terms of the failure probability) is robust against the uncertainties in the estimated statistics of noise factors. In this rock slope example, the noise factors are the strength characteristics of rock discontinuity, including cohesion c (ton/m^2) and friction angle ϕ (°). The COVs of these two noise factors (i.e., c and ϕ), as well as the correlation coefficient (ρ) between these noise factors, are considered random variables within the robust design framework. Furthermore, in addition to the variation in the geotechnical properties, the correlation between the horizontal distance of the tension crack behind crest b (m) and the ratio of the depth of water filled in the tension crack (i_w) is also considered a random variable (Low 2007; Wang et al. 2015). Other input parameters are assumed with fixed statistical characterization (e.g., fixed COV), as listed in Table 6.7 (Low 2007). The statistics of the noise factors, namely $COV[c]$, $COV[\phi]$, $\rho_{c,\varphi}$ and $\rho_{b,iw}$, are considered random variables. Based on the typical ranges reported by Lee et al. (2012), the mean of $COV[c]$ is assumed to be 0.20, and its COV is assumed to be 17%. Similarly, the mean of $COV[\phi]$ is assumed to be 0.14, and its COV is assumed to be 12%; the mean of $\rho_{c,\varphi}$ is assumed to be

TABLE 6.7
Statistical Information of Input Parameters of the Sau Mau Ping Slope Design Problem in Design Example 3

Random variable	Probability distribution	Mean	Std. dev.
c	Normal	10 ton/m^2	2 ton/m^2
φ	Normal	35°	5°
b	Normal	15.3 m	4.2 m
i_w	Exponential with mean 0.5, truncated to (0, 1)		
α	Exponential with mean 0.08, truncated to (0, 0.16)		

Source: Low, 2007.

−0.50, and its COV is assumed to be 25%; the mean of $\rho_{b,iw}$ is assumed to be 0.50, and its COV is assumed to be 25%. The target reliability index (β_T) is set at 2.5 in the safety evaluation of rock slope, which is equivalent to a target failure probability p_T of 0.00621 (Low 2008).

Based on the typical ranges for design parameters of rock bolt systems, the design force (*T*) ranges from 0 to 600 ton/m, and the inclination angle (θ) ranges from 0° to 80° (Wyllie and Mah 2004; Low 2007). The design force (*T*) is taken at an increment of 30 ton/m, while the inclination angle (*θ*) is rounded to the nearest 1°. Based on the combination of the discrete values of design force (*T*) and inclination angle (θ), there are 1,701 possible designs in the design space. For each design in the domain, the mean and standard deviation of the failure probability of the rock slope, denoted as μ_p and σ_p, can be evaluated using the PEM integrated with the first-order reliability method (FORM). Figure 6.17 illustrates the results of μ_p and σ_p for selected designs. It is found that μ_p and σ_p generally decrease with the increased force of rock bolts (*T*). However, the increase in the inclination angle of rock bolts θ will initially cause decreases in μ_p and σ_p, and the trend will reverse with very high

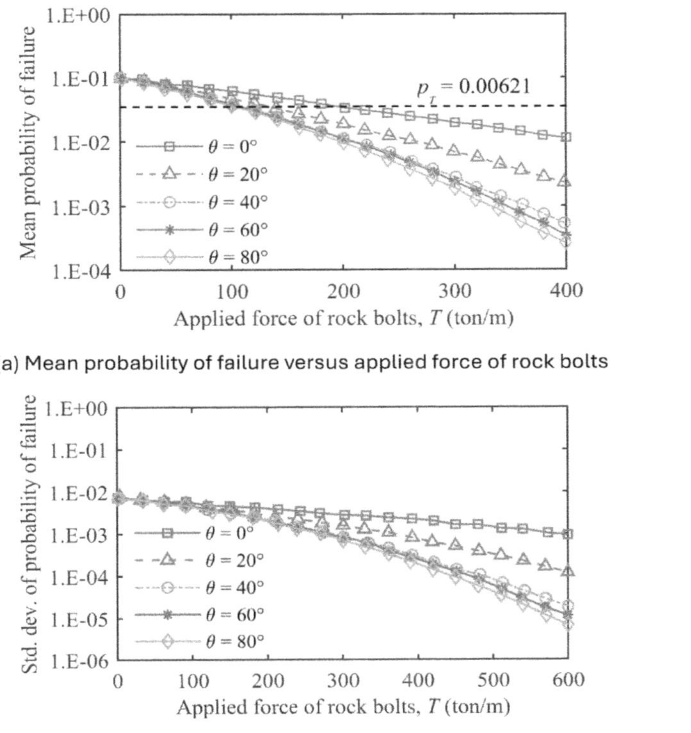

FIGURE 6.17 Mean and standard deviation of the failure probability for selected designs in Design Example 3 (modified from Figure 2 in Wang et al. 2015, with permission from American Society of Civil Engineers).

inclination angles. The cost of rock bolts, in this robust design example, is approximately represented using the average length of rock bolts per unit longitudinal direction of the slope (Chen 2003).

The robust design optimization can be conducted based on the analysis results of μ_p and σ_p for all designs. The optimization is set up to maximize the design robustness (by minimizing the standard deviation of the failure probability σ_p) and cost efficiency (by minimizing the cost of rock bolts), while subjected to the safety constraint that the mean of the failure probability (μ_p) is less than the target failure probability. The optimization results in a converged Pareto front shown in Figure 6.18, which consists of 71 unique designs that are optimal for both design robustness and cost efficiency. There exists an obvious trade-off relationship between cost and robustness. Furthermore, the feasibility robustness of each design on the Pareto front can be estimated. The FI or confidence probability is expressed as follows.

$$\text{FI}=\Pr[p_f - p_T < 0] \geq P_0$$

where p_f is the computed failure probability; p_T is the target failure probability (0.00621) based on a target reliability index of 2.5 (Low 2008); $Pr\left[\left(p_f - p_T\right) < 0\right]$ is the confidence probability that the target failure probability requirement is satisfactory; and, P_0 is the target level of the confidence probability. Figure 6.19 depicts the relationship between cost and confidence probability (representing feasibility robustness) for all 71 non-dominated designs on the Pareto front. The design decision can be chosen based on different target confidence probability requirements, as shown in Table 6.8. For example, for a required confidence probability of 90%, a final design with $T = 210$ ton/m and $\theta = 47°$ can be chosen as the final design, which provides an easy-to-use metric in the design decision-making.

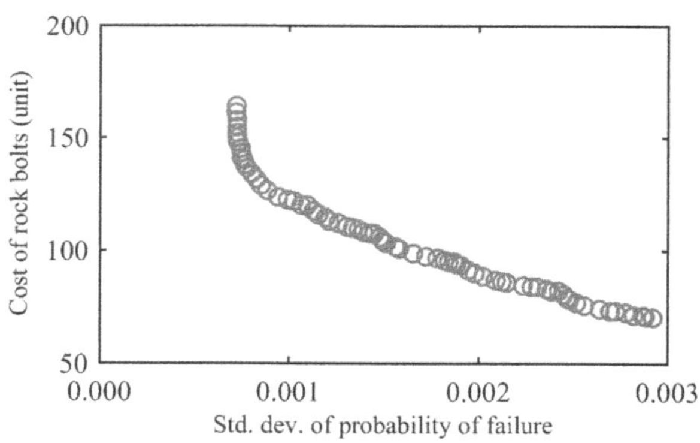

FIGURE 6.18 Pareto front based on two design objectives of design robustness and cost efficiency in Design Example 3 (modified from Figure 3 in Wang et al. 2015, with permission from American Society of Civil Engineers).

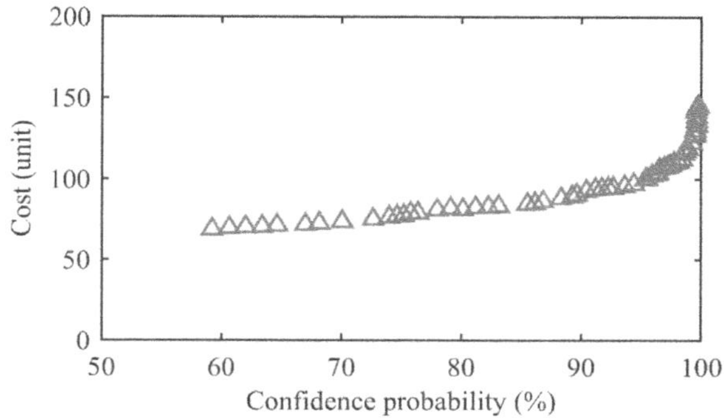

IGURE 6.19 Cost versus confidence probability for all designs on the Pareto front in Design xample 3 (modified from Figure 4 in Wang et al. 2015, with permission from American ociety of Civil Engineers).

ABLE 6.8
esigns Selected Based on Different Target Feasibility Levels in Design xample 3

P_0	*T* (ton/m)	θ (°)	Cost (units)
0%	180	38	74.6
0%	210	30	83.0
0%	210	47	94.0
5%	240	40	100.9
9%	300	36	122.7

.4 SUMMARY AND REMARKS

his chapter presents the feasibility robustness concept and its implementation in the GD. Accurate characterization of the uncertainties or statistical information of geo-echnical parameters is a prerequisite for the RBD. However, with budget constraints f site investigation, there exists no sufficient site-specific data that can accurately haracterize the statistical information of geotechnical parameters. The feasibility obustness concept is introduced to address the uncertainties involved in such statis-ical characterization. Here, feasibility robustness is employed to evaluate the confi-ence probability that the reliability requirement for the design is satisfactory. Then, he FI-based RGD is formulated, which aims to reduce the effect of the uncertainties n the statistical characterization of geotechnical parameters and achieve a certain evel of design robustness, in addition to satisfying safety and cost requirements. hrough the multi-objective optimization framework, a Pareto front consisting of ptimal designs can be obtained.

The obtained Pareto front and associated feasibility robustness can aid in th design process to make a more informed decision. Three geotechnical design exam ples, including the shallow foundation, drilled shaft, and rock slope, are analyze to demonstrate the effectiveness and significance of the feasibility robustness-base RGD approach.

REFERENCES

Akbas, S.O., and Kulhawy, F.H. (2009a) Axial compression of footings in cohesionless soils. Load–settlement behavior. *Journal of Geotechnical and Geoenvironmental Engineerin* 135(11), 1562–1574.

Akbas, S.O., and Kulhawy, F.H. (2009b) Axial compression of footings in cohesionless soil II: Bearing capacity. *Journal of Geotechnical and Geoenvironmental Engineerin* 135(11), 1575–1582.

Akbas, S.O., and Kulhawy, F.H. (2011) Reliability based design of shallow foundations cohesionless soil under compression loading: Serviceability limit state. In *Proceedin of georisk 2011: Geotechnical risk assessment & management (GSP224)*. America Society of Civil Engineers (ASCE), Reston, VA, USA, 616–623.

Amundaray, J.I. (1994) *Modeling geotechnical uncertainty by bootstrap resampling*. Ph.I thesis, Purdue University, West Lafayette, IN.

Bourdeau, P.L., and Amundaray, J.I. (2005) Non-parametric simulation of geotechnical vari bility. *Géotechnique*, 55(1), 95–108.

Chen, M. (2003) *Optimal design for permanent anchors in the c-φ soils*. Master thesis, Depar ment of Civil Engineering, National Central University, Jhongli, Taiwan.

Chen, W., Allen, J.K., Mistree, F., and Tsui, K.-L. (1996) A procedure for robust design: Mi imizing variations caused by noise factors and control factors. *Journal of Mechanic Design*, 118(4), 478–485.

Coduto, D.P. (2010) *Foundation design: Principles and practices* (2nd edition). Prentice Ha New Jersey.

Deb, K., Pratap, A., Agarwal, S., and Meyarivan, T. (2002) A fast and elitist multiobjecti genetic algorithm: NSGA-II. *IEEE Transactions on Evolutionary Computation*, 6(2 182–197.

Duncan, J.M. (2000) Factors of safety and reliability in geotechnical engineering. *Journal Geotechnical and Geoenvironmental Engineering*, 126(4), 307–316.

Gong, W., Tien, Y.M., Juang, C.H., Martin, J.R., and Luo, Z. (2017) Optimization of si investigation program for improved statistical characterization of geotechnical propert based on random field theory. *Bulletin of Engineering Geology and the Environmen* 76, 1021–1035.

Hara, T., Honjo, Y., Otake, Y., and Moriguchi, S. (2011) Application of reliability-based desig (RBD) to Eurocode 7. In *Proceedings of the 3rd international symposium on geotechn cal safety and risk*, N. Vogt, B. Schuppener, D. Straub, and G. Bräu, eds. Bundesansta für Wasserbau, Karlsruhe, Germany, 489–496.

Hoek, E. (2006) *Practical rock engineering. Chapter 7: A slope stability problem in Hon Kong; and Chapter 8: Factor or safety and probability of failure*. www.rocscience.com education/hoeks_corner.

Hoek, E., and Bray, J. (1981) *Rock slope engineering* (3rd edition). Institution of Mining an Metallurgy, London.

Juang, C.H., and Wang, L. (2013) Reliability-based robust geotechnical design of sprea foundations using multi-objective genetic algorithm. *Computers and Geotechnics*, 48 96–106.

Juang, C.H., Wang, L., Atamturktur, S., and Luo, Z. (2012) Reliability-based robust and optimal design of shallow foundations in cohesionless soil in the face of uncertainty. *Journal of GeoEngineering*, 7(3), 75–87.

Juang, C.H., Wang, L., Liu, Z., Ravichandran, N., Huang, H., and Zhang, J. (2013) Robust geotechnical design of drilled shafts in sand – new design perspective. *Journal of Geotechnical and Geoenvironmental Engineering*, 139(12), 2007–2019.

Kulhawy, F.H. (1991) Drilled shaft foundations. In *Foundation engineering handbook* (2nd edition), H.Y. Fang, ed. Van Nostrand Reinhold, New York, 537–552.

Kulhawy, F.H., and Mayne, P.W. (1990) *Manual on estimating soil properties for foundation design* (No. EPRI-EL-6800). Electric Power Research Institute, Palo Alto, CA, USA; Cornell University, Ithaca, NY, USA. Geotechnical Engineering Group.

Kulhawy, F.H., Trautmann, C.H., Beech, J.F., O'Rourke, T.D., McGuire, W., Wood, W.A., and Capano, C. (1983) *Transmission line structure foundations for uplift – compression loading*. Rep. No. EL-2870. Electric Power Research Institute, Palo Alto, CA.

Lee, Y.F., Chi, Y.Y., Juang, C.H., and Lee, D.H. (2012) Reliability analysis of rock wedge stability- a Knowledge-based Clustered Partitioning (KCP) approach. *Journal of Geotechnical and Environmental Engineering*, 138(6), 700–708.

Low, B.K. (2007) Reliability analysis of rock slopes involving correlated nonnormals. *International Journal of Rock Mechanics and Mining Sciences*, 44(6), 922–935.

Low, B.K. (2008) Efficient probabilistic algorithm illustrated for a rock slope. *Rock Mechanics and Rock Engineering*, 41(5), 715–734.

Luo, Z., Atamturktur, S., Cai, Y., and Juang, C.H. (2012a) Simplified approach for reliability-based design against basal-heave failure in braced excavations considering spatial effect. *Journal of Geotechnical and Geoenvironmental Engineering*, 138(4), 441–450.

Luo, Z., Atamturktur, S., and Juang, C.H. (2012b) Bootstrapping for characterizing the effect of uncertainty in sample statistics for braced excavations. *Journal of Geotechnical and Geoenvironmental Engineering*, 139(1), 13–23.

Mayne, P.W., and Kulhawy, F.H. (1982) K_o-OCR relationships in soil. *Journal of the Geotechnical Engineering Division*, 108(6), 851–872.

Most, T., and Knabe, T. (2010) Reliability analysis of bearing failure problem considering uncertain stochastic parameters. *Computers and Geotechnics*, 37(3), 299–310.

Nikulin, M.S. (1974) Chi-square test for continuous distributions with shift and scale parameters. *Theory of Probability & Its Applications*, 18(3), 559–568.

Orr, T.L.L., and Farrell, E.R. (1999) *Geotechnical design to Eurocode 7*. Springer, Berlin.

Parkinson, A., Sorensen, C., and Pourhassan, N. (1993) A general approach for robust optimal design. *Journal of Mechanical Design*, 115(1), 74–80.

Phoon, K.K., Kulhawy, F.H., and Grigoriu, M.D. (1995) *Reliability based design of foundations for transmission line structures*. Rep. TR-105000. Electric Power Research Institute, Palo Alto, CA.

R.S. Means Co. (2007) *R.S. Means building construction cost data*. Kingston, MA.

Taguchi, G. (1986) *Introduction to quality engineering: Designing quality into products and processes*. Quality Resources, White Plains, New York.

Vesić, A.S. (1975) Bearing capacity of shallow foundations. In *Foundation engineering handbook*, H. Winterkorn and H.Y. Fang, eds. Van Nostrand Reinhold, New York.

Wang, L., Gong, W., Luo, Z., Khoshnevisan, S., and Juang, C.H. (2015) Reliability-based robust geotechnical design of rock bolts for slope stabilization. In *International foundations congress & equipment exposition 2015*. ASCE Geotechnical Special Publication (GSP 256), 1926–1935.

Wang, L., Hwang, J.H., Juang, C.H., and Atamturktur, S. (2013) Reliability-based design of rock slopes – a new perspective on design robustness. *Engineering Geology*, 154, 56–63.

Wang, Y. (2011) Reliability-based design of spread foundations by Monte Carlo Simulations. *Géotechnique*, 61(8), 677–685.

Wang, Y., Au, S.K., and Kulhawy, F.H. (2011) Expanded reliability-based design approach for drilled shafts. *Journal of Geotechnical and Geoenvironmental Engineering*, 137(2), 140–149.

Wang, Y., and Kulhawy, F.H. (2008) Economic design optimization of foundations. *Journal of Geotechnical and Geoenvironmental Engineering*, 134(8), 1097–1105.

Wu, T.H., Tang W.H., Sangrey, D.A., and Baecher, G.B. (1989) Reliability of offshore foundations – state-of-the-art. *Journal of Geotechnical Engineering*, 115(2), 157–178.

Wyllie, D.C., and Mah, C.W. (2004) *Rock slope engineering: Civil and mining* (4th edition). Spon Press, London.

Zhang, J., Zhang, L.M., and Tang, W.H. (2011) Reliability-based optimization of geotechnical systems. *Journal of Geotechnical and Geoenvironmental Engineering*, 137, (12), 1211–1221.

Zhao, Y.G., and Ono, T. (2000) New point estimates for probability moments. *Journal of Engineering Mechanics*, 126(4), 433–436.

Index

M

N

O

P

R

S

U

V

Y

For Product Safety Concerns and Information please contact our EU
representative GPSR@taylorandfrancis.com
Taylor & Francis Verlag GmbH, Kaufingerstraße 24, 80331 München, Germany

www.ingramcontent.com/pod-product-compliance
Ingram Content Group UK Ltd.
Pitfield, Milton Keynes, MK11 3LW, UK
UKHW020158200926
475702UK00004B/521
* 9 7 8 1 0 3 2 9 1 4 2 4 4 *